POLYMER SCIENCE AND TECHNOLOGY
Volume 4

RECENT ADVANCES IN POLYMER BLENDS, GRAFTS, AND BLOCKS

POLYMER SCIENCE AND TECHNOLOGY

Volume 1 • STRUCTURE AND PROPERTIES OF POLYMER FILMS
Edited by Robert W. Lenz and Richard S. Stein • 1972

Volume 2 • WATER-SOLUBLE POLYMERS
Edited by N. M. Bikales • 1973

Volume 3 • POLYMERS AND ECOLOGICAL PROBLEMS
Edited by James Guillet • 1973

Volume 4 • RECENT ADVANCES IN POLYMER BLENDS, GRAFTS, AND BLOCKS
Edited by L. H. Sperling • 1974

POLYMER SCIENCE AND TECHNOLOGY
Volume 4

RECENT ADVANCES IN POLYMER BLENDS, GRAFTS, AND BLOCKS

Edited by
L. H. Sperling
Associate Professor of Chemical Engineering
and Senior Staff Member
Materials Research Center
Lehigh University
Bethlehem, Pennsylvania

PLENUM PRESS • NEW YORK AND LONDON

Library of Congress Cataloging in Publication Data

Main entry under title:

Recent advances in polymer blends, grafts, and blocks.

(Polymer science and technology, v. 4)
"Proceedings of the symposium . . . sponsored by the Polymer Division of the American Chemical Society, held in Chicago, August, 1973."
Includes bibliographical references.
1. Polymers and polymerization—Congresses. I. Sperling, Leslie Howard, 1932- ed. II. American Chemical Society. Division of Polymer Chemistry.
QD380.R42 547'.84 74-3439
ISBN 0-306-36404-2

Proceedings of the symposium on Polymer Blends, Grafts, and Blocks, sponsored by the Polymer Division of the American Chemical Society, held in Chicago, August, 1973

A Division of Plenum Publishing Corporation
227 West 17th Street, New York, N.Y. 10011

United Kingdom edition published by Plenum Press, London
A Division of Plenum Publishing Company, Ltd.
4a Lower John Street, London W1R 3PD, England

Printed in the United States of America

In memory of my mother

ROSE A. SPERLING

who loved books and

encouraged learning all her life

PREFACE

Polymer blends, grafts, and blocks, broadly defined, encompass all of the ways in which two or more kinds of polymer molecules can be mixed and/or joined. Because these materials exhibit non-linear and often synergistic properties, they have found increasing application in our technology. Their multifarious uses have, in turn, spurred new research efforts, to find yet different ways of joining two kinds of polymer molecules, with novel physical and/or mechanical behavior patterns.

In August, 1973, the Polymer Division of the American Chemical Society sponsored a symposium at its meeting in Chicago on Polymer Blends, Grafts, and Blocks. This book collects the papers presented at that symposium. Yet, it is more than just a collection of papers, for we here display the thinking and efforts of a number of top-ranking American and foreign scientists in one of the world's more active research areas.

The symposium emphasized the interrelationships among synthetic detail, morphology, and physical and mechanical properties. Several novel syntheses were presented. These include oxidation resistant thermoplastic elastomers (Holden), a graft copolymer based thermoplastic elastomer (Kennedy and Smith), a cationic graft copolymer (Kennedy, Charles, and Davidson), an AB crosslinked copolymer (Bamford and Eastmond), an interpenetrating polymer network (Donatelli, Thomas, and Sperling), and simultaneous interpenetrating networks (Frisch, Klempner, Frisch, and Ghiradella).

Most polymer blends, grafts, and blocks exhibit two phases. The theory of microdomain structure was discussed (Helfand). The different ways that the two molecules can be joined together was examined (Kenney), and their topology was explored (Sperling). Mechanical behavior of urea-urethane block polymers was examined (Work). The temperature dependence of blend properties was investigated (Kaplan and Tschoegl), and their optical properties were discussed (Reed, Bair, and Vadimsky). The orientation properties of segmented polyurethanes were examined (Seymour, Estes and Cooper), and physical properties and applications of block copolymers were discussed (Kraus and Railsback).

Phase separation in nearly compatible materials was studied (Robeson, Matzner, Fetters and McGrath), and the properties of blocks and blends were compared (McGrath, Robeson and Matzner).

In addition to the papers presented at the symposium, this volume also offers two invited review papers. Kennedy reviews the synthesis of graft copolymers, while Deanin discusses processing of polymer blends.

The book is conveniently divided into three sections, each representing a half day portion of the symposium. The three session chairmen, Sperling, Krause, and Kenney, have kindly consented to highlight and introduce their respective groups of papers. Because it is natural to do so, the two review papers will be included as a portion of Section I.

The editor wishes to take this opportunity to thank each of the authors for their splendid contributions, and also to thank all of their secretaries for diligent typing and neat arrangement of figure and table materials.

L. H. Sperling

Bethlehem, Pennsylvania
January 2, 1974

CONTENTS

SECTION I. INTRODUCTION

L. H. Sperling

SECTION II. INTRODUCTION

S. Krause

SECTION III. INTRODUCTION

J. F. Kenney

SECTION I. INTRODUCTION

This section opens with two invited review papers, by Dr. J. P. Kennedy and by Dr. R. D. Deanin. Dr. Kennedy considers the chemistry of block and graft copolymer synthesis, and discusses free radical, anionic, and cationic mechanisms. Dr. Deanin, in the following paper, devotes his attention to physical and mechanical behavior, and to processing. These two fine reviews provide the reader with much important background material.

We then have three papers devoted to a combination of theory and review. Dr. E. Helfand examines the phase boundary characteristics present in block copolymers, a sequel to an earlier theory on polymer blend phase boundary characteristics. In the earlier theory, Dr. Helfand predicted a density lower than otherwise expected at the phase boundary. Dr. J. F. Kenney reviews the morphology and mechanical behavior of a number of blends, blocks, and grafts, and makes a comparison to random and alternating materials of similar mer composition. Dr. Kenney concluded with a section of his own original studies. Dr. L. H. Sperling presents a new theory of polymer blend topology, making use of group theory concepts. He expounds a systematic way of arriving at novel or previously unrecognized structures, and classifies a number of known blend, block, and graft copolymers.

Lastly, we present two experimental papers. The cationic graft copolymers of Drs. J. P. Kennedy, J. J. Charles, and D. L. Davidson should be read in context with the invited review paper of Dr. Kennedy (above), and the paper by Drs. Kennedy and Smith included in Section III. Thus, Dr. Kennedy and coworkers display three papers in this book, an indication of their productivity. Drs. Bamford and Eastmond present a paper on AB cross-linked copolymers. Their topology involves grafting in such a way that all of the polymer I chains are joined to polymer II chains, and vice versa, to form a single netowrk. The Bamford and Eastmond topology might profitably be compared with the IPN and SIN structures presented in Section III by Drs. Donatelli, Thomas, and Sperling, and Drs. Frisch, Klempner, Frisch, and Ghiradella, respectively. In the latter two papers, two more or less separate networks are formed, which interpenetrate on a supermolecular or molecular level. It should be pointed out that Drs. Bamford and Eastmond, from England, were guests of the American Chemical Society.

Depending upon the source of information, between half and two-thirds of all chemists and chemical engineers in the United States are currently engaged in polymer-related activities. Since most polymeric materials are not ultimately consumed as pure homopolymers, it follows then that a very large proportion of chemists and chemical engineers are engaged in polymer blend and/or composite activities. This category includes major sections of the elastomer and plastics industries, as well as the paint and adhesive technologies. Much of the original work remains unpublished, or is of an empirical experimental nature. Few people have attempted to summarize our growing measure of knowledge, and even fewer have set about to reduce the many accumulated facts to well-understood theories and principles. It is in this light that we proudly present the following papers.

L. H. Sperling

AN INTRODUCTION TO THE SYNTHESIS OF BLOCK- AND GRAFT COPOLYMERS*

J. P. Kennedy

Institute of Polymer Science

The University of Akron, Akron, Ohio 44325

CONTENTS

*Invited paper. This chapter was not presented in the Symposium.

This paper, by Dr. J. P. Kennedy, and the following paper by Dr. R. D. Deanin, are invited review papers. This material will serve to introduce the reader to the remainder of the book by reviewing first the chemistry, and then the physical properties and technology of polymer blends, grafts, and blocks.

The Editor

I. The Scope of This Chapter, Definitions

This chapter concerns the synthesis of sequential copolymers, i.e., copolymers comprising at least two reasonably long sequences of uniform repeat units. Sequential copolymers are generally subdivided into block and graft copolymers. Random, statistical and alternating copolymers will not be considered in this chapter. A few simple molecular formulae help to visualize the structure of these materials:

AAAAAAAAAAA	homopolymer
ABBABABAABBB	random or statistical copolymer
ABABABABAB	alternating copolymer
AAAAAABBBBB	(two) block copolymer
AAAAAAAA with side chain –B–B–B–B attached at the fifth A	graft-copolymer

A block copolymer comprises two or more linear sequences of a homogeneous chemical composition. A graft copolymer consists of a backbone sequence to which a second, different sequence of branch-polymer is attached.

Graft copolymers are described in terms of the backbone (or trunk) polymer and the branch polymer. Obviously, the backbone or branches may themselves be complex systems, for example, random copolymers. One could graft a random copolymer of, say, styrene/α-methylstyrene onto another random copolymer of, say, butadiene/acrylonitrile. The idealized structure of this system would then be:

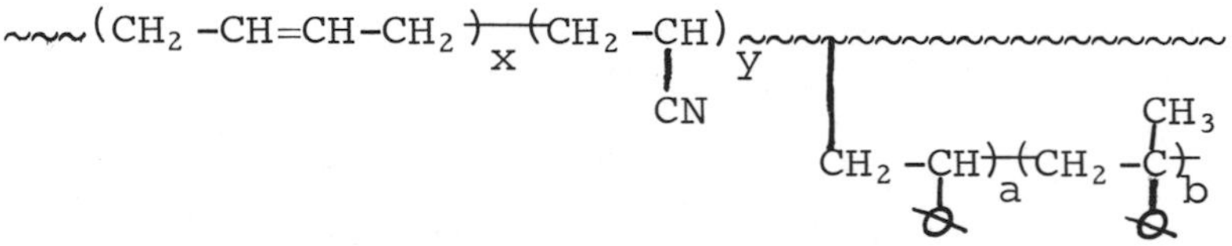

$$\sim\!\sim(CH_2-CH{=}CH-CH_2)_x(CH_2-\underset{CN}{CH})_y\sim\!\sim\!\sim\!\sim\!\sim\!\sim\!\sim\!\sim$$
$$|$$
$$CH_2-\underset{\phi}{CH})_a(CH_2-\underset{\phi}{\overset{CH_3}{C}})_b$$

Sequential polymers contain chemically combined polymeric sequences and therefore are different from polymer blends or alloys which represent physical mixtures of polymer sequences. However, to facilitate the discussion of the synthesis of sequential polymers and because of the close phenomenological relationship between sequential copolymers and polymer blends, when appropriate, the latters will also be considered.

High scientific and technological interest in sequential copolymers exists because of the attractive and varied physical properties they offer at reasonable cost. The physical properties of sequential copolymers more often than not are quite different from those of physical blends of corresponding homopolymers. Indeed, a justification for research in sequential copolymers resides in their ability to lead to properties unavailable by physical polymer blends.

Some sequential copolymers that acquired technological-commercial significance are:

1. Impact resistant polystyrenes
2. ABS resins

3. Thermoplastic elastomers
4. Cellulose derivatives
5. Polyurethanes
6. Pluronics and Tetronics of the Wyandotte Co.
7. Hevea Plus (Natural rubber-g-PMMA)
8. Irradiated wood-MMA systems

The complete characterization of sequential copolymers is an enormously complex and largely unsolved problem. At the present the polymer scientist must be content by obtaining a fraction of the desired characterization information:

1. Overall composition (sequence ratio).
2. Overall molecular weight and molecular weight distribution.
3. Molecular weight and molecular weight distribution of sequences.
4. Number of sequences per molecule.
5. Compositional distribution (homogeneity of composition).
6. Microstructure of the individual sequences.
7. Microarchitecture of the overall polymer (sequence arrangement).

Research directed toward the characterization of sequential polymers *per se* will not be considered.

A list of recent books and significant review articles concerning sequential copolymers has been assembled at the beginning of References.

II. Nomenclature of Sequential Copolymers

The architecture of block and particularly graft copolymers may be extremely varied and complex. Consequently the naming of these materials presents a considerable problem. A very useful nomenclature has been proposed by Ceresa in his 1962 book Block and Graft Copolymers, which has been adopted and is heartily recommended by this writer. This nomenclature has also been adopted by the Encyclopedia of *Polymer Science and Technology* and by Battaerd and Tregear in their book on

Graft Copolymers. (The following four paragraphs are quoted from Ceresa's book (Book ref. 1)):

"The system is simple yet it allows for expansion when further amplification of structure is required. It only becomes cumbersome when dealing with complex copolymers, but this difficulty is to be expected with any classification system. In the system to be described the Chemical Abstracts nomenclature for homopolymers is used, e.g. poly(methyl methacrylate), polystyrene, poly (vinyl acetate) etc. Random copolymers are indicated by the prefix -*co*-, e.g. poly(butadiene-*co*-styrene) and poly(vinyl chloride-*co*-vinyl acetate). Alternating copolymers can be differentiated where necessary by replacing -*co*- with -*alt*-, as in poly(ethylene-*alt*-carbon monoxide).

"To differentiate between block and graft copolymers it is necessary to introduce two additional prefixes, -*b*- for linear block copolymer segments, and -*g*- for grafted block copolymer segments. In this nomenclature the first polymer segment named corresponds to the homopolymer or copolymer which is prepared in the first stage of the synthesis. (For a graft copolymer this is of necessity the backbone polymer or copolymer.) The following exemplify the classification system: poly (styrene-*b*-methyl methacrylate), poly(methyl methacrylate-*b*-styrene), poly(ethylene-*g*-acrylonitrile), poly-(styrene-*g*-[butadiene-*co*-styrene]), poly([butadiene-*co*-styrene]-g-[styrene-*co*-butadiene]). If the convention is accepted that in a copolymer the first-mentioned monomer is present in the copolymer in the greater proportion, then the last example represents a high styrene-butadiene copolymer grafted on a low styrene-butadiene rubber.

"Conventional prefixes to indicate diene structures can be inserted in their normal positions, e.g. natural cis-1:4-poly(isoprene-*g*-methyl methacrylate) (Heveaplus), natural cis-1:4-poly(isoprene-*b*-trans-1,4-isoprene) (from the comastication of natural rubber and gutta percha). The phenomena of branching and crosslinking can likewise be accommodated by using the prefixes [*br*] and [*c.l.*], e.g. poly(styrene-g-[*br*] vinyl acetate), poly(ethylene-*b*-[*br*]vinyl carbazole, poly([*c.l.*]styrene-*g*-acrylonitrile), poly([*c.l.*]cis-1:4-isoprene-*b*-methacrylic acid).

"Tacticity may be allowed for by using the prefixes to differentiate between isotacticity [*iso*], syndiotacticity [*syndio*] and atacticity [*a*] as in:"

```
~~(AAAA....AAAAA)—(AAAAA....AAAA)—(AAAA..AAAA)~~
                 iso              syndio      iso

      ~~(AAAAAAA.......AAAAAAAAAAAAA)~~
             |               |        iso
             A               A
             |               |
             A               A
             |               .
             A               .
             .               .
             .               A
             .               |
             A               A
             |               |
             A               A
             |               |
             A               A
             ~  a            ~  a

  ~~(AAAAAA...AAAAAAAAA)—BBBBBBBBBBBBBBBBBBBBB~~
                     syndio
```

Some examples should illustrate this nomenclature. If S signifies styrene and B butadiene then the nomenclature of sequential copolymers would be as follows:

```
SSBSBBSSBBSSS        poly(styrene-co-butadiene)

SBSBSBSB             poly(styrene-alt-butadiene)

SSSSSBBBBB           poly(styrene-b-butadiene)

SSSSSS               poly(styrene-g-butadiene)
  |
  B
  |
  B
  |
  B
  |
  B

BBBBBBSSSSSSS        poly(butadiene-b-[styrene-g-buta-
        |                                   diene])
        B
        |
        B
        |
        B
        |
        B
```

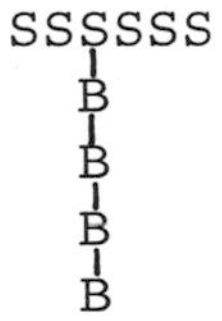

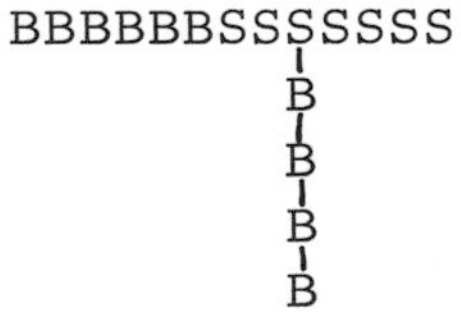

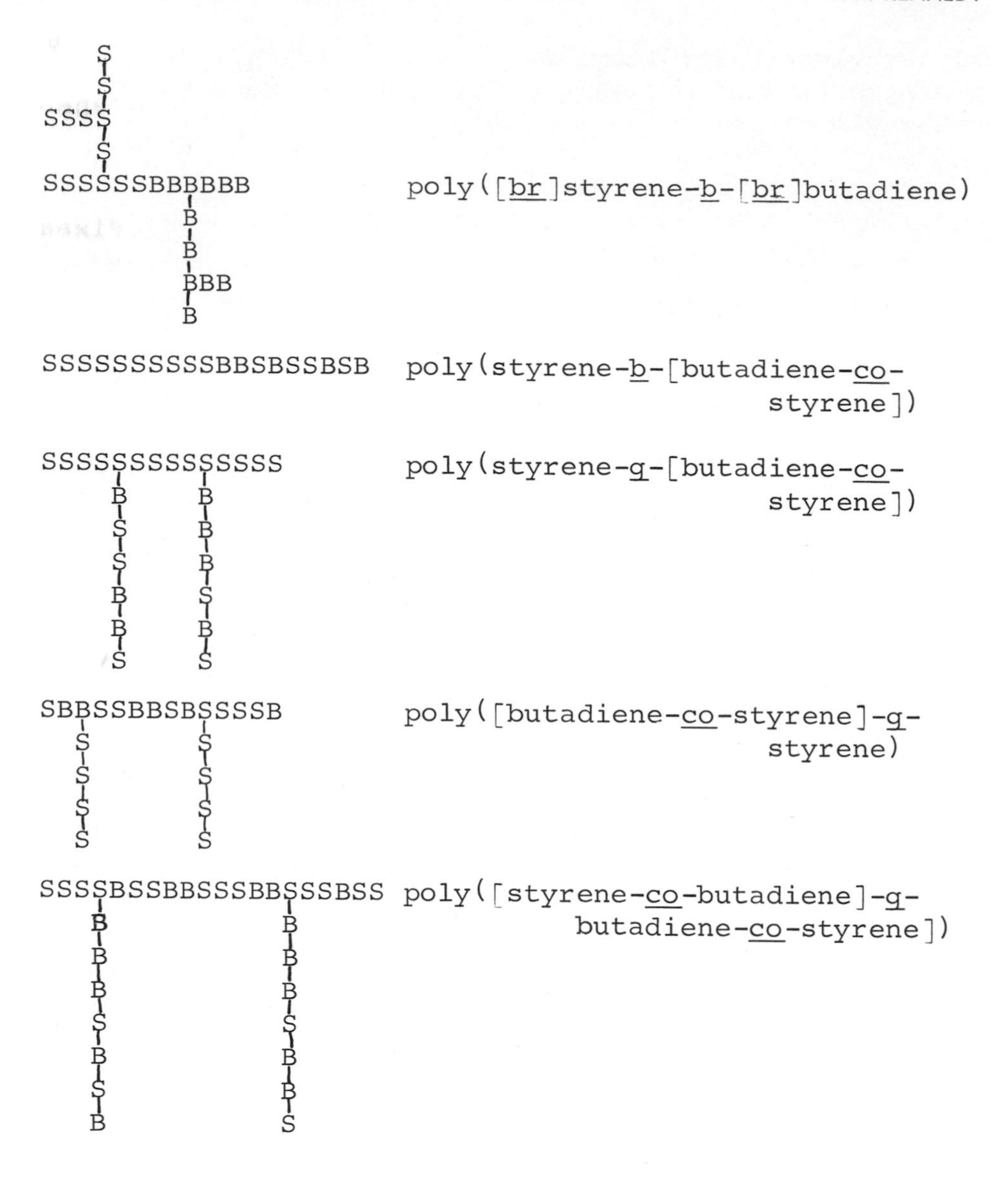

III. Synthesis of Sequential Polymers

The field of synthesis of block and graft copolymers is conveniently organized by the mechanism of the synthetic procedure. Thus block and graft copolymers may be synthesized by

1) Free Radical Mechanism
   A. Chemical Methods
   B. Photolytic Methods
   C. High Energy Irradiation Techniques
   D. Mechanochemical Methods
2) Ionic Mechanism
   A. Anionic Mechanism
   B. Cationic Mechanism
3) Coordination (Ziegler-Natta) Polymerization
4) Miscellaneous Ring Opening and Coupling Reactions

In the following discussion on the synthesis of block and graft copolymers no distinction will be made between these two species. The reason for this is that from the point of view of synthesis the only fundamental difference between the preparation of block or graft copolymers is the location of the active site: If the active site(s)is(are) at the termini of a polymer, block copolymers will be obtained; if the active site(s) is(are) somewhere along the polymer chain, graft copolymers will result.

A useful terminology concerns the synthesis of graft copolymers: If the active site is located somewhere along the backbone and it starts to propagate, and thus produce branches, we call this process "grafting from"; if, however, a growing polymer attacks another polymer along the chain, and thereby attaches a branch to a preformed trunk, we term this event "grafting onto". Schematically:

Grafting from:

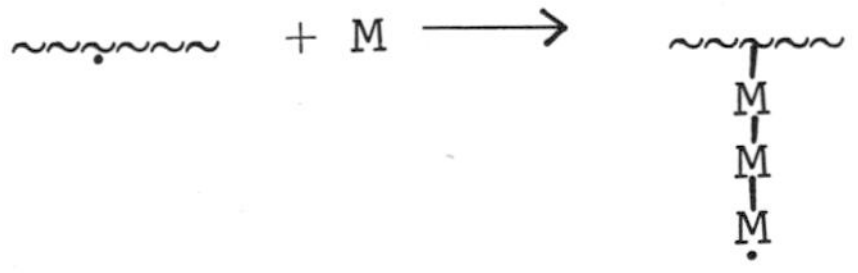

$$\sim\sim\dot{\sim}\sim\sim\sim + M \longrightarrow \begin{array}{c}\sim\sim\top\sim\sim\\ M\\ |\\ M\\ |\\ \dot{M}\end{array}$$

Grafting onto:

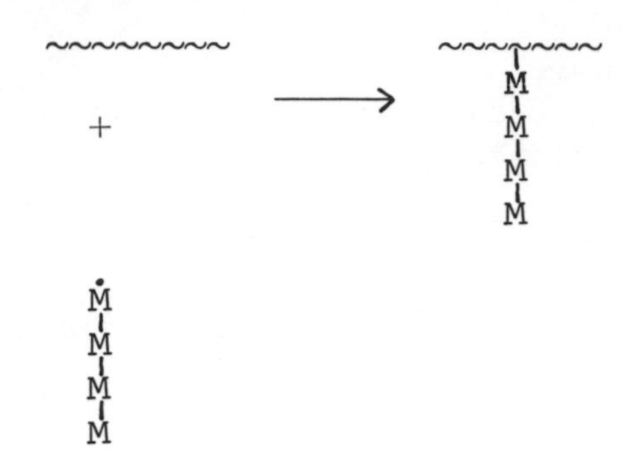

Besides the grafting from and onto methods which involve one preformed polymer (the backbone or trunk) and a monomer supply, graft and block copolymers can also be synthesized by a variety of coupling techniques. In these processes two preformed polymer sequences are coupled either directly or by the use of a separately added suitable coupling agent:

~~~~~⊕ + ⊕——— ⟶ ~~~~~———

~~~~~ + ⊕ + ——— ⟶

III.1. Synthesis by Free Radical Mechanism

There are in principle four methods for the synthesis of sequential polymers by free radical mechanisms:

A. Chemical Generation of Free Radicals
B. Photolytic Methods
C. Ionizing Radiation Techniques
D. Mechanochemical Methods

This classification is based on the method of generation of free radicals. These methods can of course be further subdivided along general characteristics.

III.1.A. Synthesis by Free Radicals: Chemical Methods

a) Chain Transfer to Polymer

This method for the synthesis of sequential polymers is illustrated by the following scheme:

$$R^{\cdot} + -CH_2-\underset{X}{CH}- \xrightarrow{(RH)} -CH_2-\underset{X}{\dot{C}}- \xrightarrow{+nM} -CH_2-\underset{X}{\overset{M^{\cdot}\!-\!(M)_{n-1}}{C}}-$$

or

$$\sim\sim\sim\sim\cdot + -CH_2-\underset{X}{CH}- \xrightarrow{(\sim\sim H)} -CH_2-\underset{X}{\dot{C}}-$$

Grafting efficiency and graft yield are strongly affected by the free radical source. For example, large amounts of graft product is obtained by using benzoyl-peroxide (BPO) initiator for grafting of methyl methacrylate in the presence of polystyrene,however, the results are much poorer with azobisisobutyronitrile (AIBN) initiator (1). Methyl methacrylate is reluctant to graft onto natural rubber with AIBN, however, gives good yields with BPO (2). The reason for the inactivity of AIBN might be due to the resonance stabilization of cyanopropyl radical $(CH_3)_2\dot{C}-CN$.

In principle these graft and/or block copolymerizations can be carried out by dissolving a polymer in the monomer to be grafted or in a monomer-solvent system, and introducing an effective free radical source.

The above grafting-from reactions involve graft-initiation by macroradicals. These reactions are governed by the reactivity and/or polarity of both the radical and the monomer and follow the Q-e scheme (3,4).

Graft copolymers can also be synthesized by introducing into a chain epoxy groups, for example, by copolymerizing small amounts of glycidyl methacrylate which can subsequently be converted with thioglycolic acid to pendant thiol groups. The transfer activity of these side groups with growing styrene or methyl methacrylate is quite high and thus can be used to prepare grafts (5):

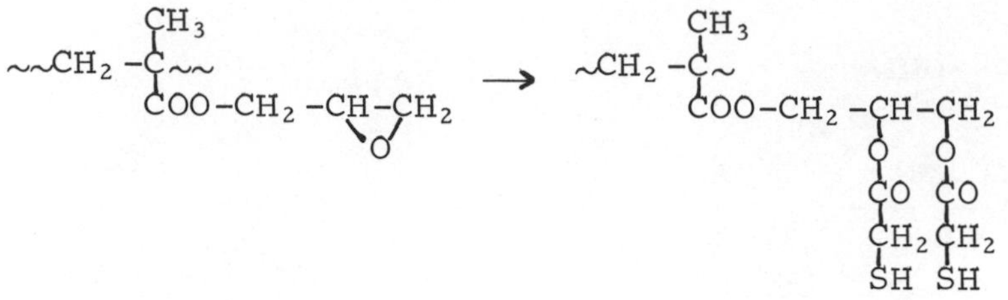

$$\sim\sim CH_2-\underset{COO-CH_2-CH\overset{O}{-}CH_2}{\overset{CH_3}{\underset{|}{\overset{|}{C}}}}\sim\sim \longrightarrow \sim CH_2-\underset{COO-CH_2-\underset{O-CO-CH_2-SH}{CH}-\underset{O-CO-CH_2-SH}{CH_2}}{\overset{CH_3}{\underset{|}{\overset{|}{C}}}}\sim$$

Block copolymers can be obtained by chain transfer mechanisms provided chain transfer activity occurs exclusively at the end-groups and does not involve statistical grafting onto other pendant groups among the chain. Poly(methyl methacrylate) contains unsaturated end groups which arise during chain breaking via disproportionation and whose transfer activity is orders of magnitude higher than that of in-chain groups. These polymers have been used to produce blocks with vinyl acetate monomer to give poly(methyl methacrylate-<u>b</u>-vinyl acetate)(3):

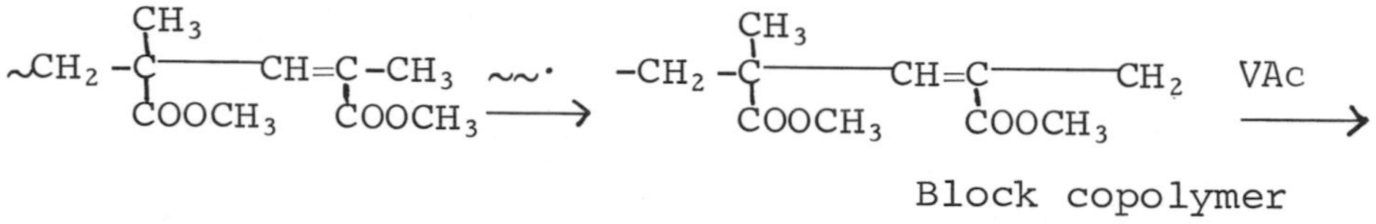

$$\sim CH_2-\underset{COOCH_3}{\overset{CH_3}{\underset{|}{\overset{|}{C}}}}-CH=\underset{COOCH_3}{C}-CH_3 \quad \sim\sim\cdot \longrightarrow -CH_2-\underset{COOCH_3}{\overset{CH_3}{\underset{|}{\overset{|}{C}}}}-CH=\underset{COOCH_3}{C}-CH_2 \xrightarrow{VAc} \text{Block copolymer}$$

An interesting method for the preparation of poly(methyl methacrylate-<u>b</u>-acrylonitrile) has been developed (6,7): It was known that tertiary amines are effective transfer agents in the polymerization of acrylonitrile, but are much less active with methyl methacrylate. Thus one can prepare, in the presence of sufficiently high amounts of a tertiary amine, poly(methyl methacrylate) molecules terminated by tertiary amine end groups, which in turn can be used as transfer sites in a polymerization of acrylonitrile. The following equations summarize these steps:

$$\text{nMMA} + N(C_2H_5)_3 \xrightarrow{\text{pzn.}} \sim CH_2-\underset{COOCH_3}{\overset{CH_3}{C}}-CH\begin{matrix} N(C_2H_5)_2 \\ CH_3 \end{matrix}$$

$$\downarrow + \text{AN}$$

$$\sim CH_2-\underset{COOCH_3}{\overset{CH_3}{C}}-\underset{CH_3}{\overset{N(C_2H_5)_2}{C}}-CH_2-\underset{CN}{\overset{\cdot}{CH}} \leftarrow \sim CH_2-\underset{COOCH_3}{\overset{CH_3}{C}}-\overset{\cdot}{C}\begin{matrix} N(C_2H_5)_2 \\ CH_3 \end{matrix}$$

$$\downarrow \text{AN}$$

block copolymer of MMA + AN

This technique of block copolymer synthesis, of course, can be successful only when the chain transfer activity of amine (or other agent) is significantly different for the two monomers involved. Also, statistical chain transfer to the polymer should be absent. Keeping these principles in mind Bamford and White first polymerized MMA in the presence of Et_3N at Et_3N/MMA ratio of 14:1 at 60^0 with AIBN; the excess base was removed and the PMMA isolated. Subsequently the PMMA was dissolved in AN and the second (block) copolymerization initiated by AIBN.

Another method for the preparation of graft copolymers by free radical chain transfer is first to prepare a backbone polymer containing halogens (halogens have high transfer activity) and to use these groups in a subsequent grafting reaction as initiating sites. For example Schonfield and Waltcher (8) first synthesized a suitable polyester which contained the following grouping:

$$\sim O-CH_2-\underset{CH_2Br}{\overset{CH_2Br}{C}}-CH_2OCO-(CH_2)_4-CO\sim$$

and used this to graft polystyrene by transfer.

Disadvantages of all these free radical methods for the synthesis of block or graft copolymers are that a) in addition to the desired grafting or block-forming reaction homopolymers are also obtained due to either chain transfer to monomer or direct initiation, or because not every preformed macromolecule enters into chain transfer reaction, and b) the possibility for main chain degradation exists. None of these side reactions can be tolerated if the goal is the synthesis of "clean" sequential structures. If homopolymerization and/or degradation occur, laborious extraction/purification procedures become necessary to obtain the homogeneous sequential product.

b) Reactions of Reactive Groups in the Chain

Sequential copolymers can be synthesized by introducing reactive groups into a preformed polymer sequence (either along its chain or at the termini) and transforming these groups into radicals capable of initiating the polymerization of a monomer different from the one which provided the preformed polymer. Suitable reactive groups are unsaturations, peroxide groups, peresters, diazonium salts, etc. These reactive groups can be introduced into the preformed polymer either by copolymerization of small but controlled amounts of a suitable monomer or by post-polymerization treatment, such as oxidation or peroxidation, etc.

b) α Unsaturation in the Chain

Since unsaturated groups in a polymer chain are reactive sites, the grafting of vinyl monomers onto polymers with unsaturated groups is discussed in this class. Indeed, the adducting of polystyrene or cis-1,4-polybutadiene or poly(butadiene-co-acrylonitrile) to natural rubber are commercially important grafting techniques for the production of high impact materials (impact polystyrene, ABS resins).

Distinguished research has been carried out on the natural rubber (or gutta percha) plus methyl methacrylate graft system by Allen et al. (2). These authors used C^{14} labelled benzoylperoxide at 60^0 in benzene to initiate the grafting of MMA to natural rubber. Theoretically the following eight possibilities exist for grafting to the simplest unsaturated backbone, that obtained with butadiene:

A. Attack on the Double Bond by Initiator Radical ($I^{\cdot}$) or Polymer Radical ($\sim P^{\cdot}$)

$60 \pm 5\%$:

$$I^{\cdot} + -CH_2-CH{=}CH-CH_2- \longrightarrow -CH_2-\dot{C}H(I)-CH-CH_2-$$

1,4 enchainment

$$I^{\cdot} + -CH_2-CH(CH{=}CH_2)- \longrightarrow -CH_2-CH(\cdot CH-CH_2I)-$$

1,2 enchainment

0%:

$$\sim P^{\cdot} + -CH_2-CH{=}CH-CH_2- \longrightarrow \sim CH_2-\dot{C}H(P\sim)-CH-CH_2\sim$$

$$\sim P^{\cdot} + -CH_2-CH(CH{=}CH_2) \longrightarrow \sim CH_2-CH(\cdot CH-CH_2-P\sim)-$$

B. Attack on the Allylic Hydrogen:

$40 \pm 5\%$:

$$I^{\cdot} + \sim CH_2-CH{=}CH-CH_2\sim \longrightarrow IH + \sim\dot{C}H-CH{=}CH-CH_2\sim$$

$$I^{\cdot} + \sim CH_2-CH(CH{=}CH_2)\sim \longrightarrow IH + \sim CH_2-\dot{C}(CH{=}CH_2)$$

0%:

$$\sim P^{\cdot} + \sim CH_2-CH{=}CH-CH_2\sim \longrightarrow \sim PH + \sim\sim\dot{C}H-CH{=}CH-CH_2-$$

$$\sim P^{\cdot} + \sim CH_2-CH(CH{=}CH_2)\sim \longrightarrow \sim PH + \sim CH_2-\dot{C}(CH{=}CH_2)$$

According to Allen et al.'s experiments, grafting is largely initiated by the $I^{\cdot}$; more in particular, $40 \pm 5\%$ of the graft is formed by an attack on the allylic hydrogen and $60 \pm 5\%$ by direct addition to the double bond. Interestingly, when AIBN was used instead of BPO, only homopoly(methyl methacrylate) formed and no graft polymer was obtained. Thus growing PMMA radicals cannot chain transfer with polybutadiene. Similar results have also been obtained by Smets (9) in experiments concerning the grafting of MMA to polystyrene with BPO and AIBN. Allen et al. explain this "Initiator Effect" by the resonance stabilization of the isobutyronitryl radical (see above).

Termination in the above polybutadiene/methyl methacrylate system may occur by either recombination to yield a network of two strands of rubber connected by a PMMA sequence or by disproportionation (the usual termination process for PMMA) of two growing rubber branches to yield essentially soluble "comb" grafts:

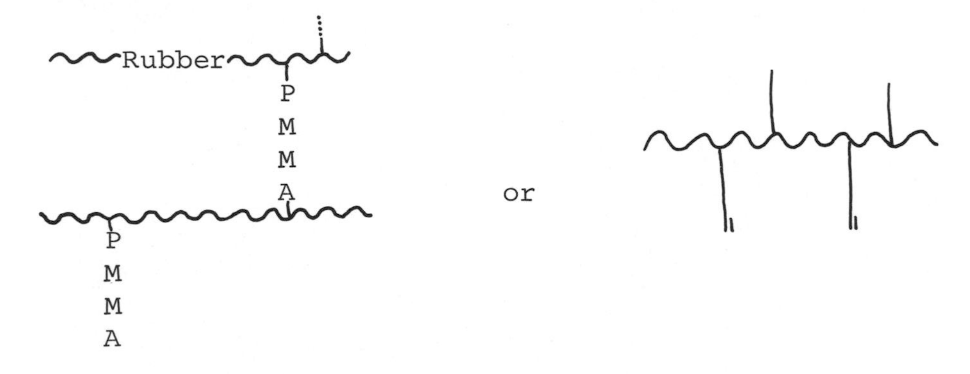

The natural rubber/styrene system, however, contains insoluble gel which consists of polyisoprene and polystyrene sequences. This is explained by the combination-termination mechanism operating in radical-initiated styrene polymerization. Impact-resistant polystyrene prepared by graft copolymerization is not completely soluble even in toluene, a good solvent for both the polydiene and polystyrene sequences (10).

Impact-resistant polystyrene is produced commercially by dissolving a suitable elastomer (natural rubber, synthetic cis-1,4-polyisoprene, cis-1,4-polybutadiene, poly(butadiene-co-styrene), etc.) in styrene monomer and initiating the polymerization-grafting of the latter by introducing an initiator such as BPO. The grafting reaction can be carried out in emulsion or in solution. Grafting in emulsion is used for the preparation of ABS resins (acrylonitrile-butadiene-styrene resins). The desired multicomponent systems are obtained by mixing the grafted rubber dispersion with a separately prepared poly(styrene-co-acrylonitrile) dispersion and coprecipitating

(11). Grafting in solution is carried out on the large scale for the preparation of rubber-modified homopolymers. During polymerization, homopolymerization of styrene and grafting to the elastomer proceed simultaneously. The ultimate physical properties of the product are affected by the ratio of elastomer/polystyrene, the nature of the elastomer, rate of stirring, molecular weight of the components, etc.

At the present time huge quantities of impact-resistant (rubber-filled) polystyrenes are produced and consumed in a variety of applications. To relieve sudden stresses arising during impact, it is necessary to have an incompatible multiphase system consisting of a continuous glassy phase and a discontinuous rubbery phase, and to establish intimate contact between the rubbery and glassy domains. The continuity of the glassy phase provides the desired stiffness and thermal form-stability, while the finely dispersed rubber phase acts as an energy absorber. During shock or impact energy is first absorbed by the hard phase and if brittle failure is to be avoided, the energy must be transferred to the embedded rubber phase where the mechanical energy is converted to heat and is rendered harmless. If energy transfer from the glassy to the rubbery phase is slow or impossible, the energy remains in the brittle phase where it may concentrate at strain points and lead to fracture.

For efficient transfer of mechanical energy from the hard to the soft phase, intimate contact, and good adhesion between phases must exist. Sufficient adhesion can be expected only when the two phases have a limited compatibility. This necessary limited compatibility can be achieved by modern graft copolymer technology.

b) β <u>Peroxide Groups</u>

One way to obtain grafts on polystyrene backbones is to introduce peroxy groups into the preformed polystyrene and then to initiate the polymerization of a second monomer by the macromolecular peroxide. Since polystyrene is difficult to peroxidize directly, styrene is copolymerized with a small amount of **p**-isopropylstyrene, which is subsequently peroxidized with oxygen.

The macromolecular peroxide readily initiates the polymerization of various monomers by heating or by treatment with ferrous salts:

$$\sim CH_2-CH\sim(C_6H_4)-C(CH_3)_2-H \xrightarrow{+O_2} \sim CH_2-CH\sim(C_6H_4)-C(CH_3)_2-O-O-H \xrightarrow{\Delta \text{ or } Fe^{++}} \sim CH_2-CH\sim(C_6H_4)-C(CH_3)_2-O^{\cdot} \xrightarrow{+M} \text{graft copolymer}$$

In this manner methyl methacrylate or vinyl acetate have been grafted to polystyrene by decomposing the peroxide with heat (12) or by a redox mechanism with Fe^{++} salts (13).

Natta (14,15) prepared poly(atactic propylene-g-methyl methacrylate) by first oxidizing atactic polypropylene with air oxygen at 70^0 (hydroperoxide content 2-4%) and subsequently using this polymer-derivative to initiate the polymerization of MMA at 70^0 in toluene solution. In addition to MMA, this author has also grafted styrene and vinyl chloride onto poly(α-olefins).

The disadvantage of the heat-initiated grafting of peroxidized polymers is that currently with the macromolecular radical a separate $OH^{\cdot}$ radical is also formed which then leads to homopolymerization. An advantage of redox initiation is that the formation of small radicals is avoided:

$$\sim\!O\!-\!O\!-\!H \xrightarrow{\Delta} \sim\!O^{\cdot} + OH^{\cdot}$$

$$\sim\!O\!-\!O\!-\!H \xrightarrow{+ Fe^{++}} \sim\!O^{\cdot} + OH^{-} \; (+F^{+++})$$

In addition to direct oxidation, ozonization can also be used to introduce peroxy groups into a polymer (16). In principle ozonization can be used with any polymer containing unsaturations and, depending on the position of the double bonds, grafts or block copolymers can be obtained: Pendant unsaturations would lead to grafts and terminal or in-chain double bonds to blocks. In fact the ozonization technique has been used for the preparation of many sequential copolymers by French (17) Japanese (18) and Russian (19) investigators.

Peroxy groups can also be introduced in a polymer by direct copolymerization of a peroxide containing monomer. For example, Smets et al. (20) copolymerized methyl acrylate and t-butyl peracrylate. The random copolymer which contained ~2% of the per-compound was heated in the presence of styrene monomer and a graft copolymer was obtained: poly(methyl acrylate-g-styrene).

Another way to introduce peroxy (perester) groups into polymers is by converting polymeric acid chlorides as follows:

~CH_2-CH~ (with pendant CO–Cl) —tBuOOH→ ~CH_2-CH~ (with pendant CO–O–O–tBu)

~CH_2-CH~ (with pendant CO–Cl) —COOOH–φ→ -CH_2-CH- (with pendant CO–O–O–CO–φ)

The t-butyl perester and perbenzoate of polyacrylic acid were found to be active in initiating the polymerization of styrene vinyl acetate (21) or methyl methacrylate (22).

In a somewhat similar method, neighboring poly (acryl chloride) groups have been reacted with perbenzoic acid to give polyperanhydrides and used to initiate the graft copolymerization of other monomers e.g. methyl methacrylate (23):

$$-CH_2-\underset{\underset{Cl}{|}}{\underset{CO}{|}}{CH}-CH_2-\underset{\underset{Cl}{|}}{\underset{CO}{|}}{CH}- \xrightarrow{\text{perbenzoic acid}} -CH_2-\underset{\underset{\diagdown O-O \diagup}{}}{\underset{CO}{|}}{CH} \overset{CH_2}{\diagup\;\diagdown} \underset{CO}{|}{CH}-$$

Still another technique to introduce labile peroxy bonds into the main chain of a polymer is by the use of polymeric phthalyl peroxides. Smets et al. (24) initiated the polymerization of styrene by the peroxide:

$$-O\cdot CO-C_6H_4-CO-O-O-CO-C_6H_4-CO\cdot O-$$

and obtained a polystyrene which contained terminal or in-chain peroxide groups. When solutions of this polymer were heated in the presence of methyl methacrylate, ethyl chloroacrylate etc., the residual peroxide groups ruptured giving rise to macroradicals of the first polymer which initiated the polymerization of the second monomer (24,25).

Since oxygen sometimes behaves as a diradical, it may be incorporated into a growing polymer chain. This "copolymerization of oxygen" leads to a peroxy linkage in the main chain. If these macromolecular peroxides are subsequently heated in the presence of available monomers, block copolymers may be obtained:

$$CH_2=\underset{X}{\underset{|}{C}}H + O_2 \longrightarrow \sim CH_2-\underset{X}{\underset{|}{C}}H-O-O-CH_2-\underset{X}{\underset{|}{C}}H-\sim$$

Ceresa (26) polymerized in the bulk methyl methacrylate in the presence of oxygen then swelled the polymer with styrene and heated the system to 70^0. The heat ruptured the peroxide links, initiated the polymerization of styrene and thus led to the formation of block copolymers. Various block copolymers obtained by Ceresa by this technique are reproduced in Table 1.

According to Orr and Williams (27,28) macrohydroperoxy groups useful for the synthesis of block copolymers can be synthesized by initiating the polymerization

Table 1

Vapor phase block copolymerization with peroxidic copolymers (26)

Block copolymer	Composition of the polymerization product			Poly-B in block copolymer
	Poly-A* %	Poly-B %	Block AB %	%
Poly(methyl methacrylate-b-styrene)	45	Nil	55	23.1
Poly(methyl methacrylate-b-acrylonitrile)	38	Nil	62	25.4
Poly(methyl methacrylate-b-vinylidene chloride)	57	Nil	43	29.6
Poly(methyl methacrylate-b-vinyl acetate)	48	23	29	35.6
Poly(styrene-b-methyl methacrylate)	27	Nil	73	32.2
Poly(styrene-b-acrylonitrile)	--	--	100 (crosslinked)	19.9
Poly(vinyl acetate-b-styrene)	--	--	100 (crosslinked)	22.2

*In this table poly-A refers to the polymer which is swollen with monomer B.

of a vinyl monomer by a m-diisopropylbenzene dihydroperoxide/Fe^{++}-pyrophosphate system. The following equations convey the gist of this method:

$$HOO\text{-}R\text{-}OOH + Fe^{++} \longrightarrow HOO\text{-}R\text{-}\dot{O} + OH^{\ominus} + Fe^{+++}$$

$$\downarrow +M$$

$$HOO\text{-}R\text{-}O\dot{M}$$

$$\downarrow +M$$

$$HOO\text{-}R\text{-}O\text{-}(M)_n\text{-}O\text{-}R\text{-}OOH \longleftarrow HOO\text{-}R\text{-}OMMM\dot{M} \text{ (in oil phase)}$$

b γ) <u>Diazo Groups</u>

Graft copolymers can be synthesized by converting aromatic amino groups to diazonium salts and decomposition by iron salts to phenyl radicals:

$$\text{percursor} \longrightarrow -CH_2\text{-}\underset{\underset{NH_2}{|}}{\overset{}{CH}}\text{(}C_6H_4\text{)}\sim \xrightarrow{\text{diaz.}}$$

$$\sim CH_2\text{-}CH(C_6H_4N^{\oplus}{\equiv}N\,Cl^{\ominus})\sim \xrightarrow{Fe^{++}} \sim CH_2\text{-}CH(C_6H_4^{\bullet})\text{-} + N_2 + Fe^{+++}$$

This method was first described by Valentine and Chapman (29) and later modified by Hahn and Fisher (30) who first produced the acetylamino derivative, converted it with NOCl to the N-nitroso-N-acetyl amino derivative which on decomposition also gave rise to phenyl radicals in the chain:

$$-CH_2-\underset{\underset{NH_2}{|}}{CH}(C_6H_4)- \longrightarrow -CH_2-CH(C_6H_4NHCOCH_3)- \longrightarrow\longrightarrow \text{NOCl} \quad -CH_2-CH(C_6H_4N(NO)COCH_3)- \longrightarrow -CH_2-CH(C_6H_4^{\cdot})-$$

The latter authors used this method to prepare poly (styrene-g-acrylonitrile) from poly(p-aminostyrene). A similar method was also used by Richards (31) and Simonescu et al. (32) who grafted various vinyl monomers onto cellulose derivatives.

b δ) Macromolecular Redox Initiators

A series of organic reducing agents e.g., alcohols, thiols, glycols, aldehydes, can be converted to radicals by one-electron transfer to ceric (IV) ion. For example, for alcohols:

$$Ce^{4\oplus} + R\text{-}CH_2\text{-}OH \rightleftharpoons [\text{ceric/alcohol complex}] \longrightarrow$$
$$Ce^{3\oplus} + H^{\oplus} + R\dot{C}HOH \quad (\text{or } RCH_2O^{\cdot})$$

Thus the reducing agent provides a free radical and may be utilized to initiate vinyl polymerization. In the absence of a suitable reducing agent, ceric salts (nitrate, sulfate) initiate the polymerization of acrylcnitrile but only after a long induction period. Mino and Kaizerman (33) employed first the ceric system to synthesize various graft copolymer systems of polyvinylalcohol backbone with polyacrylamide, polyacrylonitrile and poly(methyl acrylate) branches. These authors also mentioned the use of natural polymers like cellulose, starch, dextrane, etc. for backbones. The ceric ion technique leads to relatively clean graft copolymers since the free radical is formed only at the prepolymer so that homopolymerization is minimized. The chemistry of graft copolymerization of cellulose with redox systems has recently been summarized (34,35).

b ω) Trapped Radicals

Free radical polymerizations may sometimes lead to highly viscous and/or crosslinked products or to precipitated phases. Due to the diminished mobility of polymer sequences in such systems radicals may be trapped and when undisturbed, these radicals may survive for long periods of time. If now a second monomer is admitted, the still active radicals may restart the polymerization giving rise to block (or graft) copolymers.

This two step polymerization principle has been utilized by Sashoua and Van Holde (36) for the synthesis of graft copolymers. These authors first prepared a microgel of styrene containing a small amount of divinylbenzene, and after long aging used this gel to initiate the polymerization of acrylonitrile.

Trapped radicals also survive in "popcorn" polybutadiene which can initiate the polymerization of styrene (37) or methyl methacrylate (38).

III.1.B. Photolytic Methods

Chemical reactions can be initiated by light directly provided the irradiated molecules absorb optical energy. Selective absorption by well defined chemical groups of electromagnetic radiation in the visible and UV region may result in bond clevage and consequently in radical formation which in turn may lead to polymerization initiation. Block or graft copolymers can be synthesized by this general principle if radicals can be generated on preformed polymers by irradiation.

If none of the bonds in a molecule absorb light, indirect photolysis may be used. In these methods photosensitizers are added which absorb the light and are able to transfer light energy to other species in the system. Photosensitizers may also degrade into polymerization-initiating radicals or into radicals which are able to abstract atoms from other molecules converting the latters into initiating entities.

a. Direct Methods

According to Norrish et al. (39,40,41) aliphatic ketones are useful photosensitizers and they photodegrade by either of two mechanisms:

$$\text{Type I:}\quad R\text{-}CO\text{-}R' \xrightarrow{h\nu} R^{\cdot} + {}^{\cdot}COR' \longrightarrow R^{\cdot} + CO + R'^{\cdot}.$$

$$\text{Type II:}\quad R\text{-}CH_2CH_2CH_2\text{-}CO\text{-}R' \xrightarrow{h\nu} R\text{-}CH{=}CH_2 + CH_3\text{-}CO\text{-}R'$$

Later Norrish et al. (39) extended this basic chemistry for the preparation of graft copolymers. These authors photolyzed at 3130 Å poly(methylvinyl ketone) in the presence of other monomers e.g. methyl methacrylate (40) and acrylonitrile (41) and obtained a mixture of graft and homopolymers:

$$-CH_2\text{-}\underset{\underset{CH_3}{|}}{\underset{CO}{|}}{CH} - CH_2\text{-}\underset{\underset{CH_3}{|}}{\underset{CO}{|}}{CH}- \quad \xrightarrow{h\nu}$$

$$-CH_2\text{-}\underset{\dot{C}O}{|}{CH}\text{-}CH_2\text{-}\underset{\underset{CH_3}{|}}{\underset{CO}{|}}{CH}- + \dot{C}H_3 \xrightarrow{+\,M} \text{graft and homopolymer}$$

$$-CH_2\text{-}\dot{C}H\text{-}CH_2\text{-}\underset{\underset{CH_3}{|}}{\underset{CO}{|}}{CH}- + CH_3\dot{C}O \xrightarrow{+\,M} \text{graft and homopolymer}$$

Carbon-halogen bonds can also function as the labile site. The polymerization (thermal or UV) of styrene in the presence of CCl_3Br or CBr_4 give rise to polystyrenes with halogen-containing end-groups, for example:

$$CCl_3Br \xrightarrow[\text{or UV}]{\Delta} CCl_3^{\cdot} + Br^{\cdot} + n\ S \longrightarrow \begin{array}{l} Cl_3C\text{-}SSSS\text{-}Br \text{ or} \\ Br\text{-}SSSS\text{-}Br \text{ or} \\ Cl_3C\text{-}SSSS\text{-}CCl_3 \end{array}$$

Irradiation of the reaction product in the presence of methyl methacrylate then resulted in a mixture of homopolystyrene and poly(styrene-b-methyl methacrylate)(42).

Random bromination of polystyrene results in tertiary bromides in the chain which can readily photolyze by UV light. Thus Jones (43) prepared a polystyrene chain containing ~3% bromine and subsequently irradiated this macromolecular bromide in the presence of methyl methacrylate in benzene solution. In this manner he obtained an average of 1.2-5.9 poly(methyl methacrylate) branches per polystyrene backbone.

Photolyzable end-groups can be introduced to polystyrene by heating the monomer with certain sulfur compounds, e.g., tetraethyl thiuram disulfide (44):

$$(Et_2N-C(=S)-S)_2 + \text{Styrene} \longrightarrow$$

$$Et_2N-C(=S)-S-CH_2-CH(\phi)------$$

When the residual disulfide is removed and the polymer is irradiated in the presence of methyl methacrylate, graft copolymer, poly(styrene-g-methyl methacrylate), can be obtained.

b. Indirect Methods:

As mentioned above, photosensitizers upon irradiation may give rise to radicals which in turn may interact with a polymer in the system by removing an atom or group of the chain and thus ultimately provide radicals on polymer chains. Oster et al. (45) utilized this concept to graft acrylamide to natural rubber by benzophenone as sensitizer. Cast natural rubber films containing benzophenone were irradiated in the presence of aqueous acrylamide solution. The side of the film which was exposed to the acrylamide solution was found to be hydrophilic due to the formation of poly(cis-1,4-isoprene-g-acrylamide).

Cooper et al. (46) have studied the grafting of methyl methacrylate branches to natural rubber latex by UV radiation in the presence of a variety of sensitizers e.g., benzil, benzoin, 1-chloroanthraquinone, xanthone, etc. These authors defined important reaction variables for the preparation of poly(cis-1,4-isoprene-g-methyl methacrylate).

Grafting vinyl monomers to natural rubber latex in the presence of various photosensitizers e.g., ferric ion pairs ($FeCl_3/Fe(OH)_3$), hydrazine plus cupric ion, have been studied by Menon and Kapur (47).

Stannett et al. (49) synthesized grafts of cellulose and cellulose-derivatives by adsorbing photosensitive dyes (e.g., anthraquinone-2,7-disulfonic acid sodium salt) on these substances and irradiating them in the presence of monomers such as acrylamide. According to these authors, the photoexcited dye molecule abstracts a hydrogen atom from the cellulose and thus forms a free radical which initiates the polymerization:

$$^{\ominus}O_3S\text{-(anthraquinone)-}SO_3^{\ominus} \xrightarrow{h\nu} \text{(semiquinone diradical)} \xrightarrow{\text{cellulose}} \text{(}\dot{O}\text{, OH semiquinone)} + \text{cell}^{\cdot}$$

$$\xrightarrow{+M} \text{graft}$$

By this method these authors grafted a variety of branches e.g., acrylamide, acrylonitrile, methyl methacrylate, styrene, etc., onto cellophane and other polymers (49).

III.1.C. High Energy Irradiation Methods

High energy radiation produced by ^{60}Co sources or Van der Graff accelerators has been extensively explored in polymer science, both as a method for polymerization-initiation and polymer-derivatization. High energy irradiation methods can be used to produce sequential polymers by

a) Direct or Mutual Irradiation
b) Preirradiation

These, in principle only little different techniques can be further subdivided depending on the medium of irradiation, e.g., in vacuo or in the presence of oxygen, in emulsion or in solution, etc.

a) Direct or Mutual Irradiation Techniques

This technique involves the direct irradiation of polymer/vinyl monomer mixtures. The polymer is either dissolved or swollen by the vinyl monomer or sometimes the vinyl monomer is in the vapor phase. Graft copolymerization starts at the radical sites generated along the polymer backbone. The great advantage to this technique is its simplicity, however, disadvantages are simultaneous homopolymerization and radiation damage of the exposed preformed polymer.

The G values (the number of free radicals formed per 100 eV absorbed per gram) obtained experimentally for vinyl polymerizations can be qualitatively correlated with the G values of potential polymer backbones, although in some cases the G values calculated for polymers from homopolymerization data are much below the expected values. For example, the G value for natural rubber was found to be ~3 from crosslinking experiments, however, in the presence of styrene this value dropped to ~0.26. This "protective effect" could be due to energy transfer between the rubber and styrene monomer (50).

Diffusion of monomer into a polymer importantly affects sequential copolymer formation by the direct irradiation technique. Complications due to diffusion might arise when the polymer is crosslinked upon irradiation. By increasing the dose rate, simultaneous grafting and crosslinking could occur and monomer diffusion to the reactive site may become rate limiting. Or, the rate of grafting might increase autocatalytically due to the Tromsdorf effect, i.e., reduced rate of termination in viscous media.

Grafting is favored in amorphous regions or on the surface of materials where monomer can easily penetrate. For example, the grafting of styrene onto the surface of polytetrafluoroethylene, Teflon, an insoluble, unswellable, inert polymer, can be achieved by irradiating by γ-rays Teflon films immersed into styrene monomer (51).

Hoffman et al. (52) studied the grafting of styrene onto high and low density polyethylene and observed that grafting progressed long after irradiation was stopped. This "post-effect" is indication of the (generation and) survival of occluded active radicals in polymer matrices. This post-effect is useful in increasing grafting efficiency while protecting the preformed polymer from continued radiation-exposure. Thus by the "intermittent technique" the monomer is allowed to diffuse into the polymer which is subsequently exposed to relatively short bursts of irradiation, preferentially at low temperatures (53,54).

A large amount of experimentation with the direct irradiation technique involved the preparation of various rubber/vinyl monomer grafts. Polyisoprene has a relatively large G value (~17.9) and it crosslinks upon irradiation. The first step leading toward grafting probably involves the formation of polyisoprenyl plus hydrogen radicals. In the presence of methyl methacrylate, grafting commences on the rubber backbone. For some reason, the H· does not lead to the formation of free PMMA; evidently the MMA· formed by γ-rays reacts faster with cis-1,4-polyisoprene than with its own monomer, giving rise to high grafting-efficiency. Indeed, natural rubber/methyl methacrylate systems upon exposure to γ-rays yield only insignificant amounts of homo-PMAA. This is illustrated by some data (55).

Similarly, graft copolymer is obtained when a rubber latex is swollen with MMA and exposed to γ-rays. Grafts can be obtained under similar conditions by redox systems, however, γ-ray induced grafting gives much less homopolymer and the molecular weight of the PMMA branches is higher and its film-forming properties are also superior (more homogeneous product). Evidently, the redox systems have difficulty in uniformly diffusing through the swollen rubber/MMA system and consequently grafting is concentrated on the surface of the latex particles.

γ-Irradiation of Acetone Extracted Rubber/MMA Systems (55)

Dose rate[d]	Dose x 10^{-3}	Conv.% of MMA[a]	Product Analysis, %		
			Free Rubber	Graft	Homo-PMMA
5.6×10^3	5.6	2.5	68	32	0
	19	59	40	60	0
	56	93	20	79	1
	93	96	20	79	1
	370	96	16[b]	81	3
370×10^3	56	21	50	49	1
	93	36	47	53	0
	280	84	20	76	4
	650	99	10[c]	85	5

a = rubber containing 40% MMA
b = trace gel
c = 19% gel
d = rad/hr

Direct or mutual irradiation may also be carried out with two different, intimately mixed polymers. Non-selective irradiation will result in bond breaking which in turn might result in crosslinking-grafting and/or block formation, etc.

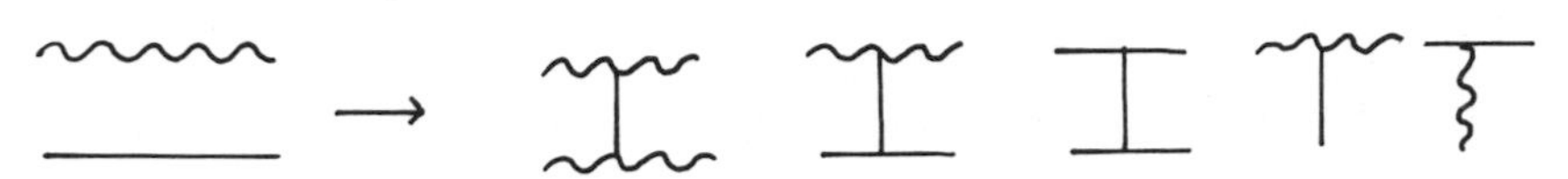

This is a very inefficient technique for the preparation of sequential copolymers and has not been studied extensively.

b) Preirradiation Techniques

Preirradiation techniques are based on the recognition that high energy irradiation is able to knock out electrons from a suitable preformed polymer and that the free radicals or peroxide groups, if preirradiation was performed in the presence of oxygen, are able to survive until polymerizable monomer is introduced in the system. Thus depending on the medium the polymer is preirradiated, we distinguish between two processes: preirradiation in air (oxygen) or in the absence of air (oxygen). Since the latter process involves the generation of immobile or trapped radicals, this method is sometimes called the trapped radicals method.

Preirradiation in the absence of air or other radical scavengers may result in the formation of trapped radicals. To maximize the generation of immobile radicals, i.e., radicals that cannot mutually destroy each other by combination, preirradiation should be preferentially performed below the Tg of the particular polymer. Grafting yield will depend on the efficiency of trapping the radical. After preirradiation and introduction of the monomer, the system can be heated to accelerate branch formation. Homopolymer formation is minimized since the monomer is not being irradiated directly.

Using this technique Bevington and Eaves (56) grafted polyacrylonitrile onto polystyrene, nylon-6,6 and poly (ethylene-g-styrene).

When polymers are preirradiated in air, due to the radical-scavenging action of oxygen, presumably, among other structures, peroxy groups are also generated:

$$-\overset{|}{\underset{|}{C}}-\overset{|}{\underset{H}{C}}-\overset{|}{\underset{|}{C}}- \xrightarrow{\gamma} -\overset{|}{\underset{|}{C}}-\overset{|}{\underset{\bullet}{C}}-\overset{|}{\underset{|}{C}}- \longrightarrow \text{other possibilities}$$

$$\downarrow +O_2$$

$$-\overset{|}{\underset{|}{C}}-\overset{|}{C}(-O-O^{\bullet})-\overset{|}{\underset{|}{C}}- \qquad \text{many other possibilities}$$

$$\xrightarrow{H\bullet} -\overset{|}{\underset{|}{C}}-\overset{|}{C}(-O-O-H)-\overset{|}{\underset{|}{C}}- \rightarrow \cdot OH \rightarrow \text{homopolymer}$$

$$\rightarrow -\overset{|}{\underset{|}{C}}-\overset{|}{C}-\overset{|}{\underset{|}{C}}- \;(O-O\text{ bridged})\; -\overset{|}{\underset{|}{C}}-\overset{|}{C}-\overset{|}{\underset{|}{C}}- \rightarrow 2\,-C-\overset{|}{C}(-O^{\bullet})-C- \rightarrow \text{graft polymer}$$

The peroxy groups are of reasonable stability, they can be stored and can be used in subsequent reactions to initiate grafting of vinyl monomers. Simultaneously with the peroxy groups, hydroperoxy groups are also formed that ultimately lead to homopolymer formation. Polypropylene is particularly prone to homopolymer formation. Evidently, the removal of the tertiary hydrogen from its backbone readily occurs under irradiation conditions. Homopolymer formation can be significantly reduced by the use of redox systems (58).

According to recent research by Sakurada et al. (59), the ^{60}Co radiation-induced grafting onto cellulose fibers by preirradiation in dry air is initiated by trapped radicals and not by peroxide groups. These results are based on the finding that the amounts (percents) of polystyrene grafts adducted to preirradiated cellulose were the same in vacuum or in air i.e., irrespective of the mode of preirradiation. ESR spectra of in vacuum or in air-irradiated cellulose were identical.

III.1.D. Mechanochemical Methods

The mechanochemical synthesis of sequential copolymers has been investigated in some detail by researchers at the British Rubber Producers Research Association and is concisely summarized in Ceresa in his book.

When a high polymer is subjected to aggressive shearing forces (mechanical, ultrasonic, etc.) nonrandom stress concentrations in parts of the molecule may exceed bond strengths and the polymer may degrade i.e., the polymer is "torn apart". Such degradations usually occur by a free radical mechanism. For example, cis-1,4-polyisoprene on milling degrades by a homolytic, radical-generating process. The free radicals formed from natural rubber during applied mechanical stress are most likely resonance stabilized allylic radicals arising by the rupture of $-CH_2-CH_2-$ bonds:

$$-CH_2-\overset{\overset{\displaystyle CH_3}{|}}{C}=CH-CH_2-CH_2-\overset{\overset{\displaystyle CH_3}{|}}{C}=CH-$$

$$\downarrow$$

$$-CH_2-\overset{\overset{\displaystyle CH_3}{|}}{C}=CH-\dot{C}H_2 \quad \dot{C}H_2-\overset{\overset{\displaystyle CH_3}{|}}{C}=CH-$$

In general, sequential copolymers can be prepared by

A) Subjecting a mixture of two or more polymers to mechanical degradation.
B) Subjecting a polymer to degradation in the presence of a polymerizable monomer.

The degradation of high polymers by free radical paths may be accomplished by a number of ways:

a) cold mastication, milling, extrusion above Tg
b) dispersing or vibro-milling below Tg of amorphous and crystalline polymers
c) ultrasonic irradiation of polymer solutions
d) high-speed stirring or shaking or forcing of polymer solutions through narrow orifices
e) the freezing-and-thawing of polymer solutions
f) the discharging of high-voltage sparks through polymer solutions
g) the swelling of crosslinked or high entangled polymers by the vapor phase

These methods will now be briefly highlighted.

Cold mastication, milling or extrusion of high polymers above their Tg and at high enough shear fields leads to homolytic degradation which manifests itself in a permanent molecular weight reduction. Degradation is initiated by nonrandom shearing, and polymer free radicals have been shown to be involved. When elastomers synthesized from conjugated dienes are masticated in the absence of radical acceptors, i.e., under pure N_2, the polymeric radicals arising by mechanical scission may attack unsaturations in the main chain and thus lead to branched products. Under certain conditions this process might even result in gelation. The rate of mechanical degradation is proportional to a) the initial molecular weight of the polymer, b) the plasticity of the system, c) the rate of shear. The limit to which degradation proceeds is dependent only upon the plasticity of the system and is independent of the initial molecular weights.

Comastication of blends of polymers e.g., natural rubber and chloroprene yields "interpolymers", polymers which contain long segments of the original components.

Similarly, comastication of polymer/monomer solutions e.g., natural rubber/methyl methacrylate systems, leads to shear degradation of the rubber and the block copolymerization of MMA by the radicals formed at the end of the degraded rubber segments.

Comastication of various rubbers and carbon black gives rise to gels of which the carbon black cannot entirely be removed. Evidently the rubber can be grafted onto carbon black surfaces by a radical process. Under masticating conditions, besides carbon black, all kinds of nonpolymerizable materials e.g., aluminum alkoxides, phenolic resins, etc. can also be bonded to rubber.

Dispersing (comminution) by vibro-milling of glassy or crystalline polymers might also result in sequential copolymer formation because free macroradicals can form at the fresh cleavage surfaces of solids. Polymers can be frozen in liquid N_2 and then exposed to vibro-milling. Inorganic materials e.g., quartz, NaCl, graphite, metals, etc., can be vibro-milled in the presence of monomers and "blocks" or "grafts" may be obtained. Polystyrene was grafted to SiO_2 in this manner.

Ultrasonic degradation of polymers under cavitation conditions leads to the generation of free macroradicals which, in the presence of monomers, can initiate the polymerization of the latters and thus give rise to sequential copolymers. Poly(styrene-g-methyl methacrylate) can be prepared by this method.

High speed stirring or shaking or forcing of polymer solutions through narrow orifices may result in polymer degradation and consequently in polymeric radicals. While it has been demonstrated that grafting may occur under these conditions there is a dearth of information in this field.

Freezing and thawing cycles of water occluded or imbibed into cavities of polymers might result in polymer degradation. For example, the freezing of aqueous solutions of starch or freezing starch swollen in water results in degradation and has been used to synthesize starch-g-polystyrene in emulsion systems.

The discharging of high-voltage sparks through polymer solutions under certain conditions may result in polymer degradation and consequently in block copolymer formation. Using this method poly(vinyl chloride-b-methyl methacrylate) was synthesized.

The swelling of crosslinked or highly entangled polymers with monomers might result in bond rupture and consequently to block copolymerization of the swelling agent.

All these mechanochemical methods in one way or another degrade preformed polymers and then build up new polymers by assembling the fragmented segments. This then is a three-step process: first, the polymerization to produce the starting polymer; second, the degradation under more-or-less controlled conditions; third, the reassembling of fragments or initiation by fragments, again under more-or-less controlled conditions. Indeed, a disadvantage of these mechanochemical methods is in the difficulty in controlling steps two and three. Consequently, the products synthesized by these processes are more often than not ill-defined mixtures and combinations of all kinds of polymer fragments.

III.2. Synthesis by Ionic Mechanism

A large amount of research has been carried out in the field of ionic block and graft copolymerization. Indeed, ionic synthesis, provide excellent methods for the preparation of sequential copolymers.

III.2. A. Synthesis by Anionic Methods

The synthesis of sequential polymers by anionic mechanisms have been reviewed and summarized recently by several authors; the interested reader should consult the following main references 60-68.

Since the discovery of "living" polymerizations (60) anionic methods have become uniquely suited for the preparation of well characterized block and graft copolymers. True, preparative difficulties of this technique are sometimes formidable, nontheless, if the aim is the synthesis of blocks or grafts of precisely defined structures and free of homopolymers, only anionic techniques should be considered. For the fundamentals of living anionic polymerizations which permit the controlled synthesis of well-defined homopolymers, the reader should study appropriate references.

The controlled synthesis of block copolymers became possible only by living polymerizations. Only after this synthesis method became available have polymer chemists been able to "tailor make" certain block (and graft) copolymers. However, a serious limitation of living anionic polymerizations is the very small number of monomers (styrene, dienes) polymerizable to well-defined structures by this technique.

There are, in principle, two ways to prepare block copolymers and two ways to prepare graft copolymers via anionic mechanism:

Block syntheses:

Sequential addition: Living polymerization of monomer A is initiated and after the complete consumption of monomer A, monomer B is added; living polymer B continues the polymerization and an A-B block copolymer is obtained.

Coupling: Living homopolymers A and B are prepared separately and are subsequently coupled by a suitable step.

Graft syntheses:

Grafting from: Anionic sites capable of initiating the polymerization of vinyl monomer B are generated on preformed polymer A chains. Subsequently monomer B is introduced which commences the polymerization on backbone polymer A.

Grafting onto: Living polymer sequence A is prepared. Polymer sequence B containing sites able to link up with living polymer A is prepared separately. Finally living polymer A and reactive polymer B are mixed to achieve coupling between the sequences.

These synthetic methods will now be surveyed.

a) Synthesis of Block Copolymers by Sequential Addition Technique

When using this method it is essential to find suitable monomer pairs so as to initiate the polymerization of one monomer with an anion generated from another monomer. For example, living polystyrene (e.g., polystyril lithium) initiates the polymerization of methyl methacrylate, but anionically growing methyl methacrylate cannot initiate the propagation of styrene. The styrene-isoprene pair is of interest since both growing anions, polystyryl or polyisoprenyl, are able to initiate the polymerization of the other monomer. A whole series of "isomeric" copolymers can be produced from these two monomers: If one polymerizes first styrene and then adds isoprene, a di-block S-I is obtained. However, if one introduces intermittently styrene and isoprene to a living system, the final product is a multi-block S-I-S-I--- structure.

It is well known that certain complexing agents, e.g. ethers, change the microstructure of butadiene polymerization initiated by organolithium compounds. This phenomenon can be exploited for the preparation of special stereoblock copolymer: One starts the polymerization of butadiene in hydrocarbon solvent and produces 1,4 enchainment; at a certain instance ether is added which changes the microstructure to largely 1,2 units. Thus one obtains a block copolymer of 1,4 and 1,2 polybutadiene sequences.

Chain transfer cannot be tolerated in the sequential addition technique. Thus polymerizations involving methyl methacrylate or acrylonitrile are suspect because these monomers have a tendency for chain transfer.

The best-initiators for the sequential technique seem to be organolithium compounds or organosodium or -potassium derivatives. The most popular monomers, particularly for the synthesis of linear sequential elastomers, are styrene, isoprene and/or butadiene because these monomers polymerize with few if any side reactions and termination can be completely eliminated.

The first step in the synthesis of a linear elastomeric triblock structure with a monofunctional initiator is the initiation of styrene (hard segment) with an organolithium compound e.g., sec.-BuLi in hydrocarbon medium. This reaction is rather slow, probably because of the association of organolithium molecules in hydrocarbon solution. The second step, after the polymerization of styrene is over, is the introduction of the diene monomer (soft or rubbery segment). The crossover from the polystyryl anion to the dienyl anion is fast. The final step is the addition of styrene. The crossover from the polydienyl anion back to the styryl anion is slow but it can be accelerated by the addition of small amounts of ethers e.g., tetrahydrofuran.

"Difunctional" initiators such as sodium naphthalene or organic dilithium compounds can be used advantageously for the preparation of linear triblock elastomers. In this technique a difunctional initiator induces the polymerization of a monomer, say isoprene, and after its consumption, the second monomer is added whose polymerization is initiated by the bifunctional active first polymer segment. By this technique only one sequential addition is needed to produce a triblock structure so that adventitious termination by impurities is minimized.

Initiation in these systems can be visualized by the following equations:

$$Na + C_{10}H_8 \longrightarrow \dot{C}_{10}H_8^{\ominus}\,Na^{\oplus} \quad + \xrightarrow{CH_2=CH-\phi}$$

$$C_{10}H_8 + \dot{C}H_2-\underset{\phi}{\overset{\ominus}{C}H}\;Na^{\oplus} \xrightarrow{comb.}$$

$$Na^{\oplus}\;\underset{\phi}{\overset{\ominus}{C}H}-CH_2-CH_2-\underset{\phi}{\overset{\ominus}{C}H}\;Na^{\oplus}$$

b) Synthesis of Block Copolymers by Coupling

The bulk of research in this field has been directed toward the synthesis of linear elastomeric triblock structures. Numerous coupling possibilities for anionic chain ends have been uncovered. One of the simplest but most efficient coupling reaction for the linking of poly(styrene-b-butadienyl) anion is the reaction with a small organic dihalide e.g., methylene chloride or bromide:

$$(PS)_x\!-\!(PBd)_m^{\ominus}\ Li^{\oplus} + Br\!-\!CH_2\!-\!Br + Li^{\oplus}\ {}^{\ominus}(PBd)_n\!-\!(PS\ )_y \longrightarrow$$

$$(PS)_x\!-\!(PBd)_{m+n}\!-\!(PS)_y + 2\ LiBr$$

Since the residual $-CH_2-$ in the triblock contributes an insignificant amount to the overall chain, one may view the product as a true block copolymer.

Besides methylene halides a variety of coupling agents have been described e.g., epoxides, phosgene, anhydrides, carbon monoxide, etc. In case dianions are coupled, this technique gives rise to multiblock structures:

$$^{\ominus}A\!-\!B\!-\!A^{\ominus} + R \rightarrow {}^{\ominus}A\!-\!B\!-\!A\!-\!R\!-\!A\!-\!B\!-\!A^{\ominus}$$

where R = coupling agent, and A and B are suitable polymeric sequences.

Triblock copolymers, for example poly(styrene-b-butadiene-b-styrene), have recently become commercial commodities. These products behave as crosslinked rubbers up to $\sim 80^0$ and find use in toys, shoe soles, etc. (thermoplastic elastomers). In these materials the glassy polystyrene domains fulfill two functions: they act as physical crosslinks and as reinforcing fillers. In other words the elastic network is held together by the glassy polystyrene domains which act as crosslink points (crosslinking by "physical" forces) and as finely dispersed reinforcing agents. These blocks should be monodisperse. Commercially available Kraton (Shell Chemical Co.) is a polystyrene-polydiene-polystyrene triblock copolymer probably made by this coupling technique.

c) Synthesis of Graft Copolymers by "Grafting From" Technique

In this technique one introduces reactive anionic sites into a polymer chain and initiates the polymerization of subsequently added vinyl monomers. A large amount of work has been spent on the exploitation of this principle. Convenient techniques have been described for the preparation of reactive site-carrying polymers. Most methods involve metalation by complexed organolithium compounds of hydrocarbon polymers. For example, in one of the most recent investigations Halasa (69) metalated polybutadiene or polyisoprene by nBuLi·TMEDA (tetramethylethylenediamine complex of n-butyllithium) and used the lithiated rubbers to initiate the polymerization of styrene. This author also metalated poly(butadiene-*co*-o- or p-chlorostyrene) and was able to control the amount and site of metalation by directing it exclusively to the aromatic ring. (The exchange of aromatic Cl for Li is a facile reaction). In an experiment the original copolymer contained 3-5% of o-chlorostyrene and when this random copolymer was lithiated and used to initiate the anionic polymerization of styrene essentially pure graft copolymer formed.

An interesting technique for grafting from -CN group-containing polymers was described by Greber (68). This author reacted polyacrylonitrile or its copolymers with BuMgCl and was able to graft from the addition product acrylonitrile, vinylpyridine and methyl methacrylate:

$$\sim\text{CN} + \text{BuMgCl} \longrightarrow \text{Bu}\overset{\sim}{\text{C}}\text{=N-MgCl} + \text{monomers} \longrightarrow \text{Bu-}\overset{\sim}{\text{C}}\text{=N-Polymer-MgCl}$$

d) Synthesis of Graft Copolymers by "Grafting Onto" Technique

By extending the principle of the synthesis of star-shaped homopolymers, one can easily envision synthetic routes for the preparation of star-shaped copolymers, e.g. graft copolymers. Star polymers are macromolecules having more than one branch originating from a single site:

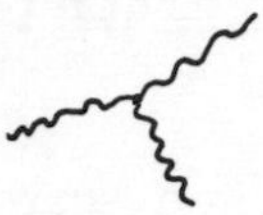

or

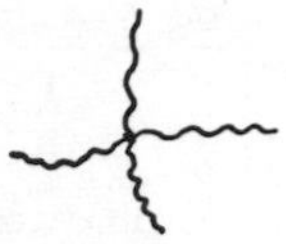

Star polymers can be synthesized by the use of multi-functional initiators, or by the multiple coupling of living polymers with a multi-functional coupling agent. For example, several living polystyrenes can be collectively joined with

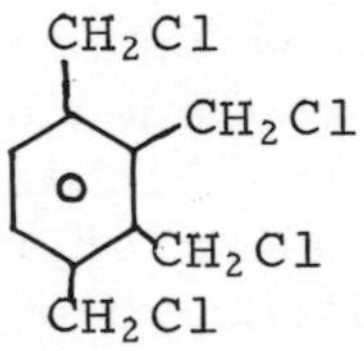

CH_2Cl CH_2Cl CH_2Cl CH_2Cl

Another possibility is to join living polystyrene chains with phosgene (ClCOCl) and then reacting the polymeric ketone with another growing chain:

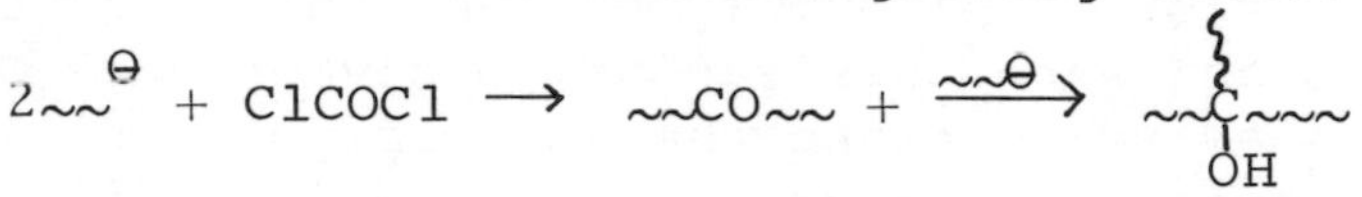

$$2\sim\sim^{\ominus} + \text{ClCOCl} \longrightarrow \sim\sim\text{CO}\sim\sim + \xrightarrow{\sim\sim\ominus} \sim\sim\text{C}(\text{OH})\sim\sim\sim$$

The OH group can be converted to bromine; subsequent coupling with another growing living polymer produces a four-armed star-shaped polymer.

Star-polymers with dissimilar branches could perhaps be synthesized by collectively joining various living polymers (e.g., living polystyrene and polydienes) with a multifunctional coupling agent.

Comb-shaped copolymers e.g., structures in which the branches ("teeth" of the "comb") are regularly spaced along the backbone, can also be prepared by this technique:

$$\ominus\sim\sim\sim\ominus + ClCOCl \longrightarrow \sim CO\sim CO\sim CO\sim \xrightarrow{\sim\sim\sim\ominus} \sim C\sim C\sim C\sim$$

Conventional graft copolymers e.g., grafts with randomly spaced branches, become available by interacting growing anions with suitable backbone sequences. For example, a living polystyrene chain may be reacted with a preformed polymethyl methacrylate or polyvinyl chloride:

$$-CH_2-\underset{COOCH_3\ (\text{or Cl})}{\overset{CH_3}{C}}- + -CH_2-\overset{\ominus}{C}(\phi)\ K^{\oplus} \longrightarrow -CH_2-\overset{CH_3}{C}(-CO-CH(\phi)-CH_2-)- + CH_3OK\ (\text{or KCl})$$

Experiments have shown that while all the preformed polymers became grafted, 20-40% of the polystyrene did not interact.

III.2. B. Synthesis by Cationic Methods

An advantage of cationic polymer synthetic techniques over those of anionic methods is the very large number of potentially available monomers. Cationic techniques have been explored in the past for the synthesis of sequential copolymers but only very recently have there been reports according to which essentially pure blocks or grafts could be prepared. Among cationic polymerizations, only oxonium ion-initiated polymerizations of cyclic ethers have led to block copolymers and block copolymers of hydrocarbons by cationic synthesis is yet unknown. In contrast, only hydrocarbon graft copolymers synthesized by carbenium ion techniques have been described and cationic grafting of cyclic ethers has not yet been studied.

a) <u>Synthesis of Block Copolymers by Oxionium Ion Techniques: Sequential Addition</u>

Living cationic, more precisely oxonium ion-initiated polymerizations have been described and have been applied for the synthesis of block copolymers. Since no living carbenium ion-initiated polymerizations are yet known, it is not too surprising that block copolymers of hydrocarbons prepared by cationic techniques have not yet been studied.

The first report of a sequential addition synthesis of a poly(tetrahydrofuran-b-3,3-bischloromethyloxetan) was described by Saegusa et al. (70). This material was prepared by first polymerizing THF with

$BF_3 \cdot CH_2-CH-CH_2Cl$ (with an O bridge between CH_2 and CH) in n-heptane at 0^0 and after complete polymerization of the THF charge the second monomer

$ClCH_2$ $ClCH_2$ (oxetane ring with O) was added.

b) <u>Synthesis of Block Copolymers by Oxonium Ion Techniques: Coupling</u>

Berger et al. (71) coupled polytetrahydrofuran cations with polystyryl anions and obtained poly(THF-b-S). Similarly, by coupling a di-initiated living polystyrene with living polytetrahydrofuran oxonium ions these authors prepared poly(THF-b-S-b-THF). The products were characterized by fractional precipitation. An improved method of the same principle was used by Yamashita et al. (72) to produce multiblock poly(THF-S) elastomers.

c) <u>Synthesis of Graft Copolymers by Carbenium Ions: "Grafting Onto"</u>

The most recent research in which the principle of cationic chain-transfer-to-the-polymer has been utilized was carried out by Overberger and Burns (73). These authors first produced the backbone poly(2,6,-dimethoxystyrene) and used it as a transfer agent in a styrene polymerization initiated by $SnCl_4$ in nitrobenzene solvent.

While the formation of poly(2,6-dimethoxystyrene-g-styrene) has been demonstrated, grafting efficiencies were low, i.e., 15-40%.

d) Synthesis of Graft Copolymers by Carbenium Ions: "Grafting From"

Recently Kennedy (74) developed a synthesis which gives rise to essentially pure graft copolymers. This method is based on the discovery that certain alkylaluminum compounds i.e., Et_2AlCl, Et_3Al, initiate the polymerization of cationically active monomers only in conjunction with purposely added alkyl halides e.g., t-BuX. According to this technique, suitable polymeric halides can initiate the polymerization of various monomers e.g., isobutylene, styrene, dienes, etc. In a typical synthesis lightly (~3%) chlorinated poly(ethylene-co-propylene) is stirred with styrene monomer and Et_2AlCl is added. Grafting starts immediately and in ~30 mins. the reaction is complete. Besides this graft of poly[(ethylene-co-propylene)-g-styrene], a series of other grafts have also been prepared, e.g., poly[(isobutylene-co-isoprene)-g-styrene], poly(vinyl chloride-g-isobutylene), etc.

Among the interesting materials that can be produced by this technique are thermoplastic rubbers comprising poly(ethylene-co-propylene) (EPM rubber) and poly(isobutylene-co-isoprene) (butyl rubber) backbones with polystyrene branches. Graft formation can be readily demonstrated by selective elution. For example, polystyrene is insoluble in n-pentane but it is soluble in acetone; butyl rubber represents the opposite situation: it is soluble in n-pentane but insoluble in acetone. Mixtures of homopolymers of polystyrene and butyl rubber can be readily separated by these two solvents. On the other hand butyl rubber-g-polystyrene is insoluble in acetone and homopolystyrene can be readily removed by this technique. The same is true for EPM rubber-g-polystyrene) obtained by cationic synthesis.

Interestingly these polymers are soluble in n-butanone (MEK) which is an excellent solvent for polystyrene but a nonsolvent for polyisobutylene or poly (ethylene-co-propylene) i.e., the backbone polymers of the above two grafts. Evidently, the solubility of the polystyrene branches in MEK is so high that the branches are able to "pull into solution" the insoluble backbones. The structure of these species in MEK solution is visualized to consist of extended polystyrene branches attached to "quasi-precipitated" backbones:

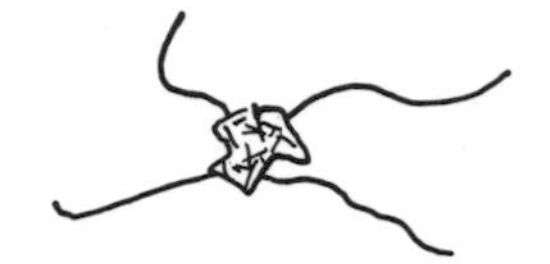

On the opposite side, these graft copolymers are also soluble in n-pentane, i.e., in the solvent only the backbone is soluble but not the branches. In this case the graft is being kept in solution by the n-pentane soluble backbones and the n-pentane insoluble polystyrene branches are "pulled into solution". In this case the branches are in a "quasi-precipitated" state:

The viscosity/temperature behavior of these species is remarkable.

The physical properties of solid sequential polymers may also be determined by the manner they have been isolated from solution. A polymer chain in solution may be extended or coiled depending on the "goodness" of the solvent. In the examples mentioned, in acetone the polystyrene chains are extended and the polyisobutylene chains are coiled. When the acetone solvent is removed by slow evaporation, the continuous solid phase will be the rigid polystyrene and therefore the properties of the dry solid will resemble this plastic. In contrast, when a n-pentane solution is evaporated, the continuous phase which precipitates last will be the rubbery polyisobutylene, and the properties of the dry solid will resemble that of a rubber rather than a rigid plastic. This phenomenon was first observed by Merrett (75) with a natural rubber-g-polymethyl methacrylate system. This graft copolymer is soluble in benzene and when benzene solutions are evaporated, rubbery products are recovered (rubber chains extended). However, when the graft is precipitated with methanol or petroleum ether a rigid material obtains (PMMA chains extended). The intrinsic viscosity of the product in benzene is strongly reduced (from ~1.5 to ~0.2) upon increased amounts of methanol additions.

The papers by Kennedy and Smith (76) and Kennedy, Charles and Davidson (77) give further details on this technique and the materials that can be obtained.

III. 3. Coordinative Methods

Stereospecific catalysts can give stereoblock copolymers. For example, one can start the polymerization of butadiene with an organolithium compound in hydrocarbon solvent and produce largely cis-1,4-enchainment. Upon the addition of ether e.g., THF the steric course of the propagation changes and a mixture of linkages, mainly 1,3, but also trans-1,4 and 1,2, are obtained. Stereoblock polypropylene has been described. This product, obtained by a typical Ziegler-Natta type catalyst e.g., $Et_3Al/TiCl_3$, contains isotactic and heterotactic sequences.

An ingenious method has been developed by Greber (66) for the synthesis of grafts onto poly(styrene-co-butadiene) and other polymers containing pendent vinyl groups using Ziegler-Natta coordination catalyst. This author first reacted diethylaluminum hydride with the potential backbone obtained and a macromolecular trialkylaluminum:

$$Et_2AlH + \sim CH{=}CH_2 \longrightarrow \sim CH_2{-}CH_2{-}Al(Et)_2$$

This trialkylaluminum was then employed in the preparation of a Ziegler-Natta catalyst with $TiCl_4$ or $TiCl_3$, etc. and used in the polymerization of ethylene, propylene or other α-olefins. If the Ti was alkylated not only by the Et groups but also by the polymer, one can visualize the grafting of poly-α-olefin branches onto the preformed macromolecule:

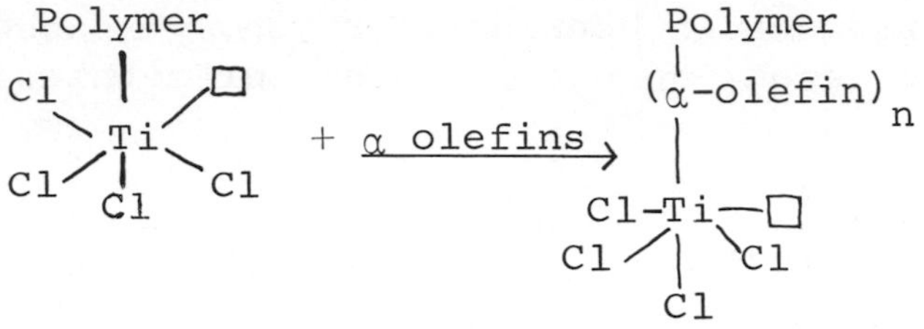

$$\mathrm{Cl_3Ti(Polymer)}\square \ \xrightarrow{+\ \alpha\ \text{olefins}}\ \mathrm{Polymer{-}(\alpha\text{-olefin})_n{-}TiCl_3}\square$$

Grafting was demonstrated by extraction techniques. Among the various grafts synthesized by this technique were poly[styrene-co-butadiene)-g-ethylene or isotactic propylene].

By using molecules with two terminal vinyl groups, this basic technique can also be used for the synthesis of block copolymers. For example, Greber described the following process:

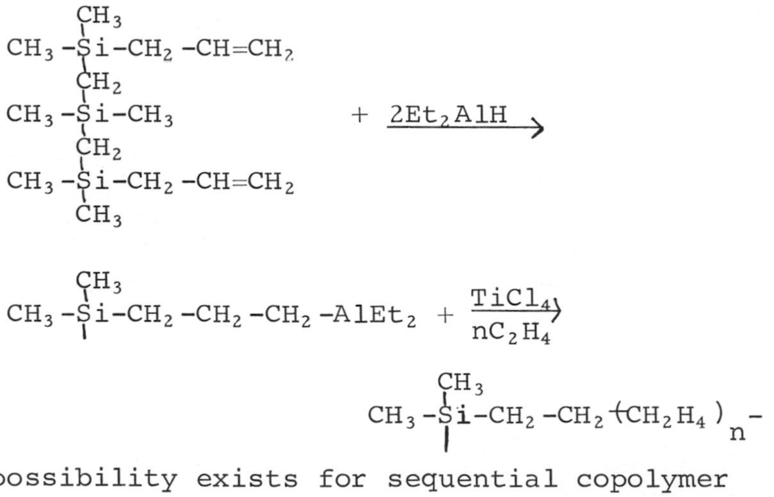

$$\mathrm{CH_2{=}CH{-}CH_2{-}Si(CH_3)_2{-}CH_2{-}Si(CH_3)_2{-}CH_2{-}Si(CH_3)_2{-}CH_2{-}CH{=}CH_2} + \xrightarrow{2Et_2AlH}$$

$$\mathrm{CH_3{-}\overset{CH_3}{\underset{|}{Si}}{-}CH_2{-}CH_2{-}CH_2{-}AlEt_2} + \xrightarrow[nC_2H_4]{TiCl_4}$$

$$\mathrm{CH_3{-}\overset{CH_3}{\underset{|}{Si}}{-}CH_2{-}CH_2{-}(CH_2H_4)_n{-}}$$

The possibility exists for sequential copolymer synthesis also with the recently discovered tungsten-based transalkylidination catalysts (78).

III.4. Miscellaneous Ring Opening and Coupling Reactions

Polymers containing a variety of acidic or basic groups, e.g., -OH, -COOH, $-NH_2$, -SH, etc. can be used as potential backbones for the synthesis of sequential copolymers. Block copolymers will be obtained if the functional groups are at the termini, whereas grafts will be produced if they are randomly distributed along the chain.

III.4. A. Ring Opening Reactions

Polymers containing active hydrogens e.g., -OH, $-NH_2$, etc. may initiate the polymerization of ethylene oxide, propylene oxide, β-propiolactone, ε-caprolactam. For example, styrene is polymerized with a small amount (~2%) of vinyl acetate and the acetate groups in the product are hydrolyzed. The -OH groups of this copolymer are capable of initiating the polymerization of ethylene oxide (79):

$$\sim CH_2-\underset{\substack{|\\ C_6H_5}}{CH}-CH_2-\underset{\substack{|\\ O-(CH_2CH_2O)_n}}{CH}-$$

A series of interesting block and star-shaped copolymers have been developed commercially. These block copolymers were synthesized by initiating the polymerization of propylene oxide by a base so that both the head-group and the end-group become hydroxy units. This double-headed hydrophobic sequence is then used to initiate the polymerization of ethylene oxide to obtain a hydrophilic sequence. The product is a triblock copolymer poly(ethylene oxide-b-propylene oxide-b-ethylene oxide):

$$HO(EtO)(PO)(EtO)H$$

These materials are nonionic detergents and are commercially available (Pluronics by Wyandotte Chemical Co.).

The polymerization of propylene oxide can also be initiated by ethylene diamine:

$$NH_2-CH_2CH_2-NH_2 + \underset{\diagdown O \diagup}{\overset{CH_3}{CH}-CH_2} \longrightarrow$$

$$\begin{matrix} H-(PO) \\ H-(PO) \end{matrix} \rangle N-CH_2-CH_2-N \langle \begin{matrix} (PO)-H \\ (PO)-H \end{matrix}$$

Further, the terminal hydroxyl groups of this four-branch star-shaped molecule can initiate the polymerization of ethylene oxide giving rise to a four-branch star-shaped block copolymer:

$$\begin{matrix} H-(EtO)-(PO) \\ H-(EtO)-(PO) \end{matrix} \rangle N-CH_2-CH_2-N \langle \begin{matrix} (PO)-(EtO)-H \\ (PO)-(EtO)-H \end{matrix}$$

These materials are also available commercially (Tetronics by Wyandotte Chemical Co.).

Polyethylene oxide branches can be readily attached to nylons since the hydrogen atoms of the amide nitrogen are easily removed.

$$\sim\sim CONH \sim\sim \xrightarrow{+\ n\ \nabla_O} \sim\sim CO-\underset{\underset{(CH_2CH_2O)_n-H}{|}}{N}\sim\sim$$

Preparatively (80) bulk nylon and liquid EtO are heated together in a bomb since nylon solutions are reluctant to react. Only a fraction (0.1-0.4) of the amide groups react and the DP of the polyethylene oxide branches is quite low 3-17.

Polyacrylamide can also initiate the grafting of ethylene oxide (81).

The functional groups in cellulose and polyvinyl alcohol are also able to initiate the grafting of ethylene oxide, β-propiolactone and ethylene imine. The cellulose-g-polyethylene oxide is a commercial product (Ethylose of the Rayonier Co.). The branches in these materials, however, are very short so that they are rather modified polymers than graft copolymers.

β-Propiolactone can be polymerized by functional groups containing active hydrogens and as such this cyclic lactone can be grafted onto a number of polymers, e.g., cellulose fiber, wool, poly(acrylonitrile-co-acrylic acid) (82).

$$\sim\!\!\sim CH_2-\underset{CN}{CH}-CH_2-\underset{COOH}{CH}\sim\!\!\sim \; + \; \text{(β-propiolactone)} \; \rightarrow \; \sim\!\!\sim CH_2-\underset{CN}{CH}-CH_2-\underset{\substack{| \\ CO \\ | \\ O \\ | \\ CH_2CH_2-COO\sim\sim}}{CH}\sim\!\!\sim$$

III.4. B. Miscellaneous Couplings

For completeness sake a few miscellaneous coupling methods for the preparation of sequential copolymers are also mentioned. For details concerning the chemistries of polyurethanes, polyepoxides and copolyesters the reader is referred to the very voluminous available literature, the surveying of which is outside the scope of this chapter.

Sequential copolymers can be obtained by coupling two or more polymers carrying suitable functional groups. If the linking-up groups are terminal, blocks are formed; if they are in-chain, grafts are obtained. For example, isocyanates R-N=C=O readily react with a variety of groups containing active hydrogens e.g., -OH, -COOH, $-NH_2$, -NH-, etc. Thus isocyanates can be linked to a large number of materials and by the use of diisocyanates O=C=N-R-N=C=O these materials may be coupled. The coupling of -OH containing polymers results in polyurethanes:

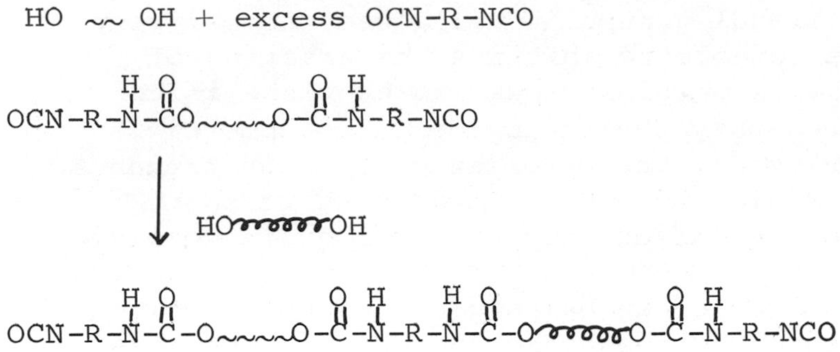

$$HO \sim\sim OH + \text{excess } OCN\text{-}R\text{-}NCO$$

$$OCN\text{-}R\text{-}\overset{H}{N}\text{-}\overset{O}{\overset{\|}{C}}O\sim\sim\sim O\text{-}\overset{O}{\overset{\|}{C}}\text{-}\overset{H}{N}\text{-}R\text{-}NCO$$

$$\Big\downarrow \; HO\text{∿∿∿}OH$$

$$OCN\text{-}R\text{-}\overset{H}{N}\text{-}\overset{O}{\overset{\|}{C}}\text{-}O\sim\sim\sim O\text{-}\overset{O}{\overset{\|}{C}}\text{-}\overset{H}{N}\text{-}R\text{-}\overset{H}{N}\text{-}\overset{O}{\overset{\|}{C}}\text{-}O\text{∿∿∿}O\text{-}\overset{O}{\overset{\|}{C}}\text{-}\overset{H}{N}\text{-}R\text{-}NCO$$

Commercially available polyurethanes are often based on medium molecular weight (mol. wt. 1000-3000) polyether glycols obtained from polypropylene oxide or copolymers of propylene oxide plus ethylene oxide. Polyurethane elastomers are prepared from polyesters obtained, for example, of ethylene glycol and adipic acid:

$$HO\text{—}CH_2CH_2\text{-}O\text{-}\left[\overset{O}{\overset{\|}{C}}\text{-}CH_2CH_2CH_2CH_2\text{-}\overset{O}{\overset{\|}{C}}\text{-}O\text{-}CH_2CH_2O\right]\text{-}H$$

which are then reacted with diisocyanates.

Isocyanate couplings are very versatile techniques for the synthesis of all kinds of block and graft copolymers. For example, coupling of isocyanates with amine groups leading to urea linkages can be used to prepare a large family of sequential copolymers. The $-NH_2$ group can be obtained as a head-group in polystyrene by initiating the polymerization of styrene by $NaNH_2$; the isocyanate groups can be introduced into vinyl polymers by copolymerizing vinyl monomers with

$$CH_2{=}\overset{CH_3}{C}\text{—}COO\text{-}CH_2CH_2\text{-}N{=}C{=}O \quad (83):$$

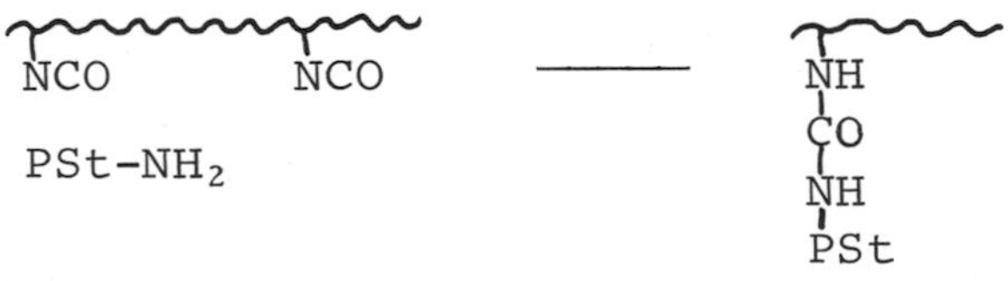

$$\underset{NCO}{\sim}\sim\sim\sim\sim\underset{NCO}{\sim}\sim\sim \quad \xrightarrow{PSt\text{-}NH_2} \quad \underset{\substack{NH\\ |\\ CO\\ |\\ NH\\ |\\ PSt}}{\sim}\sim\sim$$

This technique was used to graft polystyrene branches onto copolymers of methyl methacrylate plus β-isocyanate ethyl methacrylate (83).

Epoxy resins containing terminal (and/or in-chain) epoxy groups and secondary hydroxy groups (prepared, for example, by the condensation of epichlorohydrin with 2,2'-bis (p-hydroxyphenyl) propane) may react with a variety of functional groups in polymers capable of opening epoxy rings, to form linear or crosslinked products. Thus epoxy resins can be cured with polyamides containing $-NH_2$ or -NH- groups, with phenol-formaldehyde resins containing $-CH_2OH$ groups, with partially hydrolyzed poly(vinyl chloride-co-vinyl acetate) containing -CH-OH groups, etc., etc., to give a wide variety of materials, way beyond the scope of this chapter:

$$\left.\begin{matrix}\sim\sim SH\\ \sim\sim OH\\ \sim\sim COOH\end{matrix}\right. + \underset{\diagdown O \diagup}{CH_2-CH}-CH_2-O-C_6H_4-\overset{CH_3}{\underset{CH_3}{\overset{|}{\underset{|}{C}}}}-C_6H_4-O\sim\sim\sim \longrightarrow$$

graft or block copolymer

Block copolyesters can be prepared by coupling two different polyester sequences containing reactive terminal groups:

$$HO-(CH_2)_{10}-OCO-C_6H_4-COCl + ClCO-C_6H_4-CO-O-(CH)_{10}-OH \longrightarrow$$

$$HO-(CH_2)_{10}-OCO-C_6H_4-CO-O-(CH_2)_{10}-O-CO-C_6H_4-COCl \longrightarrow \text{etc.}$$

Polyethylene oxide (PEtO) blocks can be prepared by ester interchange between PEtO and suitable polyesters e.g., polyethylene terphthalate (PET) (84):

$$HO[-CH_2CH_2O]_{20-120}H + HO-CH_2-CH_2-O[-CO-C_6H_4-CO-CH_2CH_2O]_{100}H$$

$$\downarrow 275^0,\ EtOH$$

$$PET-PEtO-PET-PEtO$$

Acknowledgement: Financial support by the National Science Foundation in the form of grant GH-37985 is gratefully acknowledged.

IV. References

Books

1. "Block and Graft Copolymers" by R. J. Ceresa; Butterworths, London, 1962.

2. "Block and Graft Copolymers" by W. J. Burlant and A. S. Hoffman, Reinhold Publishing Co., New York, 1960.

3. "Graft Copolymers" by H. A. J. Battaerd and G. W. Tregear, Interscience Publishers, New York, 1967.

Review Articles, Review Volumes

1. S. Krause, J. Macromol. Sci., Reviews in Macromol. Sci., C7, 252 (1972).

2. L. Bohn, Kolloid Z., 213, 55 (1956).

3. "Block Copolymers" Edited by J. Moacanin, G. Holden, and N. W. Tschoegl; J. Polymer Sci., Part C, No. 26, 1969.

4. "Proceedings of the Symposium on Graft Polymerization onto Cellulose" Edited by J. C. Arthur; J. Polymer Sci., Part C, No. 37 (1972).

References

1. G. Smets, J. Roovers and W. van Humbeck, J. Appl. Polymer Sci., 5, 149 (1961).

2. P. V. Allen, G. Ayrey, C. G. Moore and J. Scanlan, J. Polymer Sci., 36, 55 (1959).

3. G. Henrici-Olive, S. Olive and G. V. Schulz, Makromol. Chem. 23, 207 (1957).

4. J. F. Voeks, J. Polymer Sci., 18, 123 (1955).

5. M. S. Gluckman, M. J. Kampf, J. L. O'Brien, T. G. Fox and R. K. Graham, J. Polymer Sci., 37, 411 (1959).

6. C. H. Bamford and E.F.T. White, Trans. Faraday Soc., 54, 278 (1958).

7. C. H. Bamford, A. D. Jenkins and E.F.T. White, J. Polymer Sci., 34, 271 (1959).

8. E. Schonfeld and I. Waltcher, J. Polymer Sci., 35, 536 (1959).

9. G. Smets, J. Appl. Polymer Sci., 5, 149 (1961).

10. D. J. Angier and E. M. Fettes, Rubber Chem. Tech., 38, 1164 (1965).

11. H. Willersinn, Makromol. Chem., 101, 296 (1967).

12. W. Hahn and H. Lechtenböhmer, Makromol. Chem., 16, 50 (1955).

13. D. J. Metz and R. B. Mesrobian, J. Polymer Sci., 16, 345 (1955).

14. G. Natta, J. Polymer Sci., 34, 531 (1959)

15. G. Natta, E. Beati and F. Severini, J. Polymer Sci., 34, 685 (1959).

16. Y. Landler and P. Lebel, French Patent 1,101,682 (1955).

17. Y. Landler and P. Lebel, J. Polymer Sci., 48, 477 (1960).

18. M. Imoto, T. Otsu, and T. Ito, Chem. Abst., 59, 7649 (1963).

19. V. A. Kargin, V. P. Shibaev and N. A. Plate, Vysokomolek. Soed, 3, 299 (1961).

20. G. Smets, A. Poot, M. Mullier and J. P. Bex, J. Polymer Sci., 34, 287 (1959).

21. T. Saegusa, M. Nozaki and R. Oda, J. Chem. Soc., Japan, Ind. Chem. Sect., 57, 243 (1954).

22. W. Hahn and A. Fischer, Makromol. Chem., 36, 16 (1955).

23. N. G. Gaylord, Interchem. Rev. 15, 91 (1957).

24. A. E. Woodward and G. Smets, J. Polymer Sci., 17, 51 (1955).

25. G. Smets, L. Convent, and Y. Vander Borght, Makromol. Chem., 23, 162 (1957).

26. R. J. Ceresa, Polymer 1, 397 (1960).

27. R. J. Orr and H. L. Williams, J. Am. Chem. Soc., 78, 3273 (1956).

28. R. J. Orr and H. L. Williams, J. Am. Chem. Soc., 79, 3137 (1957).

29. L. Valentine and B. Chapman, Ric. Sci., 25 Supplement 278 (1955).

30. W. Hahn and A. Fischer, Makromol. Chem., 21, 77 (1956).

31. G. N. Richards, J. Appl. Polymer Sci., 5, 553 (1961).

32. C. I. Simonescu and S. Dumitriu, J. Polymer Sci., Part C, 37, 187 (1972).

33. G. Mino and S. Kaizerman, J. Polymer Sci., 31, 242 (1958).

34. M. S. Bains, J. Polymer Sci., Part C, 37, 125 (1972).

35. pages 24-26 in Battaerd and Tregear's book (see above).

36. V. E. Shashoua and K. E. Van Holde, J. Polymer Sci., 28, 395 (1958).

37. G. H. Miller and A. F. Perizzola, J. Polymer Sci., 18, 411 (1955).

38. G. H. Miller and A. K. Bakhtiar, Can. J. Chem., 35, 584 (1957).

39. C. H. Bamford and R. G. W. Norrish, J. Chem. Soc., 1935 1504; 1938 1521, 1531, 1544.

40. J. E. Guillet and R. G. W. Norrish, Proc. Roy. Soc. (London) A233, 172 (1956).

41. J. E. Guillet and R. G. W. Norrish, Nature, 173, 625 (1954).

42. A. S. Dunn, B. D. Stead and H. W. Melville, Trans. Faraday Soc., 50, 279 (1954).

43. M. Jones, Can. J. Chem., 34, 948 (1956).

44. T. Otsu, J. Polymer Sci., 26, 236 (1957).

45. G. Oster and O. Shibata, J. Polymer Sci., 26, 233 (1957).

46. R. W. Cooper, G. Vaughan, S. Miller and M. Fielden, J. Polymer Sci., 34, 651 (1959).

47. C. C. Menon and S. L. Kapur, J. Appl. Polymer Sci., 1, 372 (1959).

49. N Geacintov, V. Stannett, E. W. Abrahamson and J. J. Hermans, J. Appl. Polymer Sci., 3, 54 (1960).

50. D. T. Turner, J. Polymer Sci., 35, 17 (1959).

51. D. S. Ballantine, P. Colombo, A. Glines, B. Manowitz and D. J. Metz, Brookhaven Natl. Lab. Report, BNL 414, T-81 (1956).

52. A. S. Hoffman, E. R. Gilliland, E. W. Merril and W. H. Stockmayer, J. Polymer Sci., 34, 461 (1959).

53. A. Chapiro, J. Polymer Sci., 34, 481 (1959).

54. J. Zimmerman, J. Polymer Sci., 49, 247 (1961).

55. D. Angier and D. Turner, J. Polymer Sci., 28, 265 (1958).

56. J. C. Bevington and D. E. Eaves, Nature, 178, 1112 (1956).

57. D. S. Ballantine, D. J. Metz, J. Gard and G. Adler, J. Appl. Polymer Sci., 1, 371 (1959).

58. L. Minnema, F. A. Hazenberg, L. Callaghan and S. H. Pinner, J. Appl. Polymer Sci., 4, 246 (1960).

59. I. Sakurada, T. Okada, and K. Kaji, J. Polymer Sci., Part C, 37, 1 (1972).

60. M. Szwarc, Nature, 178, 1168 (1956).

61. L. J. Fetters, J. Polymer Sci., Part C, 26, 1 (1969).

62. M. Morton and L. J. Fetters, Macromol. Rev. 2, 71 (1967).

63. L. J. Fetters, J. Elastoplastics 3, 35 (1972).

64. R. Zelinski and C. W. Childers, Rubber Chem. Tech., 41, 161 (1968).

65. P. Rempp and E. Franta, Pure Appl. Chem., 30, 229 (1972).

66. G. Greber, Makromol. Chem., 101, 104 (1967).

67. J. Heller, Polymer Eng. Sci., 11, 6 (1971).

68. D. Braun, W. Neumann and G. Arcache, Makromol. Chem., 112, 97 (1968).

69. A. Halasa, Polymer Preprints 13, 678 (1972).

70. T. Saegusa, personal communication, New London, N. H. 1969.

71. G. Berger, M. Levy and D. Vofsi, J. Polymer Sci., B, 4, 183 (1966).

72. Y. Yamashita, K. Nobutoki, Y. Nakamura and M. Hirota, Macromol., 4, 5481 (1971).

73. C. G. Overberger and C. M. Burns, J. Polymer Sci., Part A-1, 7, 333 (1969).

74. J. P. Kennedy, XXIII Int. Congr. Pure Applied Chem., Abstract 1, 105 (Boston) 1971.

75. F. M. Merrett, J. Polymer Sci., 24, 467 (1957).

76. J. P. Kennedy and R. R. Smith, this volume.

77. J. P. Kennedy, J.J. Charles and D. L. Davidson Polymer Preprints, 14, 974 (1973).

78. G. Pampus, W. Josef and M. Hoffman, Rev. Gen. Caout. Plast., 47, 1343 (1970).

79. H. Mark, Textile Res. J., 23, 294 (1954).

80. G. Miller and A. Bakhtiar, Can. J. Chem., 35, 584 (1957).

81. S. R. Rafikov, G. N. Chelnokova, I. V. Zhuravleva and P. N. Gribkova, J. Polymer Sci., 53, 75 (1961).

82. T. Shiota, Y. Gota, and K. Hayashi, J. Appl. Polymer Sci., 11, 773 (1967).

83. R. Graham, J. Polymer Sci., 24, 367 (1957).

84. D. Coleman, J. Polymer Sci., 14, 15 (1954).

PRACTICAL PROPERTIES OF MULTI-PHASE POLYMER SYSTEMS

Rudolph D. Deanin, Alice A. Deanin, & Todd Sjoblom

Plastics Department, Lowell Technological

Institute, Lowell, Massachusetts 01854

INTRODUCTION

Physical Heterogeneity

Theoretical polymer chemists traditionally prefer to study simple systems which exist in a single homogeneous phase. For theoretical study of solid state properties, in addition, they prefer polymers which are linear and amorphous as well. In actual practice, however, most important polymer systems are much more complex than this.

Most pure polymers are chemically homogeneous but nevertheless exhibit physical heterogeneity on a microscopic or, more often, sub-microscopic scale. For example, linear polymers which are regular enough to crystallize, still cannot reach 100% crystallinity; they contain crystalline and amorphous areas, and often further organize these into large spherulites. Thermoset plastics are generally considered to be amorphous glasses, but electron microscopic examination of fracture surfaces shows, not the expected smooth conchoidal fracture, but rather tiny hills and valleys, or "micelles," which suggest the existence of separate strong and weak areas. Even in conventional amorphous homogeneous thermoplastics, more sophisticated study of solid state properties is beginning to reveal anomalies which may indicate the existence of tiny ordered regions, which will require more sophisticated analytical techniques to detect them with certainty.

Chemical Heterogeneity

Most commercial polymer systems are not only physically but also chemically heterogeneous. In general a continuous polymeric matrix contains a second, dispersed phase of different chemical composition. Borrowing the terminology of materials science in general, there is a growing trend to refer to them as "composites." They may be classified according to the nature of the dispersed phase, and of its modulus relative to the polymeric matrix in which it is found:

1. Dispersed phase is high modulus, such as inorganic fillers and reinforcing fibers.

2. Dispersed phase is also polymeric:

   a. Dispersed phase is high modulus, as in thermoplastic elastomers and reinforcing resins for rubber compounding.

   b. Dispersed phase is low modulus, as in rubber-modified impact-resistant rigid plastics.

3. Dispersed phase is very low modulus:

   a. Gas bubbles in plastic foams.

   b. Water droplets in water-extended polyesters.

For the purposes of the present discussion, we are concerned exclusively with (2) Multi-Phase Polymer Systems, in which the continuous matrix and the dispersed phase are both polymeric, but of different chemical composition (and generally of different modulus).

Compatibility in Polymer Blends

When the chemist mixes two liquids, their miscibility depends upon the free energy of mixing ΔF, which in turn is the result of their heat of mixing ΔH and the entropy of mixing ΔS:

$$\Delta F = \Delta H + T\Delta S$$

Since any molecule is generally more attracted to similar than to dissimilar molecules, the heat of mixing is endothermic and positive. In mixing small molecules of two different liquids, the increase in randomness and entropy is very high, and easily outweighs the endothermic heat of mixing, favoring negative free energy and thus miscibility. In mixing large polymer molecules, on the other hand, the thousands of atoms in each molecule must remain together, so that mixing cannot be nearly as random, and the gain in entropy is not nearly as high. Thus it is rarely possible to outweigh the endothermic heat of mixing, and very few pairs of polymers are miscible to form homogeneous single-phase systems.

In fact, in rare cases when such complete miscibility and homogeneity do occur, they are usually due to specific interactions between the two polymer molecules, such as polar attractions or hydrogen-bonding of specific functional groups, or complex-formation between stereoisomers.

In contrast to this rigorous thermodynamic definition of polymer compatibility, the practical polymer technologist is more apt to apply a more utilitarian criterion, the practical properties of the resulting polymer blend, to define compatibility vs. incompatibility. If he blends two polymers and obtains useful properties, he is satisfied that they are "compatible" for his purposes.

A somewhat more eclectic scheme for classifying polymer compatibility divides it into several apparently distinct categories:

1. Compatibility which results from crystalline interaction between the two polymeric components.

2. True miscible compatibility.

3. Blendability without subsequent "sweat-out," macroscopic phase separation which would make the blend of no practical utility.

For the purpose of our present discussion, we are concerned with polymer blends, and with block and graft copolymers, in which we have clear evidence of phase separation, generally on a sub-microscopic scale, resulting in some improvement in practical properties. We will assume for the present that this is a clear definition which can be applied to actual polymer systems, and defer until later the more complex systems in which the fundamental definition itself may be called into question.

PROCESSABILITY

Rubbery Melt Flow

In many cases, the practical polymer chemist develops multi-phase polymer systems in order to improve processability. For example, rubbery melt flow, particularly during extrusion and sometimes also during injection molding, is often improved by addition of a second polymer phase. In some cases, the added polymer is a "reinforcing resin" of limited compatibility and relatively high modulus, and its smoothing action may be similar to the effect of conventional fillers in rubber compounding. In other cases, the added polymer may be fairly compatible, at least at processing temperature, and act as a polymeric plasticizer, to improve the fluidity of the total melt.

Melt Strength and Ductility

Polyblending with a second polymer which is rubbery in nature and/or very high in molecular weight, can also provide great improvements in melt strength for blowing or thermoforming, and also in ductility for solid-state post-forming operations such as stretching, forging, and punching.

Thermoplastic Elastomers

One outstanding example of improved processing is the increasing tempo of current developments in thermoplastic elastomers (Figure 1). Here the inconvenience of compounding rubber with carbon black, and thermosetting it permanently in a slow cumbersome curing cycle, have been eliminated (Figure 2) by

Figure 1

Schematic Diagram of Styrene/Butadiene/Styrene Block Copolymer Used as A Thermoplastic Elastomer (1)

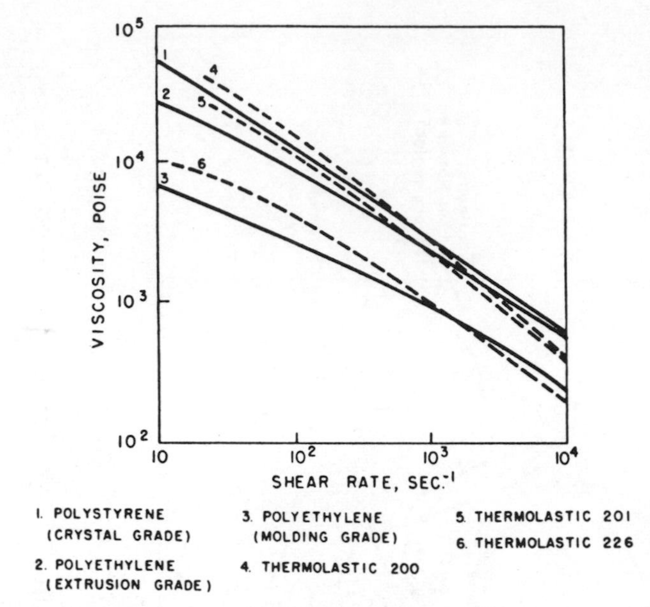

Figure 2
Melt Flow of Thermoplastics at 175°C (2)

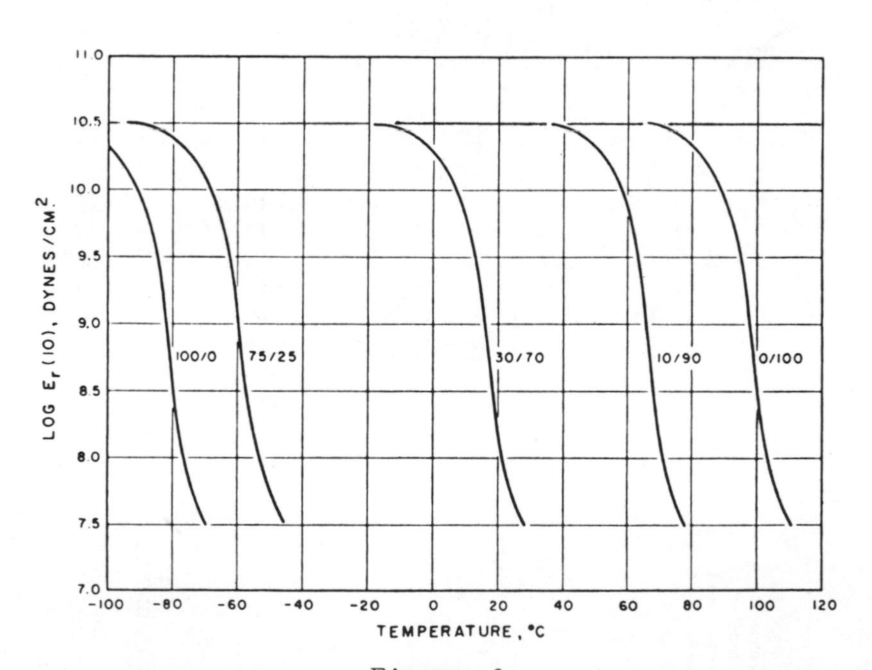

Figure 3
Modulus vs. Temperature Curves for Butadiene/Styrene Random Copolymers (3)

synthesizing A-B-A sandwich block copolymers, in which the large central rubbery block is surrounded by two (or more) "hard segments" which aggregate into firm clusters ("domains") in the finished product, acting as both reinforcing fillers and thermoplastic "cross-linking" sites.

MECHANICAL PROPERTIES

Modulus

When two polymer structures are combined in a homogeneous single-phase system, either by random copolymerization or by the more rare occurrence of a compatible polyblend, the modulus of rigidity is roughly intermediate between the two, depending fairly linearly upon the ratios of the two structures in the total composition (Figure 3). When the two polymer structures exist in separate phases, on the other hand, the relationship between composition and modulus is not nearly as simple (Figure 4). In one view, the structure which is present in largest amount should form the continuous matrix phase and play the primary role in determining the modulus; in some intermediate range, where both structures are present in fairly equal amount, there should be a rather steep transition between the two types of materials. In a more sophisticated approach, it is possible to treat the dispersed phase as a filler of higher or lower modulus, and calculate the effect of such filler content upon the modulus of the filled matrix. Using any of these theoretical techniques, the shape of the curve will also depend upon the choice of linear or logarithmic scales for the modulus, and the choice of weight, volume, or molar ratio for the two phases. Experimental data are not plentiful; and those which exist generally fall intermediate between these different theoretical approaches.

Strength

The ultimate strength of multi-phase polymer systems has not been a subject of major theoretical study. Manufacturers' literature gives the practical strength values of commercial systems, but no general analysis of such data has yet been compiled. Probably

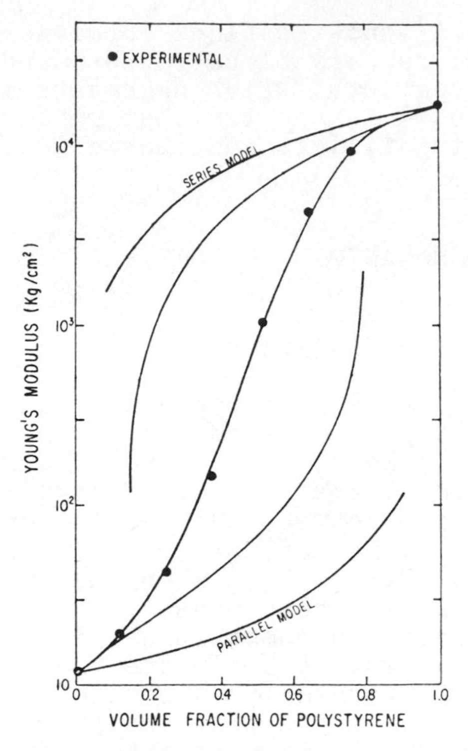

Figure 4: Theoretical & Experimental Modulus for Styrene/Butadiene/Styrene Block Copolymers (4)

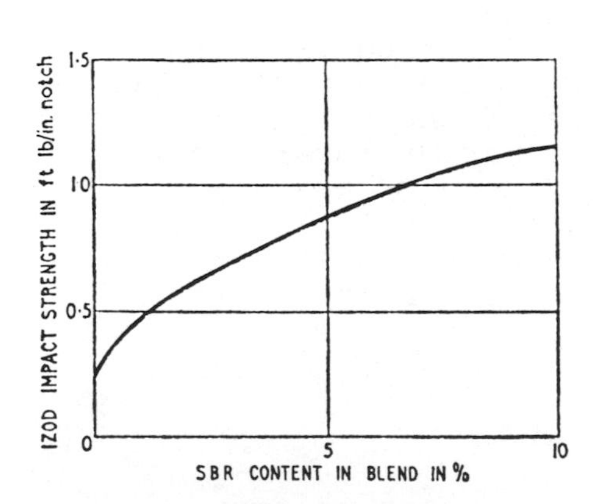

Figure 5: Effect of Rubber in Impact Polystyrene (5)

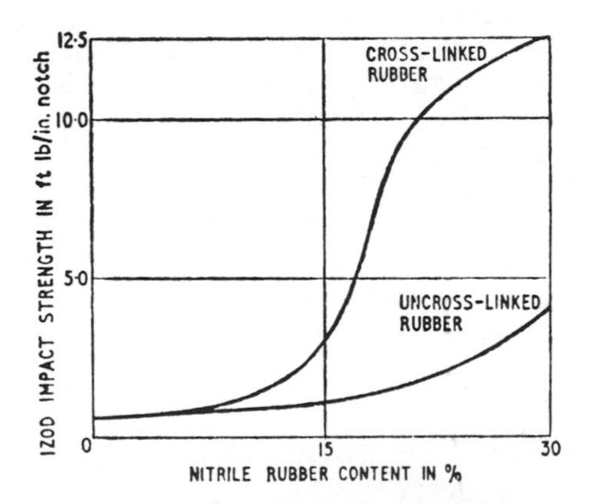

Figure 6: Effect of Nitrile Rubber in ABS (5)

Table 1

TENSILE STRENGTHS OF ELASTOMERS

Natural Rubber - Raw	300 psi
Natural Rubber - Vulcanized	3000
Natural Rubber - Reinforced & Vulcanized	4500
Thermoplastic SBS Block Copolymer - Average	3000
Thermoplastic SBS Block Copolymer - Maximum	4600

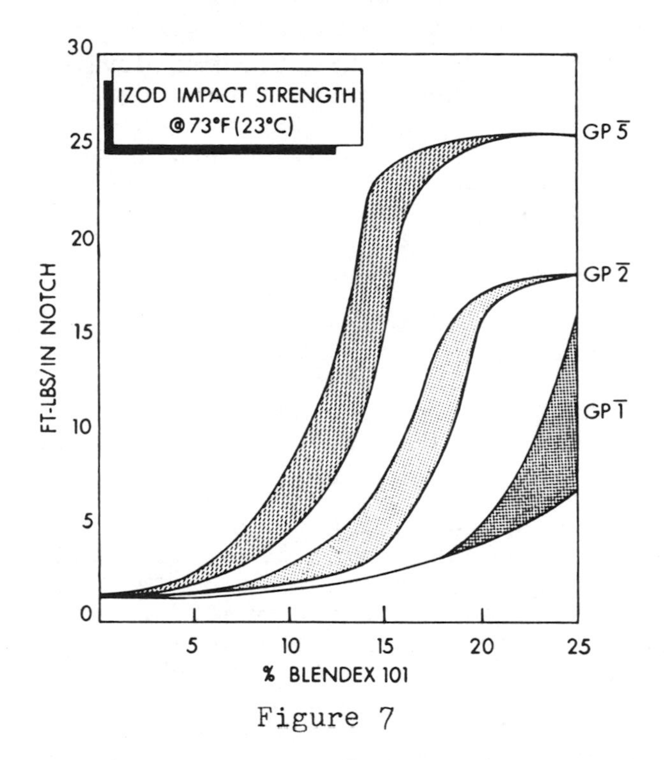

Figure 7

Effect of ABS in Rigid Polyvinyl Chloride (6)

GP5̄, GP2̄, & GP1̄ indicate high, medium, & low-molecular-weight polyvinyl chlorides

the most outstanding evidence, of course, is the strength of thermoplastic elastomers (Table 1), approaching the strength of conventional reinforced and vulcanized rubber, but relying solely on the aggregation of the terminal "hard segments" in the soft rubbery matrix, to prevent complete uncoiling, slippage, and cold flow of the elastomer molecules past each other under stress, and thus providing all the needed strength in this way.

Impact Strength

When the practical polymer chemist thinks of commercial multi-phase polymer systems, he thinks primarily of the so-called "impact plastics," impact styrene (Figure 5), ABS (Figure 6), rigid vinyls (Figure 7), and toughened polyolefins. Here a moderate concentration of tiny rubbery particles are dispersed throughout a rigid glassy plastic matrix, and produce tremendous improvement in resistance to brittle failure under high-speed impact, without any major sacrifice in other mechanical or thermal properties. Generally 5-30% rubber content is optimum in such systems.

The reasonable retention of other mechanical and thermal properties has already been considered above in the discussion of modulus. The tremendous improvement in impact strength is more problematic. Over the years, a number of distinct and/or overlapping theories have been offered to explain this phenomenon (Figure 8):

1. The crack propagates through the glassy matrix until it reaches a rubber particle. The rubber particle has the ability to relax rapidly, acting as a strain center which can elongate or deform without breaking, and it also has the ability to recover elastically after the impact shock wave has passed.

2. When a crack propagates through the glassy matrix, it produces a clean break along the crack surface. When it enters a rubber particle, on the other hand, rubber ligaments stretch across the widening crack, and their tensioning effect retards or even stops the separation of the crack.

3. The impact energy which initiates and propagates the crack is all concentrated at the apex of the crack, producing a very high force at that point. When the crack enters a rubber particle, however, the impact

A. Homogeneous Glassy Polymer

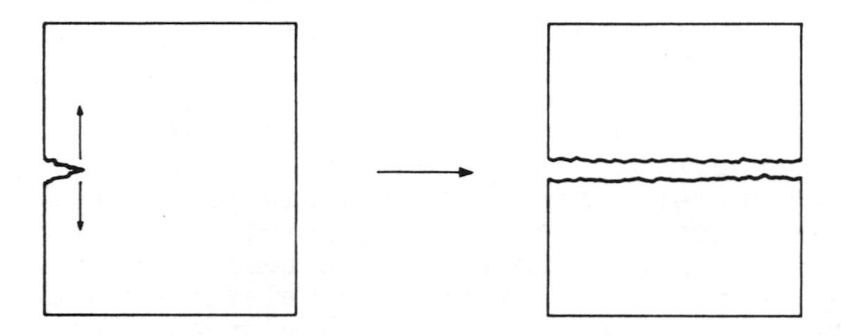

B. Rubber-Modified Glassy Matrix

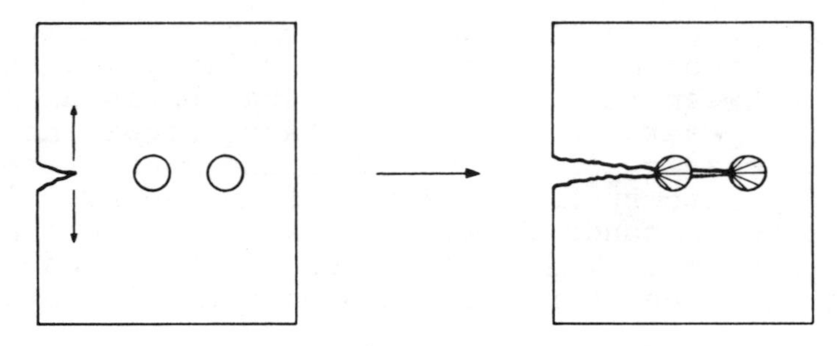

Figure 8

Effect of Rubbery Particles on Crack Propagation in a Brittle Polymer (3)

energy is dispersed throughout the particle, so that the concentration of energy at any point is quite low. When this dispersed energy reaches a surface of the particle, it is too weak to initiate a new crack in the glassy matrix at that point.

4. The mechanical energy which initiates and propagates the crack, enters the rubber particle, which is not perfectly elastic. This imperfect elasticity - internal intermolecular friction which converts mechanical energy into heat - decreases the remaining mechanical energy below the level required to propagate the crack.

5. So long as a growing crack continues to propagate through the glassy matrix, all of the energy remains concentrated at the apex, maintaining a very high stress concentration at this point. When the crack reaches a rubber particle, it tears the rubber away from the rigid matrix at this relatively weak interface, thus creating more surface area. Surface area contains surface energy, which uses up much of the incident mechanical energy, converting it into more surface energy. Thus there is not enough mechanical energy left to continue to propagate the crack.

6. Further along the same line of thought, it has been observed that minute craze-cracks form, leaving tiny voids between them. Thus such crazing is also the creation of a great deal of new surface, requiring considerable energy, and thus converting much of the impact energy into such harmless surface energy. In some cases, shear bands form, and take up the energy by similar mechanisms. This conversion of mechanical energy into surface energy is another form of hysteresis.

7. Shear band formation also represents a degree of molecular orientation. This parallel orientation of molecular segments produces an increase in strength and toughness, stopping the growth of the crack in this way.

8. "Thermally induced hydrostatic tension lowers the glass-transition temperature of the rubber particle so that the rubber particle in the composite is actually softer than one in a stress-free state. One can then postulate that the energy due to strain lowers the thermal energy requirements for reaching the glassy transition of the matrix. Above a certain strain energy, the local thermal-energy contribution is great enough to cause a yield in the matrix."

Creep

The use of "hard segment" domains in thermoplastic elastomers, in place of conventional reinforcing pigments and covalent cross-links, does suffer from the effects of long-term stress. Under such conditions, the thermoplastic "hard segment" domains do suffer gradual relaxation and permit sliding of the rubber molecules past each other, resulting in gradual creep or permanent set.

Lubricity and Abrasion

In specialized cases, the dispersed phase may be composed of fine particles of a polymer of low surface tension such as polytetrafluoroethylene or polyethylene (Table 2). To the extent that these particles are relatively incompatible in the matrix, and tend to migrate to the surface during processing and perhaps also during use, they create a surface whose surface tension is much lower than the parent matrix. This can be particularly useful for producing self-lubricating bearing surfaces and/or improving resistance to abrasive wear.

THERMAL PROPERTIES

Modulus vs. Temperature

In a conventional linear amorphous single-phase thermoplastic polymer, modulus of rigidity has a high plateau value at low temperatures where there is very little molecular mobility. With increasing temperature and atomic vibration, intermolecular spacing (free volume) increases until the simultaneous motion of chain segments of 5 or more atoms becomes reasonably probable. At this point, the material softens from rigid to leathery, and this temperature is called the glass transition temperature. With further increase in temperature, mobility of larger chain segments, such as 10-30 atoms in length, becomes highly probable, and the modulus reaches a lower "rubbery" plateau, in which the material is soft and flexible. At still higher temperatures, of course, molecular mobility becomes so high and probable that processing pressures produce melt flow during normal plastics processing technology.

In this view, the only difference between the rigid glassy plastic and the soft rubbery elastomer is

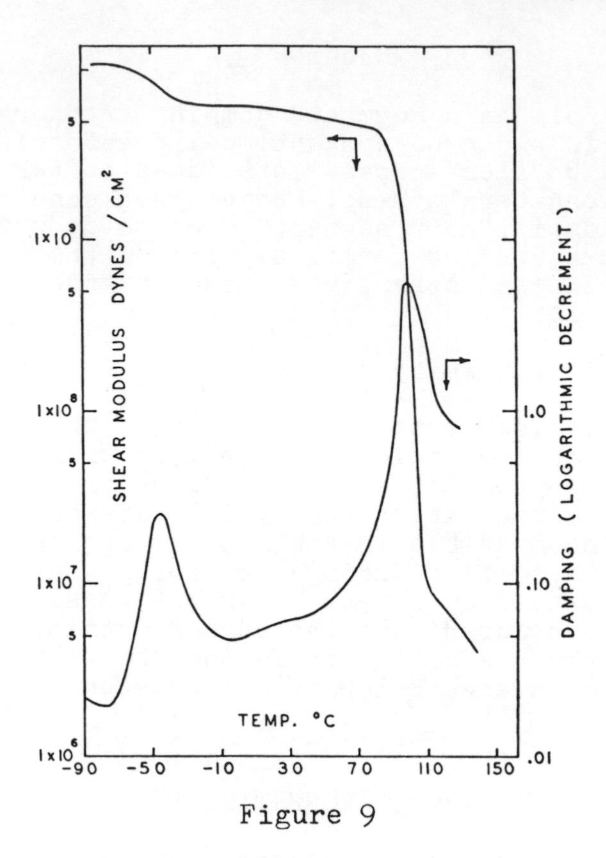

Figure 9

Modulus and Damping in a Blend of Polystyrene and Butadiene/Styrene Copolymer (3)

Table 2

FRICTION AND WEAR OF PTFE POLYBLENDS (7)

Base Resin	PTFE	Coefficient of Friction Static	Coefficient of Friction Dynamic	Wear Factor
Polycarbonate	0%	0.31	0.38	2500
Polycarbonate	15	0.09	0.15	15
Acetal	0	0.14	0.21	65
Acetal	20	0.07	0.15	17
Nylon 66	0	0.20	0.28	200
Nylon 66	20	0.10	0.18	12

the glass transition temperature at which molecular mobility appears. When two molecular structures are combined in a single homogeneous single-phase system, either by random copolymerization or by the more rare occurrence of a compatible polyblend, the leathery transition simply occurs at some intermediate temperature, depending more or less upon the average effect of the concentrations of the two structural components (Figure 3). The transition may also be somewhat broader and more gradual than it was for either pure structure alone.

When the two structures exist in separate phases, on the other hand, each phase is relatively free to exhibit its own glass transition temperature, producing a 2-step modulus vs. temperature curve (Figure 9). At very low temperature, both phases are glassy, and the entire polymer is rigid and brittle. With increasing temperature, the phase containing the more flexible molecules reaches its glass transition temperature and becomes rubbery. When the rubbery phase is dispersed in a glassy matrix, this produces an impact-resistant rigid plastic. When the "hard" phase is dispersed in a rubbery matrix, this produces the thermoplastic elastomer. Beyond this first glass transition, in either case, is where we find the most useful properties of the multi-phase polymer system.

Continuing on to higher temperatures, we eventually reach the glass transition of the phase containing the stiffer molecular structure. At this point, both phases have become soft and rubbery. In rigid impact-resistant plastics, we reach the heat deflection temperature, beyond which the plastic product distorts and is no longer useful. In thermoplastic elastomers, we reach the melt flow region in which the "hard" regions slide and flow and the rubber suffers loss of strength and high permanent set. Thus the plateau between the glass transition temperatures of the two phases defines the useful range for each multi-phase system. The temperature span of this plateau defines the useful temperature range. The level of the plateau depends upon the relative concentration and continuity of the two phases, as discussed earlier under the subject of modulus, and determines whether the material is primarily a rigid impact-resistant plastic or a thermoplastic elastomer (Figure 10).

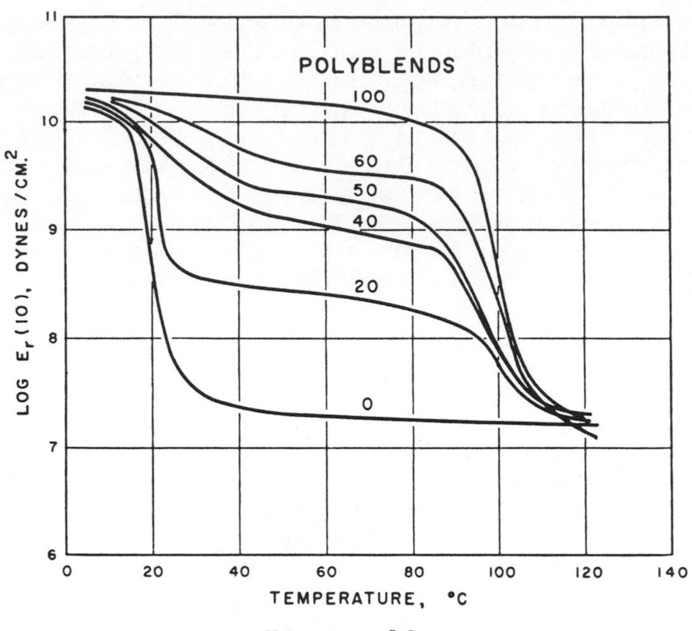

Figure 10

Modulus vs. Temperature for Semi-Compatible Blends of Polystyrene and 30/70 Butadiene/Styrene Copolymer (3)

Table 3

FLAME RETARDANCE OF PVC IN ABS*(8)

PVC	Sb_2O_3	Grade of Marbon ABS			
		Cycolac X-27		Cycolac HM	
		Burning Rating	Burning Rate	Burning Rating	Burning Rate
0%	0%	B	1.47	B	1.37
10	0	B	1.17	B	1.39
20	0	SE	0.40	B	1.12
30	0	NB	-	B	0.79
40	0	NB	-	NB	-
7.5	2.5	B	0.73	B	1.15
15	5	NB	-	SE/NB	0.34/-
22.5	7.5	NB	-	NB	-

*ASTM D635

Conversely, cooling such a material below the lower of its two glass transition temperatures, will cause the embrittlement of a rigid plastic, or the stiffening and eventual embrittlement of a thermoplastic elastomer.

Thermo-Analytical Techniques

A variety of thermal and electrical analytical techniques have been used in the laboratory to study the fundamental behavior of multi-phase systems. These include: (1) Dilatometry, (2) Differential Thermal Analysis, (3) Differential Scanning Calorimetry, (4) Dielectric Relaxation.

Thermal Stability and Flammability

Thermal stability of multi-phase polymer systems has not received very much study. In some cases, the component of higher thermal stability may preserve the integrity and properties of the sample reasonably well up to its own limit, even though the less stable component may have failed at considerably lower temperature. In others, the least stable component may initiate degradation of the entire sample. Such distinctions need to be defined more precisely.

Flammability of a polymer is sometimes reduced dramatically by blending it with a second flame-retardant polymer. Thus for example ABS, which burns readily and is not easy to protect, can be blended with polyvinyl chloride, which is itself non-burning (Table 3). At about equal ratios of the two polymers, depending upon the specific test employed, the resulting blend is also non-burning. Similarly chlorinated polyethylene has been blended into polyethylene to improve flame-retardance, particularly for wire and cable insulation.

OPTICAL PROPERTIES

Clarity

When two polymers, of different refractive index, are mixed and form a multi-phase polymer system, incident light which enters this system must pass through the polymer A/polymer B interface a number of times before emerging on the other side of the sample. Each time it passes through an interface it is scattered more and more. This converts two transparent polymers into a translucent-to-opaque multi-phase polymer system. Originally this was naively accepted as one way of determining compatibility or incompatibility. Practically, it means that multi-phase polymer systems in general will not be transparent, and that their lack of transparency will also affect the concentrations of pigments and colors that must be used in them, and often change the final appearance of these colors in any case.

In many cases, the polymer technologist requires a clear transparent plastic for his desired end-product. There are several ways he can achieve such transparency in a multi-phase polymer system:

1. Choose or modify his polymers to produce homogeneity instead of heterogeneity. This may however negate the benefits he hoped to achieve from a multi-phase system.

2. Reduce the particle size of the dispersed phase until the particles are smaller than the wave length of visible light. In this situation, the particles will no longer scatter visible light, and the system will be transparent. Unfortunately, the minimum particle size for best mechanical and thermal properties is generally larger than the maximum particle size for transparency to visible light.

3. Make the sample so thin that there are very few inter-phase interfaces to scatter the light as it passes through from one surface to the other. Such extreme thinness would of course limit the polymer to a few very select applications.

4. Modify the refractive indexes of the two phases until they are equal. This can generally be done by copolymerizing one or both phases with suitable comonomers of different refractive index, until the two phases match. Once this is accomplished, light passing through the interface between the phases will not be scattered, but will pass straight through, and the material will be transparent.

This last technique is the most promising, and has recently been applied to the production of impact-resistant rigid plastics of excellent clarity as well as other mechanical and thermal properties. One limitation of the technique is the differential effect of temperature upon refractive index in different polymers: as temperature changes, the refractive index of each phase changes at a different rate. Thus, though they may be matched at room temperature, they will diverge gradually on heating or cooling, producing change in color and cloudiness as they go. A number of manufacturers are currently wrestling with more sophisticated solutions to this problem.

Ultraviolet Stability

In many cases, one phase will have more ultraviolet light stability, the other less. In the blended system, the overall stability may then either increase or decrease, as compared with average expectations. In some systems, the less stable component initiates earlier and faster degradation of the whole system; this is useful for disposal of solid waste. In others, the stabler component may actually act as a stabilizer for the whole system; this is useful for long-lived outdoor structures. This difference requires much further study to understand, control, and use it effectively.

CHEMICAL PROPERTIES

Solubility

Earlier studies on block copolymers demonstrated remarkable behavior in selective solvents (Figure 11). Use of a solvent for structure A, which was not a

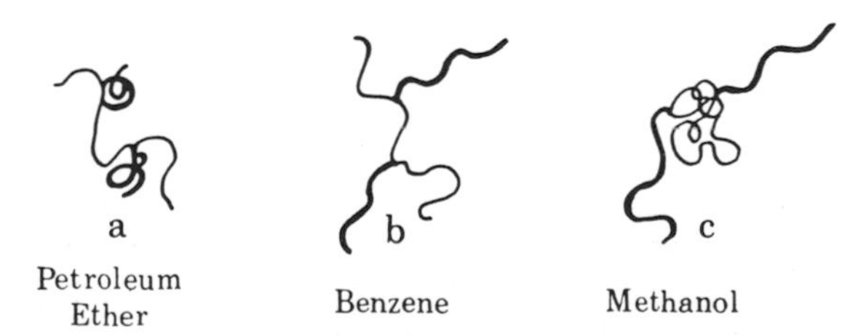

Figure 11
Effect of solvent on the conformation of chains in a graft copolymer made up of poly(methyl methacrylate) chains (heavy lines) grafted to a rubber backbone (lighter lines) (9)

solvent for structure B, caused structure A to extend and structure B to collapse into an insignificant filler particle, and made the block copolymer behave very much like polymer A alone. Conversely, a solvent for structure B, which was not a solvent for structure A, made the block copolymer behave very much like polymer B alone. These initial theoretical studies may eventually bear practical fruit in some of the recent discoveries about phase inversion which will be discussed below.

Differential solubility was one of the earliest techniques used for analyzing block and graft copolymers and polymer blends. If the chemist can find a solvent which only dissolves one component and leaves the rest of the system undissolved, or vice versa, he can begin to determine the relative amounts and structures of the different components. While this technique has been used very frequently to prove the researcher's hypothesis or goal, in many cases the results and conclusions were more teleological than logical. Proper use of such techniques requires rigorous critical logic, and extensive study of solvent and non-solvent systems on model polymers, to define precisely what is being measured during such an analysis.

Environmental Stress Crack Resistance

It has been reported that in some polymer systems, conversion from homogeneous single-phase polymers into multi-phase polymer systems, produced a significant improvement in stress crack resistance. This might be explained in a variety of ways. Since our understanding of stress cracking itself is still far from complete, perhaps it is too early to try to define the reason for its improvement in multi-phase systems.

PRICE

As a starting assumption, the price of a polyblend system should be equal to the price of the raw materials plus the processing cost. Examination of commercial price lists indicates that some polyblends are priced above this criterion, some a bit below. Complicating factors include the internal cost of the polymeric components in a multi-polymer company, and the type and affluence of the market which the polyblender picks as his target.

In some cases, polyblenders very consciously start with a high-price high-quality polymer, and blend with a lower-price polymer to reduce cost; obviously they and their customers must then balance price improvement vs. property decline and decide whether this balance is of value to them. Conversely, polyblenders often start with a low-price polymer and add moderate amounts of high-price polymer to improve properties; again they must balance price increase vs. the resulting property improvement, and hopefully not stint on a major improvement because it requires a modest increase in price.

BALANCE OF PROPERTIES IN MAJOR COMMERCIAL SYSTEMS

Polyolefins, particularly polypropylene, have long been blended with butyl rubber, and more recently with EPR, to improve impact strength, low-temperature brittleness, and stress crack resistance without serious sacrifice in rigidity, strength, or heat deflection temperature. Propylene/ethylene block copolymers have likewise been used for improvement over straight polypropylene homopolymer.

EPR Block Copolymer is one of the newest arrivals on the commercial scene. Here the ethylene/propylene random copolymer center block is soft and rubbery, while the polypropylene end blocks crystallize and bind the molecules in place, forming a saturated thermoplastic elastomer.

Styrene/Butadiene/Styrene block copolymer has proved outstanding for combining thermoplastic processability and clarity with most of the properties normally obtained only in vulcanized rubber. The styrene ends aggregate into "hard" domains which keep the rubbery polybutadiene segments from slipping past each other in normal use, and thus provide most of the expected mechanical and chemical behavior of vulcanized rubber along with improved processability and broad color range. Major weaknesses are creep, poor hot strength, and attack by oxygen and solvents. These have recently been partially remedied by replacing the butadiene middle block by EPR.

High Styrene/Butadiene copolymers and phenolic resins have been used as "reinforcing resins" in rubber compounding to increase the hardness and strength of the rubber without excessive use of vulcanizing ingredients and cycles. Polyvinyl chloride has similarly been added to nitrile rubber to increase its hardness and strength.

Impact Styrene is one of the earliest and the leading commercial example of multi-phase polymer systems. Originally a simple blend of polystyrene with rubber, of marginal properties, it is now generally a graft copolymer of polystyrene onto a rubber backbone, producing a large fraction of polystyrene homopolymer and leaving some ungrafted rubber as well. Conventional rubber contents range from perhaps 2 to 15%. As rubber content increases, impact strength increases up to 10-fold, while rigidity, strength, and heat deflection temperature decrease gradually. Clarity is sacrificed early in this series,and is the major price the polystyrene chemist must pay for high impact strength.

ABS is generally the quality brother of impact styrene, at double the price. Copolymerization with acrylonitrile improves hardness, rigidity, strength, heat deflection temperature, and chemical resistance. All these permit use of higher rubber contents, generally 20-30%, producing much higher impact strengths in return.

In recent years, ABS has been further blended with a number of other commercial polymers for specific reasons. Blends with polyvinyl chloride have high impact and flame resistance and intermediate cost. Blends with polysulfone and polycarbonate compromise between the high properties of these polymers vs. the processability and low cost of the ABS. Blends with polyurethane combine the stiffness, processability, and low cost of the ABS with the elastic recovery and abrasion resistance of the polyurethane.

Rigid Polyvinyl Chloride alone has many excellent properties; its major limitations are processability and impact resistance. These are most commonly improved by blending with ABS or acrylic graft copolymers, or occasionally with nitrile rubber, chlorinated polyethylene, ethylene/vinyl acetate copolymer, or other such "impact modifiers."

Acrylic impact plastics, all based upon a rigid acrylic matrix partially grafted onto and/or blended with an elastomeric dispersed phase, include acrylonitrile/styrene copolymer on acrylic elastomer, methyl methacrylate multipolymers grafted and blended with elastomers, and acrylonitrile/ethyl acrylate copolymer with nitrile rubber. Most of these also use matched refractive indexes to produce clarity, particularly for blown bottles.

Epoxy and Polyester resins have been toughened by adding liquid reactive nitrile rubber to them before cure. During cure, the rubber polymerizes to high molecular weight, and separates into discrete tiny rubber particles. In the final cured resin matrix, these rubber particles improve the impact strength and dynamic mechanical properties of the matrix resin.

Polyester/polyether block copolymers combine rubbery polyether segments with crystallizable polyester segments (polyoxybutylene glycol/polyethylene terephthalate) to produce one of our newest thermoplastic elastomers, actually fairly stiffly flexible plastics.

Polycaprolactone has recently been offered as a polymer of very versatile compatibility. The range of different polymers with which it can be blended is truly amazing, but the practical significance remains to be developed.

Spandex elastomeric fibers are block copolymers which contain polyurethane elastomer segments sandwiched between "hard setments" which provide enough polarity, hydrogen-bonding, and/or crystallinity to obviate much of the cross-linking normally needed to strengthen elastomers.

Phenolic resins are toughened by blending with nitrile rubber before final cure, to improve their impact strength and ductility. It is not clear whether this forms a compatible or semi-compatible polyblend, or whether there is actually copolymerization between the two by ring formation.

COMPLICATIONS

This discussion started with an idealized and simplified description of the structure and properties of multi-phase polymer systems. In actual research and commercial practice, a number of complications arise, which indicate that we are only on the threshold of beginning to understand these systems and to use them to best advantage. Some of these complications deserve mention at this point:

1. The interface between the two separate phases has always been a major problem. Since the two polymers are generally incompatible, the interface might easily be weak and fail readily under stress. In useful polyblend systems, the interface behaves as if it is strong and well-bonded. This is easily understood in block and graft copolymers (Figure 12). In polymer

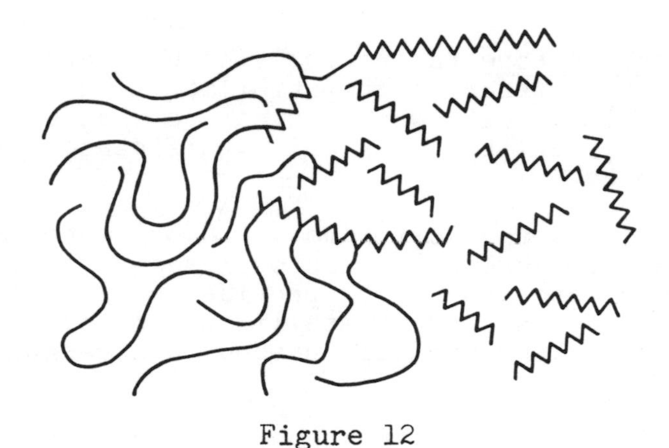

Figure 12

Interface in Block and Graft Copolymers (3)

Figure 13

Interfacial Mixing in a Polymer Blend (3)

Transparency
Glass Transition Temperature
Dynamic Modulus & Loss
Dielectric Relaxation
Electron Microscopy

Figure 14

Critical Dimensions for Compatibility by Various Tests (10)

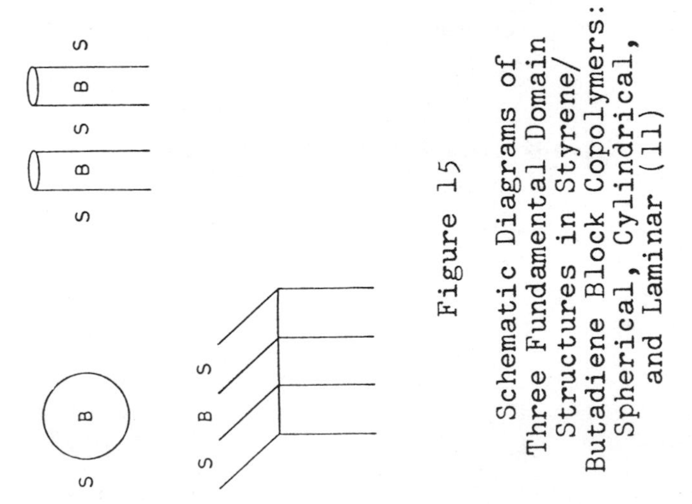

Figure 15

Schematic Diagrams of Three Fundamental Domain Structures in Styrene/ Butadiene Block Copolymers: Spherical, Cylindrical, and Laminar (11)

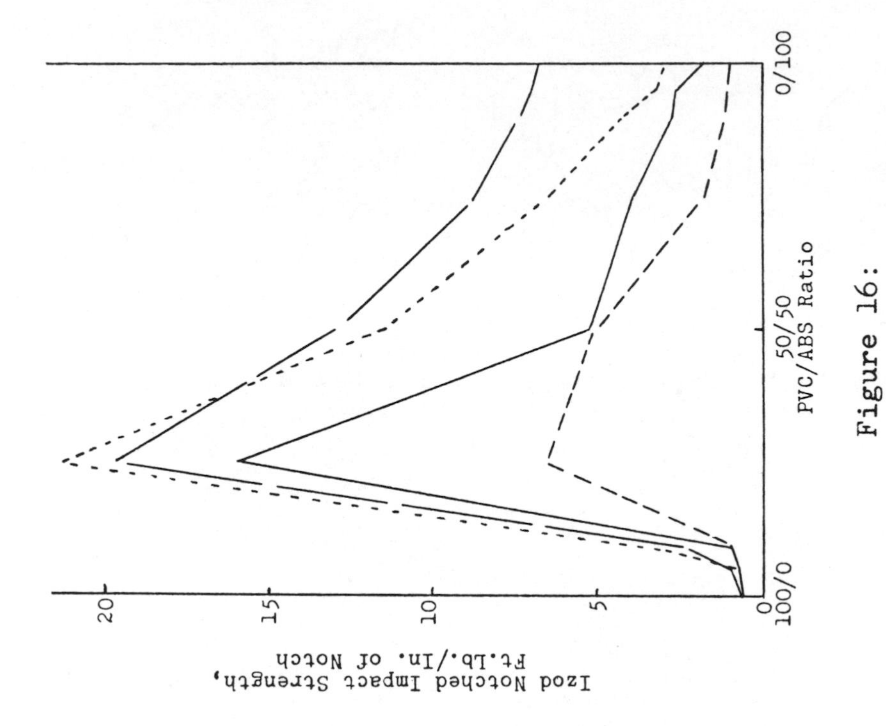

Figure 16:

Impact strength of 4 polyblend systems of polyvinyl chloride with different grades of ABS (12)

blends, it is possible that some mixing across the interface, a low degree of mutual solubility, may explain the strength of this transition region (Figure 13). In some theories of impact improvement, on the other hand, the strength of the interface actually appears unimportant or insignificant. Obviously further understanding is needed.

2. Critical size of the dispersed phase,frequently called the "domains," appears to depend upon the property being measured and the test used to measure it. A test which requires large-size domains, when applied to smaller particle sizes, will make the system appear homogeneous rather than multi-phase (Figure 14). Conversely, a test which detects very small domains may make almost all systems appear heterogeneous. Incidentally, this latter type of test may soon tell us that even pure linear amorphous polymers which we have always considered to be a single continuous phase, may actually contain very tiny ill-formed heterogeneous areas that explain some of their more sophisticated properties in this way.

3. The shape of the dispersed domains has proved to be of critical importance in thermoplastic elastomers, and this may apply to other systems as well. Depending on the polymers, their ratios, processing conditions, and past life-history of the product, these domains may be spherical, cylindrical, laminar, or even interconnecting, occasionally producing two or more continuous phases in a single polyblend system (Figure 15). This obviously has a tremendous effect on properties. Greater understanding and control of domain structure is thus of tremendous practical importance.

4. Current research in several systems has demonstrated that it may be possible to blend two "hard" semi-compatible polymers with each other and obtain impact strengths in the blends which are 5-10 times higher than in either of the two original components (Figure 16). Present multi-phase theory simply does not even begin to explain such behavior. Possibly some of the considerations throughout this discussion may apply here; but it will require considerable further study to understand such systems. Partial solubility, particles inside of particles, two continuous phases, and systems of more than two discrete phases, are a few of the possible approaches to this problem, and many more may have to be considered.

REFERENCES

General Reviews

Turner Alfrey & E. F. Gurnee, "Organic Polymers" Prentice-Hall, 1967, Ch. 7.
C. B. Bucknall, Private Communication.
E. M. Fettes & W. N. Maclay, Appl. Polym. Symp., 7, 3 (1968).
B. D. Gesner, Encyc. Polym. Sci. Tech., 10, 694 (1969).
S. Krause, J. Macromol. Sci., Rev. Macromol. Chem., C7, 251 (1972).
S. L. Rosen, Polym. Eng. Sci., 7, 115 (1967).
A. J. Yu, Adv. in Chem., 99, 2 (1971).

Symposia

S. L. Aggarwal, "Block Polymers," Plenum, 1970.
P. F. Bruins, "Polymer Blends," Interscience, 1970
N. A. J. Platzer, "Multicomponent Systems," Adv. in Chem., 99 (1971).
J. Polym. Sci., Polym. Symp., 4, 579-764 (1963).
J. Appl. Polym. Sci., Appl. Polym. Symp., 7, (1968).
Polym. Preprints, 11 (2), (1970).
J. Macromol. Sci., Rev. Macromol. Chem., C4 (2), 313-366 (1970).
J. Macromol. Sci., Rev. Macromol. Chem., C5 (2), 167-221 (1971).
J. Macromol. Sci., Rev. Macromol. Chem., C7 (2), 251-314 (1972).
Polym. Preprints, 13 (2), (1972).

Sources of Tables and Figures

1. Shell Chemical Company, Polymers Division, "Kraton 1000 Polymers for Adhesives, Sealants, and Coatings," SC:70-17.
2. M. A. Luftglass, W. R. Hendricks, G. Holden, & J. Bailey, "Thermoplastic Elastomers," 22nd ANTEC, SPE, Montreal, Quebec, March 7-10, 1966.
3. R. D. Deanin, "Polymer Structure, Properties, & Applications," Cahners, 1972.
4. L. E. Nielsen, Private Communication.
5. J. A. Brydson, "Plastics Materials," Van Nostrand, 1966.
6. Marbon Chemicals Division of Borg Warner, "Modification of PVC Compounds with Blendex ABS Resins,"Technical Bulletin B-101A.

7. J. E. Theberge & B. Arkles, ACS Org. Coatings & Plastics Preprints, 33 (2), 188 (August 1973).
8. R. D. Deanin, R. O. Normandin, & P. P. Antani, Org. Coatings & Plastics Preprints, 33 (1), 556 (April 1973).
9. M. L. Miller, "The Structure of Polymers," Reinhold, 1966.
10. F. E. Karasz, Private Communication.
11. M. Matsuo, S. Sagae, & H. Asai, Polymer (Brit.), 10, 86 (1969).
12. R. D. Deanin & C. Moshar, ACS Polym. Preprints, 15 (1), (1974), In Press.

ON THE GENERATION OF NOVEL POLYMER BLEND, GRAFT, AND IPN STRUCTURES THROUGH THE APPLICATION OF GROUP THEORY CONCEPTS

L. H. Sperling

Materials Research Center, Lehigh University

Bethlehem, Pa. 18015

Polymer blends, grafts, and blocks have achieved importance as toughened plastics and novel elastomers because of their complexity, not in spite of it. By definition, these materials are characterized as some intimate combination of two (or more) kinds of polymer molecules. Because of the very small entropy gain on mixing long polymer chains, and the usually encountered positive heat of mixing, most polymer blends, grafts, and blocks form two phases. Most strikingly in these materials, the exact mode of synthesis controls the two-phase morphological features, which in turn influences their mechanical behavior. Thus a one-to-one relationship exists between synthetic detail and potential application.

At the present time, the scientific and patent literature details over 150 ways of organizing two kinds of polymer molecules in space. Surprisingly, most of these are broadly classified as "graft copolymers", immediately pointing out the inadequacy of the term. The order of the synthetic steps, the presence of crosslinking, etc., are rarely included in the standard nomenclature. In undertaking the application of group theory to polymer blends, grafts, and IPN's, three relevant questions may be posed:

(1) How can the known organizations of two kinds of polymer molecules be best classified? In what fundamental

---

The author wishes to acknowledge the generous support of the National Science Foundation through Grant GK-13355, Amendment No. 1.

ways are the several joining modes interrelated?

(2) Based on topological[1,2] graph theory,[3] and particularly group theory[4-8] considerations, how many distinguishable structures exist? Does their number form an open or closed set?

(3) Most importantly, can the group theory be used to predict yet unknown or unrecognized structures?

The present paper will outline a qualitative and a quantitative approach to organizing possible structures. While emphasis will be placed on the more general problems, explicit examples of some of the more important and/or unusual structures will be noted. The quantitative application of group theory allows not only a superior classification scheme, but also a powerful, general, predictive theory for the generation of novel structures.

A SIMPLE CHART

As illustrated in Figure 1, some of the more important types of blends,[9,10,11] grafts,[12-15] and blocks[16,17] can be organized in the form of a chart. Thus, we may consider some of the more important distinguishable ways of organizing two kinds of mers in space. Since the one-phase random[19] and alternating[20,21] copolymers are beyond the scope of the present discussion, they will not be considered further. For simplicity, we consider only two main categories: grafts and blends, depending on whether or not the two polymers are joined by chemical bonds. The graft copolymers are subdivided into occasional,[12] regular,[22,23] and surface[24,25] grafts. The occasional grafts contain all materials where grafting is irregular, haphazard, and/or incomplete. In such compositions most of the material actually forms simple homopolymers, with grafting being restricted to portions near phase boundaries.[25a] The most important subclass of the occasional grafts are the so-called solution graft copolymers, wherein a linear polymer I is dissolved in monomer II, followed by polymerization of II, usually with agitation. High impact polystyrene is commonly prepared in this manner.[26,27] The technologically important ABS materials[26] are commonly prepared as latex grafts. Interpenetrating polymer networks,[28,29] IPN's, and simultaneous interpenetrating networks,[30] SIN's, are distinguishable examples of cases wherein both polymers form infinite networks within the bulk materials. When considering combinations of a linear and a crosslinked polymer, two kinds of semi-IPN's may be distinguished: semi-IPN's of the first kind where the first synthesized polymer is in network form, and the second synthesized is linear,[31] and semi-IPN's of the second kind where

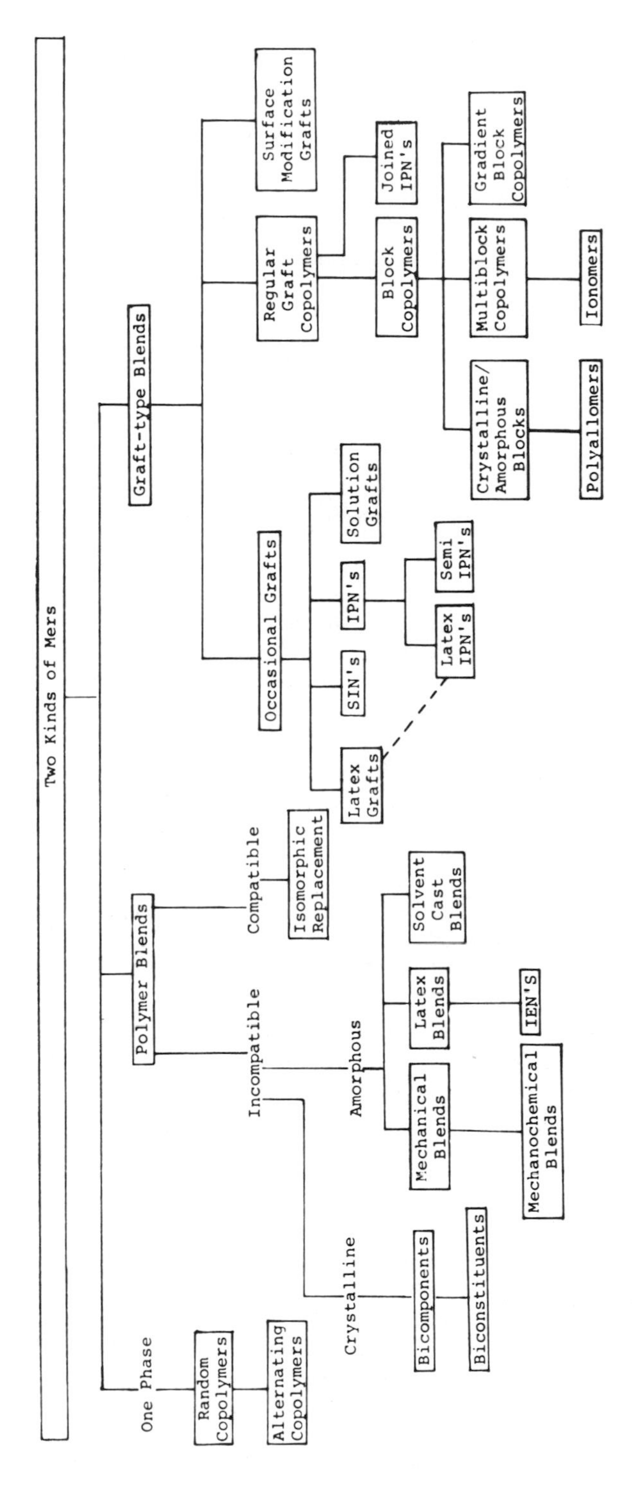

Figure 1. A schematic classification scheme for polymer blends, blocks and grafts.

the first is linear and the second is crosslinked.[32] Vastly different morphologies and properties are developed in what would otherwise be expected to be nearly identical materials.

The regular graft copolymers, by difference, include all the ways in which the joining of the two polymers is completed by specified chemical reactions to the greater or lesser exclusion of homopolymer formation. Block copolymers [16,17,18] constitute the most important subclass of the regular grafts, considered as a subclass (somewhat arbitrarily) because grafting is constrained to occur solely at the ends of the molecules. While recognizing that the ionomers also form a subclass of the random copolymers, their two-phase nature[33-36] suggests classification as a block copolymer wherein the block length is reduced to one mer. In the joined IPN's, polymer I is grafted to polymer II to form a single, continuous network.[37,38,38a]

The blends, by and large, concern mixtures of the two molecules wherein no grafting takes place.[11,39] Mechanical blends constitute the most important subclass, usually between a plastic as the major component and an elastomer as the minor component.[9,13] Latex blends are formed by mixing two species of latex, followed by coagulation or film formation.[27,40,41] Materials designated as interpenetrating elastomeric networks, IEN's, are subjected to a later crosslinking reaction.[42,43] The bicomponents and the biconstituents form distinguishable classes of oriented, fibrous materials.[44,45] Figure 1 thus enables us, in a qualitative way, to establish relationships among the numerous previously prepared materials. In the following, group theory will be applied to establish quantitative relationships.

SYMMETRY OPERATIONS

It is the custom to describe graft-copolymers with the notation poly(A-g-B), where the first mentioned polymer serves as the backbone, and the second as the side chain. While this nomenclature has met the requirements of many investigators, two criticisms may be posed:

(1) The nomenclature does not adequately differentiate among the many more complicated materials described in the literature, and

(2) Because of its essentially qualitative nature, the nomenclature system does not readily suggest operations to allow subtle changes to be distinguished.

During the course of the remainder of this paper, a number of exotic blends and grafts will be mentioned which will justify the first criticism. Let us explore the second comment at this time. As a simple case, suppose we have synthesized poly(A-g-B), and wish to deduce the structure of its inverse, poly(B-g-A). Group theory provides a number of inversion centers and symmetry operations, through which one spatial arrangement of objects (such as a crystal lattice) can be transformed into a different, but related spatial arrangement. Figure 2 portrays the application of the as yet undescribed inversion center and operations, to transform poly(A-g-B) to poly(B-g-A).

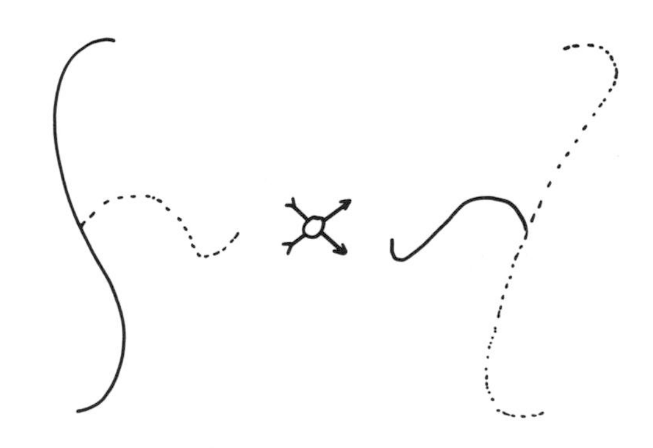

Figure 2. An inversion center as a symmetry operation.

It is helpful to imagine the inversion center as a special type of mirror. Group theory also provides for a mathematical language of "words" to describe operations in a quantitative way. The following sections will outline the application of group theory to polymer blends and grafts.

A GROUP THEORY APPROACH TO POLYMER BLEND CLASSIFICATION

Following the above descriptive approach, the need for a concise, mathematical formulation of polymer blend classification became obvious. Making use of certain simple concepts of group theory,[4-8] a preliminary effort at such classification will now be outlined.

The basic elements will be designated Polymer 1 and Polymer 2, to represent two distinguishable kinds of polymer molecules. As the basic operations, we designate the following:

Symbol	Operation
M	Addition of monomer
P	Polymerization
G	Grafting
C	Crosslinking

Polymerization will be assumed to be linear unless the operation C is specified.

In order to form a group, four conditions must be met: (1) The set must be closed: i.e., the combination of two operations also yields a member of the set. (2) The operations must include an identity, which leaves the molecules unchanged. (3) The associative law of combination must hold. (4) Every operation must have an inverse.

Examples may be helpful in the application of the four group theory conditions to polymer chemistry. For instance, condition (1) requires that if we crosslink a polymer, and then graft it, we generate a crosslinked, grafted polymer. The identity is interpreted as "no reaction". Two cases are important: (a) A man puts a beaker of styrene over an unlit Bunsen burner, or more significantly, (b) he polymerizes it, and then degrades it carefully back to monomer. An example of condition (3) requires that addition of monomer, polymerization, and crosslinking may be grouped in two identical ways: (a) The addition of monomer and polymerization, followed by crosslinking, or (b) the addition of monomer, followed by polymerization and crosslinking. Most importantly, however, the present group is non-commutative, and the time sequence of events must be preserved. Thus, the order of the operations must be maintained. One of the key findings to be explored below is that we must increase our sensitivity to the order of the operations in more complex cases, and not assume, for example, that crosslinking followed by grafting is identical to grafting followed by crosslinking.

The fourth condition is the most interesting, because it indicates we should consider a series of inverse operations. Thus, P^{-1} is introduced as depolymerization, G^{-1} as degrafting, C^{-1} as decrosslinking. In this context, C and C^{-1} are inverses. Thus, $CC^{-1} = I$, where I represents the mathematical identity. Also, $P_1P_1^{-1} = I$, etc.. In more chemical terms, the reaction was reversed. The quantity M_1M_1 would indicate two sequential additions of the same monomer. One should not assume, in general, that the effect will be identical as only one larger addition. Also, the sequence ...C...C^{-1}...C... should not be assumed to be equal to just C_1.* For example, the operations $M_1(P_1C_1)M_2P_2$ leads to network I being strained. If the subsequent operations are $C_1^{-1}C_1$, the polymer I network may no longer be strained, with concomitant changes in morphology and/or mechanical or swelling behavior. The operation M_1^{-1} indicates removal of monomer. As emphasized below few investigators have included the inverse operations in their polymer blend research in a systematic way, although a number of novel topologies are theoretically possible. More symbols will be introduced below to represent refinements. The symbols will be written from left to right, and in general will be non-commutative; i.e., the order of the operations is important.

Simultaneous operations will be placed in parentheses. Thus, parentheses will be used to distinguish operations which happen at the same time, from those that occur sequentially in time. Square brackets will be employed to distinguish separate polymerizations, of which their products are later mixed and/or reacted. In a formal sense, the parentheses and brackets constitute operations in their own right.

The polymer molecule will be associated with the designated operation with the use of the subscripts 1 and 2. Thus, P_1 indicates polymerization of polymer 1 in a linear form. The operation G will be specified by both subscripts, the order of the subscripts indicating respectively the backbone and the branch side chain. Post-polymerization grafting between equal partners will be designated by the subscript 3. Unless otherwise specified, G_{12} and G_{21} will indicate T-junctions, while G_3 will indicate X-junctions. A T-junction indicates one side chain only at the graft location, and an X-junction indicates continuity of both chains beyond the graft site. In the present context, the operation G will be employed as commonly designated in the graft-copolymer literature,[12] regardless of the true extent of chemical grafting.

A few examples are in order. The combination

$$\left[M_1P_1\right]\left[M_2P_2\right] \quad (1)$$

*However, $C_1C_1^{-1}C_1$, with no intervening steps, may be the same as just C_1, but should not in general be considered so. Likewise, $C_1C_1^{-1} \neq C_1$, in general.

represents a mixture of two linear polymers, as in a simple mechanical blend.[9,11,26] The solution-type graft-copolymer referred to above[12,14] may be written

$$M_1P_1M_2(P_2G_{12}) \qquad (2)$$

The combination

$$M_1(P_1C_1)M_2(P_2G_{12}) \qquad (3)$$

refers to a simple semi-IPN of the first kind.[31] Note that the order of the operations is important. For example,

$$M_1P_1M_2(P_2G_{12})C_1 \neq M_1(P_1C_1)M_2(P_2G_{12})C^{-1} \qquad (4)$$

the left hand side indicating a post-crosslinking operation for polymer 1, and the right hand side indicating that the crosslinks in polymer 1 are removed as the last step.

The finite collections of symbols as described above are sometimes called "words".[47] If we consider the permutations of all of the operations, each considered once in a word, the number of possible structures may be arrived at. Some, of course, are physically impossible. If we add more operations, the number grows rapidly. Before trying to assess the total possible number of ways of preparing these materials, let us seek some refinements which will permit different subclasses of materials to be more clearly distinguished.

INVERSION OPERATIONS

In an earlier section, the operations necessary to convert poly(A-g-B) to poly(B-g-A) were qualitatively discussed, and in Figure 2 a novel inversion center was introduced. With the aid of the four basic operations, M, P, G, and C, a more quantitative description may be provided. The inversion in Figure 2 can be represented by $G_{12}^{-1}G_{21}$, which means the degrafting of chain 2 from 1, followed by a grafting of 1 onto 2. Table I provides a short list of possible centers of inversion.

It is important to observe that inversion $G_{12}^{-1}G_{21}$ does not necessarily lead to the same properties as $G_{21}G_{12}^{-1}$, Figure 3. The latter leads through a joined-IPN architecture, and the molecules are never completely separated. The former leads through a polymer blend situation, where the molecules are decoupled. In the decoupled state, the molecules may relax, or assume a different morphology than permitted with the joined-IPN intermediate state.

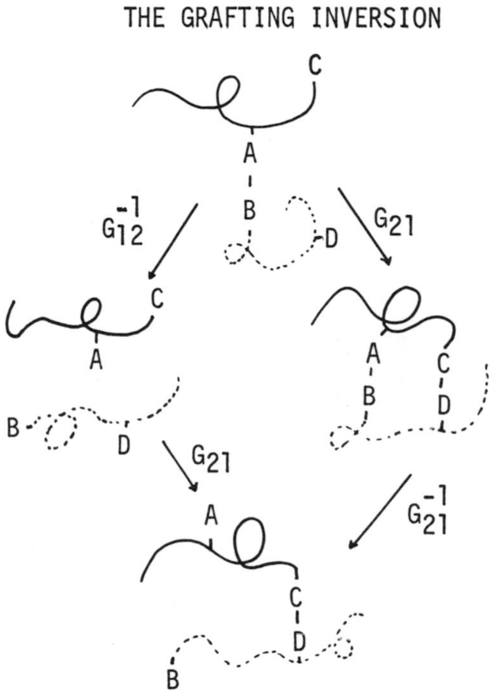

THE GRAFTING INVERSION

Figure 3. Two distinguishable ways of carrying out the grafting inversion, assuming the intramolecular mode. Site pairs A and B, and C and D are presumed reactive on a mutually exclusive basis. The intermolecular mode leads to a joined-IPN intermediate.

Table I. Some Simple Inversions

$G_{12}^{-1}G_{21}$	$G_{21}G_{12}^{-1}$
$C_1^{-1}\ C_2$	$C_2\ C_1^{-1}$
$P_1^{-1}\ P_2$	$P_2\ P_1^{-1}$
$M_1^{-1}\ M_2$	$M_2\ M_1^{-1}$
$(C_1G_{12})^{-1}(C_2G_{21})$	$(C_2G_{21})(C_1G_{12})^{-1}$
$(P_1C_1)^{-1}(P_2C_2)$	$(P_2C_2)(P_1C_1)^{-1}$

STATE OF THE POLYMER

Polymers may be either amorphous or crystalline, and separately they may be glassy or rubbery. Since the state of the polymer influences the mechanical behavior, it is important to designate these conditions. Thus, we choose the superscripts

A Amorphous
K Crystalline
V Glassy (Vitreous)
R Rubbery

The combinations are to be listed choosing either A or K, and either V or R, as appropriate. For example

$$P_1^{AV} \tag{5}$$

represents polymer 1 in the linear, amorphous, glassy state. Similarly,

$$M_1P_1^{KR}\ M_2(P_2C_2G_{12})^{AR} \tag{6}$$

represents a linear, partly crystalline, rubbery polymer grafted with a crosslinked, amorphous, rubbery polymer. An example is polyethylene, to which has been added a multifunctional acrylic monomer, followed by polymerization of the acrylic.[32] A complication in the nomenclature may arise if monomer II is added to glassy polymer I, followed by devitrification of I.

POLYMERIZATION MODE

The mode of polymerization is sometimes important; for example, some materials are prepared as latexes and mixed as such,[27,40,41] which is distinguishable from preparations in bulk which are subsequently mechanically blended. Thus we require the following sub back scripts:

S	Solution (graft)
E	Emulsion Latex
X	Suspension
B	Bulk

Thus,

$$\left[M_1\ {}_EP_1\right]\left[M_2\ {}_EP_2\right]\quad (C_1C_2) \tag{7}$$

represents mixing linear latex polymer 1 with linear latex polymer 2, followed by simultaneous crosslinking of polymers 1 and 2. The result is Frisch's IEN's.[42,43] The combination

$$M_1\ {}_B(P_1C_1)\ M_2\ {}_S(P_2C_2G_{12}) \tag{8}$$

represents the general designation of IPN's.[28,29] In this latter case, the deliberate introduction of crosslinks in both polymers is thought to dominate the numerically fewer grafting sites.

POLYMERIZATION ORDER

Two polymers may be synthesized sequentially, as in a simple graft copolymer. However, polymerization may also be simultaneous.[30] For example,

$$(M_1M_2)(P_1G_{12}P_2G_{21}) \tag{9}$$

indicates the simultaneous polymerization of Chains 1 and 2.

BLOCK COPOLYMERS

Topologically speaking, a block copolymer is a special kind of graft copolymer where the two ends are grafted together. The specialized nature of these materials, and the practice of linking three or more blocks together demands a special nomenclature for the operation. Thus we propose the operation B to designate the special grafting reactions noted to form blocks. Thus

$$M_2(P_2B_2(2))M_1(P_1B_1(1))\ M_2(P_2B_2(2)) \qquad (10)$$

indicates a triblock copolymer formed with a diradical initiator,[12,16,17] or anion.[18]

An interesting case results from a consideration of graded block copolymer, such as the result of an anionic polymerization of two monomers where the reactivity ratios are such that most of monomer I polymerizes before there is a gradual changeover to polymerization of monomer II, to form, lastly, block II.[48] In this situation, the block copolymer cannot be considered a special case of a graft copolymer. Indeed, the writer knows of no example in ordinary graft copolymers that corresponds to the graded blocks. (Let's make one!) While the material might be grouped as block I-random copolymer-block II, the situation is more complex, and will be deferred until a later date.

Other operations that might be considered involve linear and biaxial orientation, swelling, dissolving, compression, ionization, backbone chain reactions, oligomerization, branching, crystallization mode, etc. Since some of these bear importantly on morphology rather than topology, these too will be deferred until a later date.

FURTHER LITERATURE EXAMPLES

In a most interesting 1962 U.S. Patent, Vollmert[31] gives 31 examples of graft copolymers and semi-IPN's, about half of which form distinguishable materials in the present context. A few of these examples will now be presented and classified according to the group theory nomenclature above.

In his example 1,[31] poly(butyl acrylate-co-acrylic acid) is formed, and separately poly(styrene-co-1,4 butane-diol monoacrylate) is polymerized. The two polymers are mechanically mixed and heated with the simultaneous addition of 1,4 butane-diol. This material may be described:

$$\left[M_1P_1{}^{AR}\right]\left[M_2P_2{}^{AV}\right]\ {}_B(C_1G_3) \qquad (11)$$

In example 2,[31] the poly(butyl acrylate-co-acrylic acid) is dissolved in styrene plus 1,4 butane-diol monoacrylate and 1,4 butane-diol and polymerized and heated. This structure may be written:

$$M_1P_1{}^{AR}\ M_2\ {}_S(P_2G_{12})^{AV}\ {}_B(C_1G_3) \qquad (12)$$

In example 21,[31] a crosslinked latex of poly(butyl acrylate-co-1,4 butane diacrylate) is mixed mechanically with a polystyrene latex, which may be written:

$$\left[M_1\ _E(P_1C_1)^{AR}\right]\left[M_2\ _EP_2^{\ AV}\right] \qquad (13)$$

The above structure is distinguishable from example 22[31] wherein a poly(butyl acrylate-co-diallyl fumarate) crosslinked seed latex is overcoated with linear polystyrene, which may be described:

$$M_1\ _E(P_1C_1)^{AR}\ M_2\ _E(P_2G_{12})^{AV} \qquad (14)$$

Recently, Johnson and Labana[49] patented a structure wherein a crosslinked polymer network I prepared by emulsion polymerization served as a seed latex, to linear polymer 2. After coagulation and molding, polymer 2 was crosslinked to form a continuous network, resulting in a thermoset material. This may be described:

$$M_1\ _E(P_1C_1)\ M_2\ _E(P_2G_{12})\ _BC_2 \qquad (15)$$

While the material in formula (15) has two crosslinked networks, it obviously has a different topology from the IPN's described in formula (8).

NUMBERS OF BLENDS AND GRAFTS POTENTIALLY POSSIBLE

A key question may be posed at this point: Is the set of all possible operations open or closed? Put in other words: Is the potentially distinguishable number of blends and grafts infinite or finite? We have a small number of operators and subscripts, which if taken in all possible orders (some of them chemically impossible), yield, of course, a finite collection of materials. But the operations may be repeated, perhaps indefinitely, alternately with their inverses. This obviously leads to an infinite number. The introduction of new operations also increases the size of the set. If, as a practical limit, we establish that no operation may be repeated more than twice, the number of combinations possible is large, but not unlimited.

By contrast, a systematic group theory study of the number of crystal lattice structures possible (face centered, body centered, hexagonal, etc.) led to exactly 230 possible structures.[5,6]

UNEXPLORED OR UNRECOGNIZED COMBINATIONS

On practical grounds, it was concluded in the previous section that the number of possibly distinguishable blends and grafts is finite, but large, based on the limitation of two repeats per operation. Many combinations have not yet been tried. While some differ in trivial ways from known materials, it behooves us to look for broad classes of materials that may not yet have been made.

Attention will now be focused upon the inverse operations and some of their implications. A search of the literature reveals that very few inverse operations have been employed, outside of the tangential study of polymer degradation. As a simple example, let us consider two ways of preparing an IPN, one of them employing an inverse operation. As mentioned above, during the introduction of monomer 2, polymer network I is swollen, and remains under tension after the formation of network 2. Now if network I is decrosslinked, the chains will, in general, relax. See Figure 4.

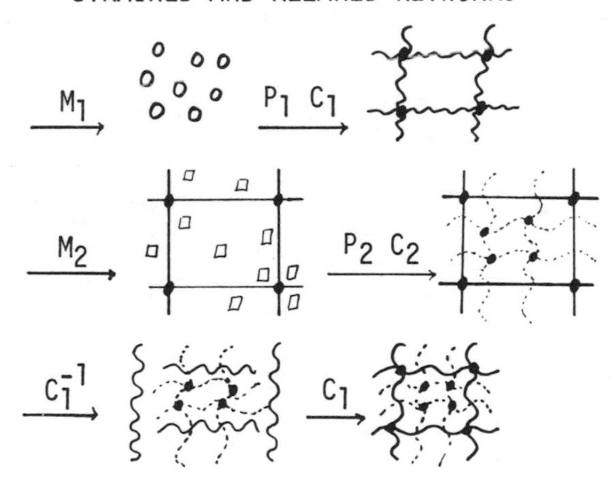

Figure 4. The synthesis of strained (straight lines) and relaxed (curved lines) networks in IPN's. The PC operations may be simultaneous or sequential.

(In Figure 4, the PC operations are left without parentheses, indicating sequential steps. Simultaneous and sequential steps should not be considered identical, as discussed in the next section.) Upon re-crosslinking of polymer I, a new IPN structure will be formed, which is distinguishable from the original IPN structure by a lack of tension in network I, with far-reaching consequences to morphological and physical behavior.

More generally, group theory contains a theorem which states that the inverse of a combination of two or more operations is equal to the combination of the inverses, in reverse order. In terms of the present context, this theorem relates to the decomposition of what was previously formed. Taking equation (3) as an example,

$$\left[M_1(P_1C_1)M_2(P_2G_{12})\right]^{-1} = (P_2G_{12})^{-1}M_2^{-1}(P_1C_1)^{-1}M_1^{-1} \quad (16)$$

which ends in the total removal of the material. Broadly interpreted, the inverse operations bear upon environmental pollution problems.

While in practice many of the polymerization, grafting, and crosslinking reactions now undertaken are difficult to reverse, it is easy to imagine model laboratory preparations capable of the required chemistry. A simple example involves the use of hydrolyzable ester groups.

TOPOLOGICALLY SIMILAR MATERIALS

It is interesting to compare a few combinations that are topologically the same, or very similar, although arrived at through different synthetic routes. A case of considerable importance involves crosslinking simultaneously with polymerization, as compared with post-curing. Many preparations involve (PC), the simultaneous polymerization and crosslinking. The use of divinyl benzene in addition polymerizations, or multifunctional condensation groups provide widely used examples. However, the vulcanization of rubber involves the post-curing of the linear polymer with sulfur, written PC, without the parentheses. More generally for blends and grafts, this may be written ....P....C.... to indicate possible synthetic steps between the polymerization and crosslinking reactions. While the two procedures result in very similar materials, the theory of rubber elasticity, more particularly the theory of thermoelasticity provides a tool for distinguishing the two materials.

A further example involves the "word"[46]

$$\left[M_1P_1\right]\left[M_2P_2\right]C_1 \quad (17)$$

which leads to a semi-IPN of the first kind with neither network strained. To the extent that grafting reactions can be ignored, the "word"

$$M_1(P_1C_1)M_2(P_2G_{12})C_1^{-1}C_1 \tag{18}$$

also leads to the same topology. The morphology, however, is probably much different.

Returning to Figure 4, it may be noted that delaying the operation C_1 until after (a) addition of M_2, (b) polymerization P_2, or (c) crosslinking C_2, leads to three more potentially distinguishable structures, all having relaxed network 1's, without going through the inverse step. Polymer II may be made linear, and subsequently crosslinked, or crosslinked simultaneously with synthesis, or if C_1 takes place after P_2, C_2 may take place last. Thus a total of 7 syntheses of IPN's may be made, all potentially distinguishable. Inverting the order of polymerization (synthesizing polymer 2 first), results in an additional 7 IPN's. Of these, 12 will have two relaxed networks, and two will have the first polymerized polymer network strained.

CLASSIFICATION BY TABLES

Although the number of blend and graft combinations known to the writer exceeds 150, and those yet unsynthesized are far more numerous, it is instructive to summarize briefly a range of examples. A systematic approach listing all the possible (and perhaps the impossible!) combinations will, unfortunately, have to await a sequel to the present paper.

Let us consider two tables. Table II details the first several blends and grafts, listed in the order of increasing complexity starting with the simplest mechanical blend. Table III samples some of the more exotic known materials, in no particular order. Although not comprehensive, the reader will understand the volume of work that remains to be accomplished.

Table II. Simple Blends and Grafts

Classification	Description	Reference*
$[M_1P_1]\,[M_2P_2]$	mechanical blend	11
$[M_1P_1]\,[M_2P_2]\,C_1$	mechanical blend	46 [I]
$[M_1P_1]\,[M_2P_2]\,C_1C_2$	mechanical blend	46 [I]
$M_1P_1M_2(P_2G_{12})$	graft copolymer	26
$M_1(P_1C_1)M_2(P_2G_{12})$	graft copolymer (Semi-IPN)	31
$M_1P_1M_2(P_2C_2G_{12})$	graft copolymer (Semi-IPN)	32
$M_1(P_1C_1)M_2(P_2C_2G_{12})$	IPN	28

* A typical recent reference is cited.

[I] Millar-type semi-IPN's and IPN's are considered, where polymers I and II are identical except for reactive groups. J. R. Millar, J. Chem. Soc., 1311 (1960).

Table III. More Complex Blends and Grafts

Classification	Description	References
$\left[M_1\ {}_{E}P_1\right]\left[M_2\ {}_{E}P_2\right]\ (C_1C_2)$	IEN	42
$\left[M_1\ {}_{X}(P_1C_1)\right]\left[M_2\ {}_{X}(P_2C_2)\right]$	mechanical blend	50
$M_1P_1^{AR}\ M_2\ {}_{S}(P_2G_{12})^{AG}{}_{B}(C_1G_3)$	graft copolymer	31
$M_1\ {}_{E}(P_1C_1)M_2\ {}_{E}(P_2G_{12})_{B}C_2$	graft copolymer	49
$M_1P_1M_2(P_2G_{12}C_{12})^{*}$	graft copolymer	37
$M_1(P_1B_1)^{AG}M_2(P_2B_2)^{KR}$	block copolymer	51
$M_1(P_1B_1)^{KG}M_2(P_2B_2)^{KR}$	block copolymer	52
$M_1(P_1B_1)^{AR}M_2(P_2B_2)^{AV}$	block copolymer	17
$M_1(P_1B_1)^{AR}M_2(P_2B_2)^{AV}$	block copolymer	26
$M_1P_1^{AV}M_2(P_2G_{12})^{KV}$	graft copolymer	53
$(M_1M_2)(P_1C_1G_{12}P_2C_2G_{21})$	SIN	30

* The C_{12} operation is introduced to describe the joined IPN topology of Bamford and Eastmond, where a single cross-linked network involves two polymers.

DISCUSSION

After developing the symbolism and format of the preceeding sections, let us re-examine the basic properties of group theory. How can group theory be best applied to polymer blends and grafts?

In an earlier section the four basic conditions to form a group were laid out. With the chemistry of polymers in mind, the identity I may be interpreted as "no reaction", or that nothing happened. Thus,

$$I\,X = X\,I = X \tag{19}$$

where X stands for the several operators previously established, M, P, C, G, and B.

All of the operations outlined have unique inverses. Once again, an inverse may be interpreted as an "undoing", or taking away, that which was done previously. From a chemical point of view, an inverse operation may follow, but not preceed, the specified operation. See Appendix I.

As mentioned previously, the present make up of the "words" is definitely not commutative. The order of the operations is important and unique. In general, then,

$$X\,Y \neq Y\,X \tag{20}$$

where X and Y are any two previously defined operations. The theory here is incomplete, and is concerned entirely with topology. However, qualitatively, it must be realized that most of the polymer blends and grafts considered have two distinct phases, and the detail of the phase structure depends not only on the final product formed, but also on the synthetic route. Thus, the order of the chemical reactions (operations) is paramount.

More serious in the consideration of group theory is the associative property. In a narrow sense, the present "word" make-up is not associative. The contents of parentheses have been taken to indicate simultaneous operations. Taking the simple semi-IPN of the first kind, for example:

$$M_1(P_1C_1)M_2(P_2G_{12}) \tag{21}$$

it must be observed that moving the parentheses:

$$M_1P_1(C_1M_2)(P_2G_{12}) \tag{22}$$

brings to mind a far different material, where polymer 1 is crosslinked simultaneously with the addition of monomer 2. Non-associative groups are defined by Baumslag and Chandler[54] as groupoids. Group theory includes groupoids as the broadest category, followed by semigroups and groups.

In a broader sense, however, the operations are associative. The first condition of a group requires that the combination of two members gives another member. The problem could be alleviated by defining new, more complex operators for all of the contents physically possible within the same parentheses. At the present, incomplete state of development of the theory, however, this appears to be an unnecessary complication.

Let us further consider the right-orientation property of the several operations. Assuming a normally conducted organic synthesis, the first operation will be the addition of monomer, M. Polymerization must follow, and never precede the addition of the appropriate monomer. (Of course, a second addition of the same monomer may follow a first polymerization.) Cases where crosslinking or grafting precede polymerization are discussed in Appendix I. Restrictions similar to right-orientation often lead to semi-group rather than group formation, in a formal sense. This difficulty can be partially obviated by including all possible combinations of operations in the mathematical analysis, and later separating out those "words" which make physical sense. Appendix I considers mathematically possible operations independent of presentation order.

APPENDIX I

If a mathematical or physical system meets the requirements of group theory, the operations written in a "word" must make sense independent of the order of presentation. (Of course, each such order of presentation will mean something different.) Two important cases arise in the present discussion.

(1) Can grafts, G, or crosslinks, C, occur before the appropriate polymerization step? Broadly interpreted, multifunctional monomers such as divinyl benzene may be considered preformed crosslink sites, even before polymerization. Thus,

$$(M_1C_1)P_1\cdot\cdot\cdot\cdot\cdot \qquad (23)$$

considers the addition of monomer and crosslinker, followed by polymerization. The operator M must now be considered to involve a difunctional, linear chain forming monomer, and P now represents polymerization, either of a linear polymer or a crosslinked material, depending on the presence or absence or crosslinker. The unrestricted placing of C clarifies and simplifies the polymerization step. We may now distinguish, in a formal manner, among crosslinks added before, during, or after polymerization with a clear and simple nomenclature.

Let us now consider grafting also as possibly occuring before, during, or after polymerization, in a context similar to the above. For example, the monomer N-methylol acrylamide may be reacted with cellulose in several ways. This monomer may be polymerized first, and then grafted to the cellulose,[55] or grafted to the cellulose first,[55] then polymerized.[56-58] For the latter, the "word" may be written:

$$P_1^{KV} M_2 G_{12} P_2^{AV} \qquad (24)$$

Note that in equation (24) P appears first. This may be easily interpreted as the direct addition of a pre-existing polymer.

(2) Can an inverse operation come before its "normal" counterpart? At the present time it appears that a much broader interpretation of the operations will be required than is desirable for this paper. Therefore, such considerations will be deferred until a later date.

REFERENCES

1. W. G. Chinn and N. E. Steenrod, "First Concepts of Topology", Random House, 1966.

2. W. B. Temple, Makromol. Chem., 160, 277 (1972).

3. O. Ore, "Graphs and their Uses", Random House, 1963.

4. D. McLachlan, Jr., "X-Ray Crystal Structure", McGraw Hill, 1957.

5. F. A. Cotton, "Chemical Applications of Group Theory", 2nd Ed., Wiley Interscience, 1971.

6. A. J. Mabis, Acta Cryst. , 15, 1152 (1962).

7. C. Hermann, Z. Kristallog., 79, 186, 337 (1931).

8. E. Artin, American Scientist, 38, 112 (1950).

9. P. F. Bruins, Ed., "Polyblends and Composites", Interscience, 1970. J. Appl. Polym. Sci. Applied Polymer Symposia No. 15.

10. H. Keskkula, Ed., Polymer Modification of Rubbers and and Plastics", Interscience, 1968. J. Appl. Polym. Sci. Applied Polymer Symposia No. 7.

11. J. E. Work, Polym. Eng. Sci., 13, 46 (1973).

12. G. E. Molau, Ed., "Colloidal and Morphological Behavior of Block and Graft Copolymers", Plenum, 1971.

13. N. A. J. Platzer, Chmn., "Multicomponent Polymer Systems", Adv. Chem. Series No. 99, ACS, 1971.

14. Graft Copolymer Symposium, V. Stannett, Chairman, American Chemical Society meeting, New York, August, 1972.

15. J. A. Manson and L. H. Sperling, "Polymer Blends and Composites, Broadly Defined", Plenum, in Press.

16. S. L. Aggarwal, Ed., "Block Polymers", Plenum, 1970.

17. J. Moacanin, G. Holden, and N. W. Tschoegl, Eds., "Block Copolymers", Interscience, 1969. J. Polymer Sci. 26C.

18. M. Szwarc, Polym. Eng. Sci., 13, 1 (1973).

19. F. W. Billmeyer, Jr., "Textbook of Polymer Science", 2nd Ed., Interscience, 1971.

20. J. Furukawa, Y. Iseda, and E. Kobayashi, Polym. J., 2, 377 (1971).

21. J. Furukawa, Angew. Makro. Chem., 23, 189 (1972); Rubber Chem. Tech., 45, 1532 (1972).

22. J. P. Kennedy and F. P. Baldwin, Fr. 1, 564, 485; Chem. Abs., 71, 102926g (1969).

23. J. P. Kennedy, presented at the XXII IUPAC meeting, Boston, Mass., July, 1971, Preprints, Vol. 1, p. 105.

24. D. J. Lyman, Revs. Macromol. Chem., 1, 355 (1966).

25. I. H. Silman and E. Katchalski, Ann. Rev. Biochem., 35, 873 (1966).

25a. S. L. Rosen, J. Appl. Polym. Sci., 17, 1805 (1973).

26. M. Matsuo, Japan Plastics, 2, 6 (July, 1968).

27. K. Kato, Japan Plastics, 2, 6 (April, 1968).

28. Volker Huelck, D. A. Thomas, and L. H. Sperling, Macromolecules, 5, 340, 348 (1972).

29. A. J. Curtius, M. J. Covitch, D. A. Thomas, and L. H. Sperling, Polym. Eng. and Sci., 12, 101 (1972).

30. L. H. Sperling and R. R. Arnts, Note, J. Appl. Polym. Sci., 15, 2371 (1971).

31. B. Vollmert, U. S. 3,005,859 (1962).

32. G. Odian and B. S. Bernstein, Nucleonics, 21, 80 (1963).

33. R. H. Kinsey, Appl. Polym. Symp., 11, 77 (1969).

34. A. Eisenberg, Macromol., 3, 147 (1970).

35. E. P. Otocka and T. K. Kwei, Macromol., 2, 244, 401 (1968).

36. A. V. Tobolsky, P. F. Lyons, and N. Hata, Macromol., 1, 515 (1968).

37. C. H. Bamford, G. C. Eastmond, and D. Whittle, Polymer, 12, 247 (1971).

38. L. H. Sperling and H. D. Sarge III, J. Appl. Polym. Sci., 16, 3041 (1972).

38a. B. Vollmert and H. Stutz, Angew. Makromol. Chemie, 20, 71 (1971).

39. M. Shen and M. B. Bever, J. Mats. Sci., 7, 741 (1972).

40. S. L. Rosen, Polym. Eng. Sci., 7, 115 (1967).

41. E. H. Merz, G. C. Claver, and M. Baer, J. Polym. Sci., 22, 325 (1956).

42. D. Klempner, H. L. Frisch, and K. C. Frisch, J. Polym. Sci., A-2, 8, 921 (1970).

43. M. Matsuo, T. K. Kwei, D. Klempner, and H. L. Frisch, Polym. Eng. Sci., 10, 327 (1970).

44. R. B. Mumford and J. L. Nevin, Modern Textiles, April, (1967).

45. J. Zimmerman, U. S. 3,393,252 (1968).

46. H. A. Clark, U. S. 3,527,842 (1970).

47. W. Magnus, A. Karass, and D. Solitar, "Combinatorial Group Theory", Interscience, 1966.

48. S. L. Aggarwal, presented before the American Physical Society meeting, February, 1969, Philadelphia, Pa.

49. O. B. Johnson and S. S. Labana, U. S. 3,659,003 (1972).

50. J. N. Weinstein, B. M. Misra, D. Kalif, and S. R. Caplan, Desalination, 12, 1 (1973).

51. A. J. Kovacs, J. A. Manson, and D. Levy, Kolloid-Z., 214, 1 (1966).

52. H. J. Hagenmeyer, Jr., and M. B. Edwards, J. Polym. Sci., 4C, 731 (1966).

53. M. Matzner, D. L. Schober, and J. E. McGrath, European Polym. J., 9, 469 (1973).

54. B. Baumslag and B. Chandler, "Theory and Problems of Group Theory." McGraw-Hill, 1968.

55. G. J. Mantell, U. S. 2,837,512 (1958).

56. J. L. Gardon, J. Appl. Polym. Sci., 5, 734 (1961).

57. J. L. Gardon, J. Polym. Sci., A, 2, 2657 (1964).

58. J. L. Gardon, U. S. 3,125,405 (1964).

EFFECTS OF MONOMER UNIT ARRANGEMENT ON THE PROPERTIES OF COPOLYMERS

JAMES F. KENNEY

M&T CHEMICALS INC.

Rahway, New Jersey 07065

Introduction

In recent years, an increasing awareness has arisen of the importance of structure as opposed to mere composition in determining the properties of copolymers (1-8). The physical properties of polymers are by far their most important attribute in regard to their practical utilization. An understanding of the relationship between the physical properties and molecular structure of polymers is essential to the polymer scientist who desires to synthesize new materials or to modify and process existing polymers. It is apparent that one can obtain unique properties in a polymeric material that can not be obtained in a homopolymer or a random copolymer by combining presently available monomers and polymers in unusual ways. There are at least four ways to formulate a polymeric material with new and useful properties (1) develop new monomers (2) develop new methods and techniques of polymerization (3) combine existing commercial monomers in such a way that the resulting material has certain superior properties and (4) combine existing polymers in an unusual way to achieve unique properties. For example, copolymers, terpolymers, and multicomponent polymers in which the monomer units are arranged in blocks, grafts, in an alternating arrangement or in some regular sequential arrangement offer opportunities to obtain unusual properties from available cheap commercial raw materials.

Monomer units in copolymers can be arranged as shown in Figure 1 in a random fashion, grafts, blocks and in an alternating fashion. In graft copolymers the backbone and side chains may both be homopolymeric or copolymeric. Methods available for producing the

ABAABABBBABABB	Random
AAAAAAAAAA B B B B B B	Graft
AAAAAAABBBBBBAAAAAAA	Block
ABABABABABABAB	Alternating

FIGURE I. MONOMER UNIT ARRANGEMENT

RANDOM
a. Free radical - solution, bulk, emulsion
b. Condensation
GRAFTS
a. Chain transfer - CH_2Br, triethylamine
b. Thiogroups - SH
c. Double bonds
d. Ceric ion
BLOCKS
a. Condensation of prepolymers
b. Melt blending of two polymers - Amide and ester interchange
c. Anionic polymerization
d. Anionic - free radical polymerization
ALTERNATING
a. Charge transfer using complex metal catalyst $ZnCl_2$, RAl halides RAlX e.g. $C_2H_5AlCl_2$.
b. Chemical reactivity of monomers, homopolymerization of one monomer does not occur.
STEREOREGULAR
Ziegler catalyst, alkyl metals, low temperature, selection of proper monomer to give isotactic or syndiotactic configuration.
POLYBLENDS - COMPATIBLE AND INCOMPATIBLE
a. Emulsion blending
b. Solution blending
c. Melt blending
d. Interpenetrating Polymeric Networks (IPN's)

FIGURE 2. SYNTHESIS METHODS

copolymeric structures illustrated in Figure I are shown in Figure 2. These techniques are well documented in the literature and will not be discussed further.

The properties of a copolymer of A units and B units are

influenced by the nature of the A and B units and by the length and arrangement of these units in the polymer chain. It is evident that various structural features possessed by units A and B which influence intermolecular interaction (van der Waals forces, hydrogen bonding, crystallinity, crosslinks, etc.) in each of the respective units will be altered to some extent in the copolymers depending not only on the molar quantities of the two units present but also on the sequence in which the units appear in the copolymer chain. Since the intermolecular interaction of A units with A units and B units with B units can be modified or altered by the presence of neighboring units of opposite kind in the interacting polymer chains, the importance of segment length and sequence becomes apparent. In general, as the length of A and B segments increases, each distinct segment of the copolymer chain patterns its individual behavior more nearly after that of its corresponding homopolymer. Block copolymers with sufficiently long blocks offer the possibility of combining the desirable properties of two different homopolymers rather than averaging them out as is frequently the case for random copolymers. Likewise, as the length of A and B segments decreases each monomer unit takes on a new behavior based on its interaction with neighboring units of opposite kind. Alternating copolymers have properties different than those of the homopolymers and random copolymers.

PRIMARY

Tensile strength, modulus, elongation, melting point, glass transition, heat deflection temperature.

SECONDARY

Impact strength, process stability, high melting point with high elongation, high elongation with high modulus, transparency, static build-up, wettability, dyeability, printability, barrier properties, soil release, etc.

FIGURE 3. POLYMER PROPERTIES

The properties of polymers can be separated into two major classes, primary and secondary properties as shown in Figure 3. Primary properties give mechanical strength to the material and allow the fabrication of a product or part whereas secondary properties enchance the primary properties and satisfies end use performance. Of practical significance is the desirability of maintaining the primary properties of a polymer while "building in" secondary properties in order to make a useful polymeric material.

This chapter attempts to correlate the effects of monomer unit arrangement on the physical and mechanical properties of copolymers in order to assess the effects of structural arrangement of a polymer on its properties. Data are presented on the physical and mechanical properties of block, graft, alternating and random copolymers, as well as polyblends.

Table I

Properties of Poly(ethylene terephthalate) Block Versus Random Copolyesters.

Copolymer	Type	Melting Point °C	Tenacity, g./den.	Elongation to break %	Modulus at 100% elongation, g./den.	Recovery %	Short-term stress decay, %
PET/PES(40/60)	Random	120	0.7	300	0.40	95	20
PET/PES(40/60)	Block	170	0.7	200	0.55	96	13
PET/PEG(40/60)	Block	200		350		98	8

PET=poly(ethylene terephthalate), PES=poly(ethylene sebacate), PEG-poly(ethylene oxide)
W. H. Charch and J. C. Shivers, Textile Research Journal, 29 536 (1959)

Discussion

Poly(ethylene terephthalate) (PET)/poly(ethylene sebacate) (PES)

The properties of random and block copolyesters received early attention. One of the first copolyesters studied was PET/PES. The polyester based on ethylene glycol and sebacic acid is amorphous and softens at about 70°C. The polyester based on ethylene glycol and terephthalic acid is crystalline and melts at about 250°C. The properties of random and block PET/PES copolyesters are given in Table I (3). The block copolymer has a melting point 50°C higher than the random copolymer and an improvement in stress decay without a loss in tenacity, modulus or elastic recovery. Through block copolymers it is qualitatively possible to combine significantly high melting points with significant elastic properties in one and the same polymer.

Poly(ethylene terephthalate) (PET/Poly(ethylene oxide) (PEG)

Further improvements in properties of PET copolymers can be achieved via block copolymers by melt blending a long chain (4000 molecular weight) poly(ethylene oxide) (PEG) with PET. This effect is also shown in Table I. There is a very substantial improvement in stress decay for the block copolymers as compared to the random copolymers. The PET/PEG block copolymer exhibits at one and the same time the highest melting point, highest elongation and the lowest stress decay.

Table II
Properties of Poly(hexamethylene adipamide)/Poly(metaphenylene) adipamide) Block Versus Random Copolymers

Type	Tenacity, g./den.	Elongation to break, %	Initial Modulus, g./den.	Relaxed cold growth %	Set	Flat-spot mils
Block(70/30)	8.7	17	100	1.4	0.56	117
Block(75/25)	7.0	10	70	2.3	0.68	127
Block(80/20)	7.1	12	75	2.6	1.04	157
Random(90/10)	7.6	13	68	3.7	1.45	190
Random(80/20)	3.1	15	46	5.7	2.3	263
Nylon-66(100/0)	8.5	16	49	3.7	1.7	210

British Patent 918,637 (1963), E. I. duPont deNemours

Poly(hexamethylene Adipamide) (Nylon-6,6/Poly(metaphenylene Adipamide)

Block copolymers of poly(hexamethylene adipamide)/poly(*meta*-phenylene adipamide) were prepared by the melt blending process and the corresponding random copolymers were obtained from the block copolymer blends containing similar proportions of the homopolymers, but after holding at high temperature long enough for randomization to take place (4). The properties of the block versus random copolymers are shown in Table II. Poly-(hexamethylene adipamide) (nylon-6,6) is included. It is clear the block copolymers possess improved modulus, cold growth, and set properties while there is no appreciable loss in other characteristics such as tenacity and elongation. More significantly, however, is the substantially reduced flat spot depth of tire cords made from the block copolymer versus the random copolymer and nylon-6,6. Furthermore, the structure of the block copolymer, as determined by X-rays, is essentially unchanged over that of poly (hexamethylene adipamide). A random copolymer of the same chemical composition shows a significantly reduced order in the structure.

Styrene (S) - Butadiene (BD)

S - BD block copolymers representing SB, SBS, BSB and SBSB sequence arrangements were synthesized by using n-butyl lithium catalyst in benzene at room temperature (2). Figure 4 shows that each structure exhibits two dynamic loss peaks E" at -80° and 110°C corresponding to polybutadiene and polystyrene, respectively. Thus a heterogeneous two-phase system is present. The stress-strain behavior is markedly affected by changes in the sequence arrangements as shown in Figure 5. SBS and SBSB are tough, as evidenced

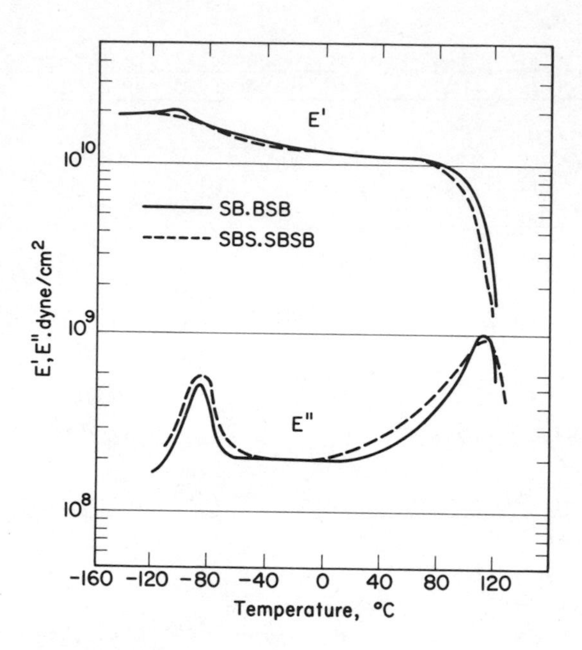

Figure 4. Temperature dependences of dynamic modulus (E') and dynamic loss (E") for styrene-butadiene block copolymer arrangements. Frequency 110 c/s. M. Matsuo, et. al., Polymer 9 425 (1968).

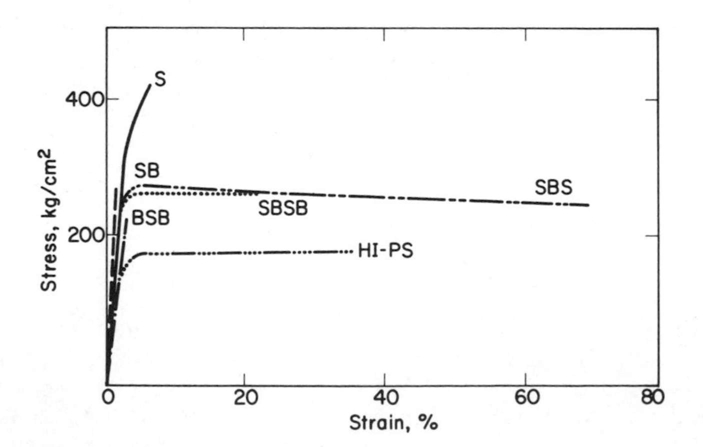

Figure 5. Stress-strain properties for styrene-butadiene block copolymer arrangements. M. Matsuo, et. al., Polymer 9 425 (1968).

Table III
Properties of Styrene-Butadiene Block Copolymers

Samples	ST/BD mole ratios (charged)	Elongn at break	Light transmittance	Charpy Impact strength $kgcm/cm^2$	Heat distort. temperature
		%	%		°C
S	100/0	7	90	2.1	97
SB	60/40	2	62	2.5	83
SBS	60/40	70	71	4.4	67
BSB	60/40	2	51	2.4	78
SBSB	60/40	23	70	3.1	55
Commercial HI-PS	---	30	0	7.0	75
Polyblend	60/40	2	0	0.5	--

M. Matsuo, et. al., Polymer 9 425 (1968).

by their high elongation at break while SB and BSB are as brittle as polystyrene. A polyblend of the same composition is also brittle. In SBS, the elongation at break increases and yield stress decreases with increasing BD sequence length. The data are summarized in Table III. SBS exhibits the highest impact strength of the sequence arrangements but does not exceed high impact polystyrene. SBS and SBSB show crazing or stress whitening under tensile stress as observed in ABS polymer. Heat distortion temperatures decrease with increasing BD ratio in SBS. The flow properties indicate easy processability, as the flow curves lie between that of polystyrene and high impact polystyrene. It is well known that two-phase systems of different refractive indices are opaque. The S-BD block-copolymers described here are transparent in contrast to the opacity of a polyblend and high impact polystyrene of the same composition. Transparency in these block copolymers is due to the small phase size. Electron micrographs of SBS show spherical PBD particles of about 300A° dispersed in the PS matrix. On the other hand, SB and BSB sequence arrangements exhibit PDB chains linked together to form irregularly shaped rod-like structures.

Acrylonitrile-Butadiene-Styrene (ABS)

The addition of rubber to a glassy polymer increases toughness without substantial loss in modulus and heat distortion temperature. Figure 6 shows the loss modulus E" and impact strength with temperature for graft and blend ABS containing the same 20 weight -% rubber (6). The E" - temperature curves for the graft and blend exhibit two loss peaks at -80°C and 110°C corresponding to polybutadiene and styrene-acrylonitrile copolymer phases, respectively. However,

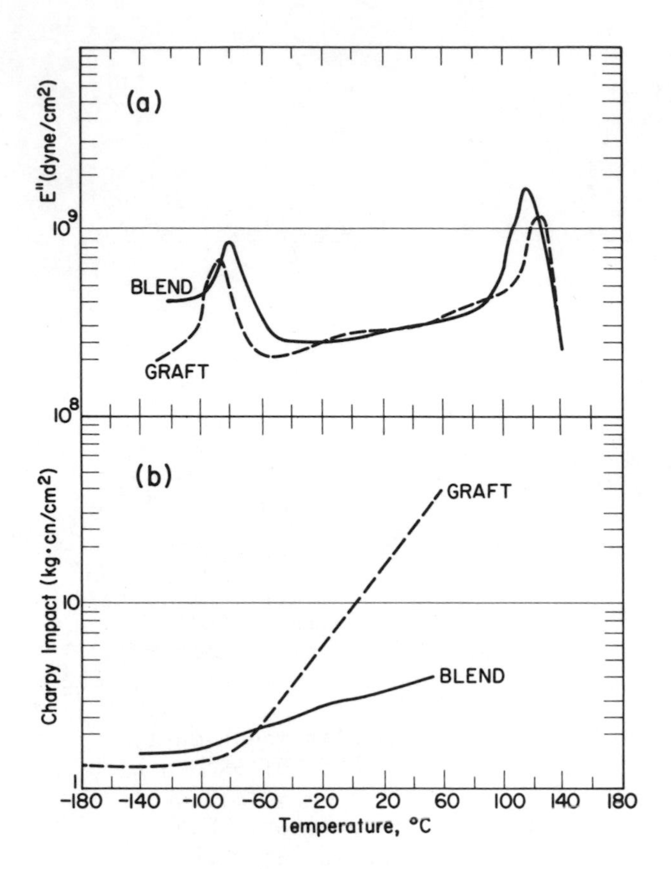

Figure 6. Temperature dependencies of dynamic loss modulus (E") and Charpy impact strength for graft-and blend-ABS. (Rubber contents are the same: 20 wt.-%). M. Matsuo, et. al., Polymer Engineering and Science 10, 253 (1970).

the impact strength of the graft ABS in the useful temperature range is far superior to the polyblend of the same composition. Grafting the SAN glassy phase to the PBD rubbery phase increases the interfacial strength or adhesion between the two phases resulting in higher impact strength.

The stress-strain curve of the two-phase graft copolymer exhibits a yield point below the ultimate tensile strength of the glass, followed by ductile deformation to failure. It is the ductile deformation after yield that results in the high impact strength of the graft copolymer. This ductile deformation is accompanied by "stress-whitening". It commences at the yield point and as the tensile strain increases beyond yield, the area of stress-whitening expands and deepens with increase in strain. A blend of polybutadiene and SAN copolymer does not possess the properties of the graft copolymer, i.e., both a high modulus and high impact strength together.

During high shear melt processing both blends and graft copolymers may suffer a very large decrease in impact strength. The two-phase graft copolymers tend to aggregate and exhibit poor dispersibility under high shear conditions during processing, resulting in poor impact strength. The loss in properties on processing can be traced to the inability of the rubber phase particles to preserve the desired size. The rubber particles not being crosslinked deform and aggregate during processing. Crosslinking the graft copolymer retards the deformation and aggregation processes. Crosslinking of rubber and glass phases concurrent with grafting lead to better preservation of morphology and improved mechanical properties. The graft copolymer must be sufficiently tightly crosslinked to resist deformation.

Crosslinking the rubber phase as well as the graft glass phase improves melt processability and reduces the amount of elastic strain energy which causes melt fracture or extrudate roughness and die swell. On processing, the crosslinked graft copolymer exhibits lower die swell, and higher extrusion rates are possible.

Crosslinking increases the yield stress which tells us how much stress a fabricated piece can withstand before undergoing a deformation or how well it can absorb an impact without fracture. A crosslinked graft copolymer exhibits a high yield stress, low die swell, reduced melt fracture and high impact strength.

Table IV

MECHANICAL PROPERTIES OF BUTADIENE-ACRYLONITRILE CURED RUBBER

Cure time 60 min.

	ALTERNATING	RANDOM
At room temperature		
Hardness	73	86
Modulus at 100% (Kg/cm^2)	43	105
Tensile strength (Kg/cm^2)	237	194
Ultimate elongation (%)	400	210
At 110°C		
Tensile strength (Kg/cm^2)	58	50
Ultimate elongation (%)	340	210
After immersion in oil		
Change in weight (%)	12	12
Tensile strength (Kg/cm^2)	147	97
Ultimate elongation (%)	300	150

J. Furukawa, et. al., Polymer Letters 7 561 (1969).

Butadiene (BD) - Acrylonitirle (AN)

Table IV summarizes Furukawa's data on BD-AN alternating and random copolymer rubbers containing 50 mole-% AN and cured with the same amount of sulfur (7). The alternating copolymer has lower hardness and modulus while its tensile strength and elongation at break are considerably greater. This means that the alternating copolymer is more flexible, stronger and has excellent elastomeric toughness. Oil resistances of both copolymers are the same. However, the alternating copolymer arrangement is superior to the random in toughness after swelling. In addition, alternating BD-AN copolymer rubber possesses a lower Tg than the random and superior rebound elasticity. The superior properties of the alternating copolymer rubber is attributed to the high regularity of alternating monomer unit arrangement in the copolymer chain.

Mukherjee (11) found that the dynamic mechanical properties of the alternating and random copolymers are nearly the same. Viscoelastic measurements show that the elastic modulus (E'), the activation energy for the glass-rubber transition (ΔH) and Tg are equivalent over a wide range of temperature and frequency. The difference in Tg value found by Furukawa (Tg-25°C) and Mukherjee (Tg-14°C) for the alternating copolymer depends on the degree of alternation in the copolymer which would have significant effects on Tg.

Stress-strain properties of the random copolymer cured with 0.8 phr dicumyl peroxide are comparable to those of the alternating copolymer cured with 1.7 phr dicumyl peroxide. Therefore, Furukawa's results indicating the superior tensile properties of the alternating copolymer could result from overcuring of the random copolymer. At the same dicumyl peroxide level, the random copolymer may be overcured resulting in a lowering of tensile strength and ultimate elongation compared to the alternating copolymer. This overcuring in the random copolymer may be due to the higher vinyl content of the butadiene than that of the alternating copolymer which contains mainly trans-1,4-units.

Stress-birefringence measurements show that the random copolymer does not exhibit any hysteresis, whereas the alternating copolymer exhibits a positive hysteresis. These data for the alternating copolymer indicate the occurrence of strain induced crystallization. Therefore, simultaneous stress-birefringence measurements indicate that the alternating copolymer crystallizes on stretching, whereas the random copolymer fails to show any crystallization even when stretched up to a level of 520%. The superior ultimate properties of the alternating copolymer result from strain induced crystallization at high strain attributed to the high regularity of alternating monomer unit arrangement in the copolymer chain.

TABLE V

PROPERTIES OF STYRENE-ACRYLONITRILE COPOLYMERS, 50/50 mole-%

	T(E")°C	TENSILE STRENGTH Kg/cm^2	IMPACT STRENGTH Kg/cm
Alternating	120	860	0.8
Random	113	990	0.9

Alternating - Higher resistance to coloration by alkali and heat.
S. Yabumoto, et. al., J. Polymer Sci., A-1 7 1683 (1969).

TABLE VI

PROPERTIES OF ACRYLONITRILE-VINYL CHLORIDE FIBERS

	Tenacity, g/denier	Elongation at break, %	Young's modulus, g/denier	Sonic modulus, g/denier
Random	4.09	22	50	100
Alternating	2.19	15	54	95

G. Wentworth and J. R. Sechrist, J. Appl, Polym. Sci., 16, 1863 (1972).

Styrene (S) - Acrylonitrile (AN)

Alternating SAN copolymer containing 50 mole-% AN was prepared in the presence of zinc chloride and its properties compared with a random copolymer of the same composition prepared by radical polymerization (8). As shown in Table V, the alternating copolymer possesses higher Tg, less color formation on heat treatment and less color formation on treatment with alkali. We are aware of the utility of high acrylonitrile polymers but they are limited by color formation on heating. The alternating acrylonitrile unit arrangement broadens the practical utility of high acrylonitrile polymers. The improvement in properties of the alternating copolymer over the random is due to the absence of long sequences of AN in the alternating S-AN arrangement. Tensile and impact strengths of alternating and random SAN copolymers are comparable.

Acrylonitrile (AN) - Vinyl Chloride (VCl)

AN-VCl fibers were prepared from an alternating copolymer and a random copolymer containing the same 50 mole -% AN (5). Tenacity of the random copolymer fiber is 4 g/denier whereas the alternating copolymer fiber has considerably lower tenacity of 2 g/denier, in spite of the presence of 50 mole-% AN. Other fiber

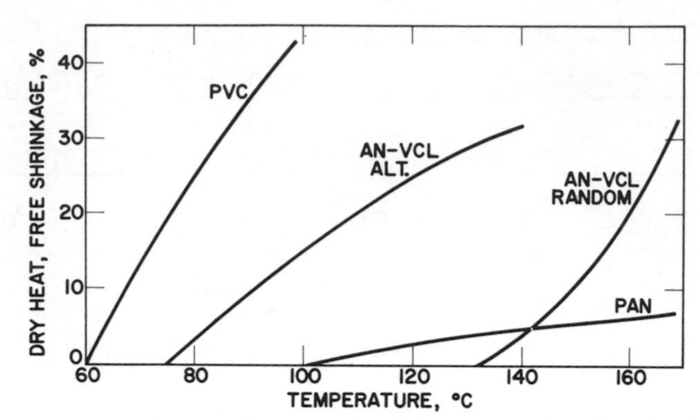

Figure 7. Effect of copolymer structure on dry-heat shrinkage properties of acrylonitrile-vinyl chloride fibers. G. Wentworth and J. R. Sechrist, J. Appl. Polym. Sci., 16, 1863 (1972).

tensile properties shown in Table VI of the alternating and random structures are similar and typical of modacrylic fibers. However, as shown in Figure 7, dry-heat shrinkage of the random copolymer fiber (an important end-use property of modacrylic fibers) is markedly superior to that of the alternating copolymer fiber at the same degree of orientation. The ANVCl random copolymer fiber exhibits rarely observed superiority over a copolymer of similar composition containing monomer units arranged in an alternating sequential arrangement. Electrostatic interactions in long sequences of AN units on dry-heat shrinkage and tenacity are evidently needed for superior fiber properties.

Vinylidene Chloride (VdCl) - Methylacrylate (MA) and Ethylacrylate (EA)

VdCl-MA and VdCl-EA copolymers exhibit considerable deviation from ideal type equations relating Tg and composition and do not show the usual minimum in Tg between the two homopolymers. These copolymers exhibit a rarely observed maximum in Tg at about 50 mole-% as shown in Figure 8 (9). The maximum is attributed to restriction of free rotation of the ester group by adjacent chlorine atoms in neighboring vinylidene chloride units. This is likely to be a maximum when there are equimolar amounts of the comonomers present and short monomer sequences.

Ethylene (E) - Chlorotrifluoroethylene (CTFE)

Figure 9 shows that the curve of melting temperature vs composition for various E-CTFE copolymers synthesized at -78°C by radical polymerization shows a maximum at 264°C for the 50-50 molar composition (10). The maximum temperature is considerably higher than the melting temperatures of both linear PE (Tm=137.5°C) and PCTFE (Tm=223°C). The very low values of both the reactivity

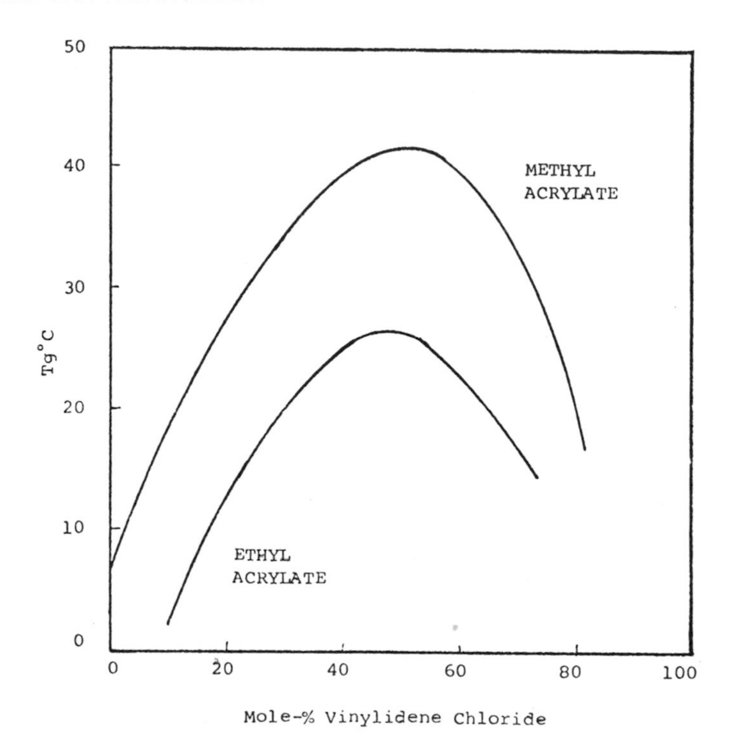

Figure 8. Glass transition temperatures of vinylidene chloride - ethyl acrylate and vinylidene chloride - methylacrylate copolymers. E. Powell and B. Elgood, Chemistry and Industry, 901 (1966.)

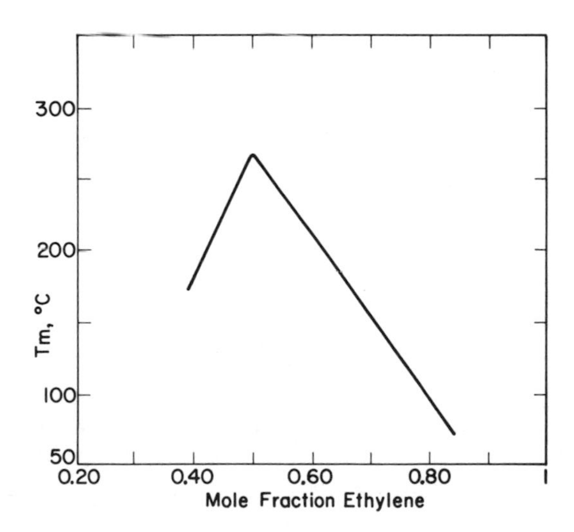

Figure 9. Melting temperatures of ethylene-chlorotrifluoroethylene copolymers, synthesized at -78°C. C. Garbuglio, et. al., European Polymer Journal 3, 137 (1967).

TABLE VII

ETHYLENE-CHLOROTRIFULOROETHYLENE COPOLYMER 50-50 MOLE-%

Pzn. Temp. °C	Product r_1r_2	M. P., °C	
60	0.00730	206	Alternation
0	0.00165	243	↓
-40	0.00058	252	increasing
-78	0.00007	264	

M. Ragazzini, et. al., European Poly. J., 3 129 (1967)

ratios of the monomers and the value of the product of the reactivity ratios as shown in Table VII is consistent with an alternating arrangement of the monomers in the chain at the 50-50 composition. The new crystalline structure formed by the alternating arrangement has a melting temperature higher than that of either homopolymer. As the temperature of polymerization is lowered from 60° to -78°C, the product of the reactivity ratios becomes nearly zero, with increasing alternation of the monomer units. The melting temperature and crystallinity increase with decreasing polymerization temperature. Commercial E-CTFE alternating copolymer has found applications as an engineering thermoplastic possessing high tensile strength, impact resistance, hardness and abrasion resistance.

New Research: Alpha-Methylstyrene (AMS) - Methacrylonitrile (MAN)

AMS-MAN copolymers were prepared by emulsion techniques. Table VIII shows a correlation for high conversion copolymer of copolymer composition, run number, sequence distribution, Izod impact and deflection temperatures under load (DTUL).

Table VIII shows results of computer calculations based on a Fortran II (IBM 1620) program written by Professor Harwood of Akron University. The run number, diad and triad distributions shown in Table VIII were calculated via this program. These data show the monomer unit arrangement for different monomer feed compositions. The run number is defined as the average number of uninterrupted monomer sequences (or runs) which occur in a copolymer chain per hundred monomer units. The run number increases with alternation of the monomer units and for a perfectly alternating copolymer is 100 and for a random copolymer is 50.

The run number can be predicted for copolymer formed initially from a given copolymerization mixture and for the azeotropic composition by (12):

Table VIII

Sequence Distribution and Properties of High Conversion AMS-MAN Copolymer

				Diad Distribution			Triad Distribution								
Monomer Charge MAN Mole-%	Conv. %	Copolymer Composition MAN Mole-%	Run No.	A-A %	A-B + B-A %	B-B %	A-A-A %	A-A-B %	B-A-B %	B-B-B %	B-B-A %	A-B-A %	ηinh	Izod impact ft.lb./in. notch	DTUL °C
100	80	100	-	-	-	100	-	-	-	100	-	-	-	0.45	93
82	91	67	40	0	40	60	0	2	98	58	34	8	1.70	0.43	113
64	93	60	75	1	75	24	0	5	95	16	47	37	1.52	0.24	124
57	84	57	83	2	83	15	0	8	92	7	40	53	1.35	0.25	129
50	57	54	86	3	86	11	0	12	88	4	32	64	1.42	0.24	123
37	48	52	88	6	88	6	2	21	78	1	20	79	1.22	0.24	121

A= Alpha-Methylstyrene

B= Methacrylonitrile

$$R = \frac{200}{2 + r_1 \frac{\% A_f}{\% B_f} + r_2 \frac{\% B_f}{\% A_f}} \qquad (1)$$

where A_f and B_f are mole-percent A and B respectively in the monomer feed. For the emulsion copolymerization of AMS and MAN, r_1 (AMS) is 0.06 and r_2 (MAN) is 0.28. For the azeotropic composition, 43 mole-% AMS and 57 mole-% MAN, R is calculated from equation (1) to be 83; compare with R_{random} = 50 and $R_{alternating}$ = 100. The run number concept which greatly simplifies the calculation of monomer sequence distribution has been coined and derived by Harwood and Ritchey (13). Table VIII shows that the run number and therefore monomer unit alternation increases with decreasing methacrylonitrile in the monomer feed. The run number allows one to rank the copolymers according to degree of alternation.

For the azeotropic copolymer the data show that 83% of the diad placements is an AMS unit followed by a MAN unit or vice versa. The triad sequence in which an AMS unit is centered between two MAN units is 92%. Note that the AMS-AMS diad sequence is only 2% and there are no AMS-AMS-AMS triad sequences in the copolymer. This is in agreement with the well known finding that AMS does not homopolymerize with free radical catalyst. The average length of a given sequence can be calculated independently of the run number by (14):

$$\overline{AMS} = r_1 \frac{\% A_f}{\% B_f} + 1 \qquad (2)$$

$$\overline{MAN} = r_2 \frac{\% B_f}{\% A_f} + 1 \qquad (3)$$

where A_f and B_f are mole-percent A and B, respectively in monomer feed. The sequence length of AMS and MAN units is calculated to be 1.04 and 1.37 respectively. These data show the highly alternating character of the AMS-MAN azeotropic copolymer and characterizes the sequence distribution quantitatively.

Glass transition temperatures of copolymers are dependent on intermolecular forces involving A and B units and on the flexibility of A-A, A-B and B-B linkages. Since sequence distribution affects chain flexibility then, Tg will be sensitive to sequence distribution as well as copolymer composition (15). Copolymer prepared at the azeotropic composition (57 mole-% MAN) which is the most homogeneous composition, has the highest DTUL (129°C), although this composition is not the most alternating (highest run number). The DTUL's can not be rationalized in terms of the sequence distribution data. The DTUL seems to depend upon compositional homogeneity in the copolymer independent of the monomer unit arrangement or sequence distribution.

The azeotropic copolymer possessed the best overall properties, however, impact strength was very low. The impact strength was improved by graft emulsion copolymerization upon a BD-MAN copolymer. Impact strength of the AMS-MAN copolymer was significantly improved on grafting upon 20-30 weight-% rubber. These data are shown in Figure 10. The impact modified copolymer possessed an Izod impact strength of 6 ft. lb/inch notch and a DTUL of 115°C.

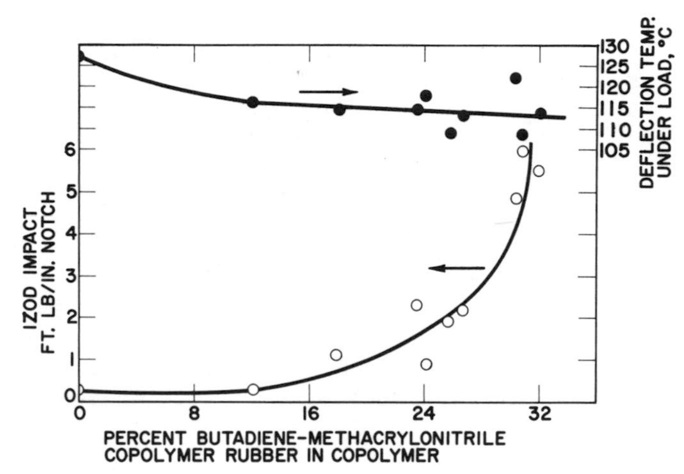

Figure 10. Effect of rubber content on impact strength and deflection temperature under load for AMS-MAN graft onto BD-MAN.

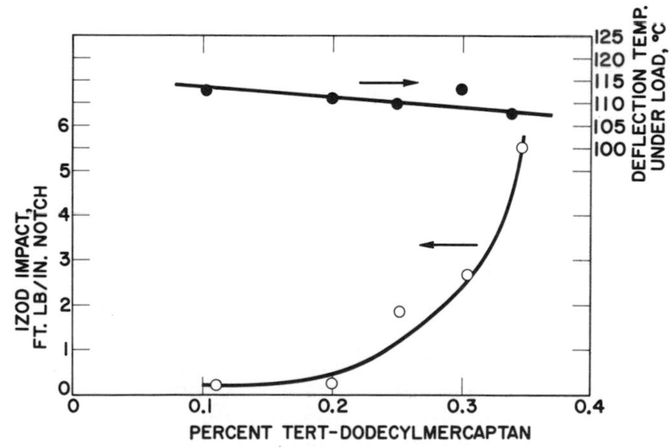

Figure 11. Effect of tert-dodecylmercaptan on impact strength and deflection temperature under load for AMS-MAN graft onto BD-MAN.

Significant improvement in Izod impact strength with a minimum lowering of DTUL was obtained when tert-dodecylmercaptan chain transfer agent was used during the graft stage copolymerization as shown in Figure 11. Increasing tert-dodecylmercaptan concentration apparently enhances grafting of AMS-MAN glass phase via chain transfer processes onto the polybutadiene copolymer substrate rubber. The following chain transfer processes are indicated. Reaction (1) shows the reaction of growing glass phase polymer with mercaptan resulting in termination of the growing chain and generation of the transfer radical RS.

$$M_x\cdot + RSH \longrightarrow M_x\text{-}H + RS\cdot \qquad (1)$$

Reactions (2) and (3) show the actual chain transfer step where radical RS• adds to the unsaturation in substrate polybutadiene rubber forming a graft site on the rubber,

$$RS\cdot + \text{-C-C=C-C-} \longrightarrow \text{-C-}\dot{C}\text{-C(SR)-C-} \qquad (2)$$

or by abstraction of hydrogen from polybutadiene.

$$RS\cdot + \text{-C-C=C-C-} \longrightarrow RSH + \text{-C-C=C-}\dot{C}\text{-} \qquad (3)$$

Reactions (4) and (5) show formation of the graft chain by glass phase forming monomer addition to the graft site on the polybutadiene rubber.

$$\text{-C-}\dot{C}\text{-C(SR)-C-} + M \longrightarrow \text{-C-C(}M\cdot\text{)-C(SR)-C-} \qquad (4)$$

or

$$\text{-C-C=C-}\dot{C}\text{-} + M \longrightarrow \text{-C-C=C-C(}M\cdot\text{)-} \qquad (5)$$

Figure 11 shows that 0.35% tert-dodecylmercaptan in the glass phase monomers results in an Izod impact strength of 6 ft. lb./inch notch and a DTUL of 115°C.

A two-phase system is readily apparent from Figure 12 with two glass transition temperatures, one for the butadiene-methacrylonitrile (62/38) copolymer rubber at -10°C and another for the glassy copolymer at around 145°C. The impact modified copolymer is opaque to translucent whereas the unmodified copolymer is transparent. The opaqueness is due, of course, to a difference in refractive index of the rubber phase and the glassy copolymer phase. Substantial impact improvement of the copolymer could not be obtained by mixing rubber and copolymer latexes, melt blending

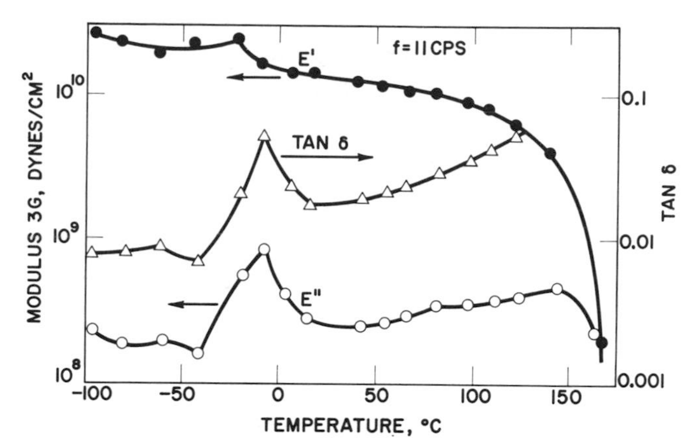

Figure 12. Modulus-temperature and tan δ temperature curve for AMS-MAN graft onto BD-MAN.

of rubber and copolymer, or melt blending with ABS.

Alpha-Methylstyrene-Methacrylonitrile (AMS-MAN) - Polyvinyl chloride (PVC) Polyblend

Blending two polymers usually leads to a class of materials whose properties are due to the presence of two phases, with two glass transition temperatures and usually opaque. Polyblends are physical mixtures of polymers and since the mixing of polymers is normally an endothermic process, heterogeneous polyblends are generally obtained. Polyblends are generally incompatible mixtures. On the other hand, if the polymers comprising the mixture have a strong enough affinity for one another, they will be compatible and mutually soluble. Such mixtures form homogeneous polyblends, have a single phase, are characterized by a single glass transition temperature and usually transparent. For compatibility, the solubility parameter of the polymers must match very closely. The degree of compatibility is determined by the size and distribution of the phases, i.e. how finely one polymer is dispersed within the other.

Blends of AMS-MAN azeotropic copolymer and PVC were prepared by melt blending at 180-200°C. Figure 13 shows the relationship between deflection temperature under load (DTUL) and composition of the polyblend. The broken line was calculated from the well-known Fox equation for random copolymers:

$$\frac{1}{DTUL} = \frac{W_1}{DTUL} + \frac{W_2}{DTUL} \qquad (4)$$

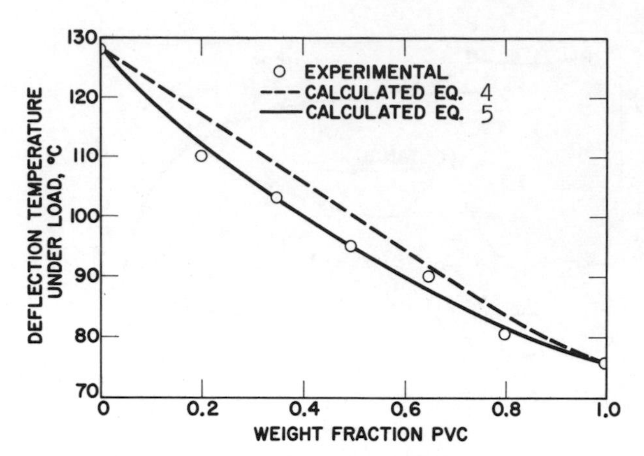

Figure 13. Dependence of deflection temperature under load on composition of AMS-MAN/PVC polyblend.

where W_1 and W_2 are the weight fractions of each of the components of the polyblend. DTUL is substituted for T_g. It is seen the polyblend approximately obeys the Fox equation for random copolymers exhibiting a DTUL about 5°C below that calculated. This behavior is not unusual even for random copolymers since volume additivity of the components is seldom ideal. Many copolymer systems which do not obey the Fox equation have been found to obey the following form of the Gordon-Taylor relationship:

$$DTUL = \frac{DTUL_1 W_1 + K\,DTUL_2 W_2}{W_1 + K W_2} \qquad (5)$$

where K is a factor which takes into consideration the thermal expansion coefficients of the polymer in the liquid and in the glassy state. If equation (5) is re-written in the form:

$$DTUL = DTUL_2 - \frac{1}{K} (DTUL - DTUL_1) \frac{W_1}{W_2} \qquad (6)$$

then it is seen that a plot of DTUL against $(DTUL-DTUL_1)\frac{W_1}{W_2}$ will have a slope $-\frac{1}{K}$ and intercept $DTUL_2$. This plot based on experimental data for AMS-MAN/PVC polyblends of different composition, results in a value of K = 0.60. Figure 13 shows that there is good agreement between the experimental points and the theoretical line calculated from equation (5) for K = 0.60. Thus, the

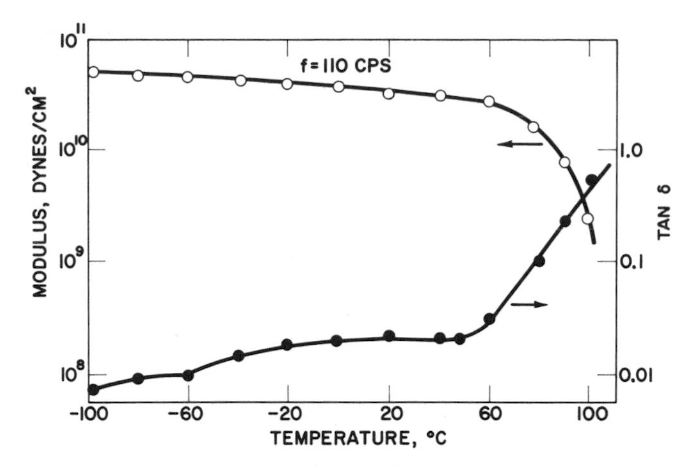

Figure 14. Modulus-temperature and tan δ-temperature curves for 50/50 wt-% AMS-MAN/PVC polyblend.

properties of this polyblend are similar to those expected from a random copolymer of the components of the polyblend.

The modulus-temperature and tan δ-temperature curves of a 50/50 weight-% AMS-MAN/PVC polyblend is shown in Figure 14. It can be seen that a single T_g at approximately 100°C exists. This unusual feature of this polyblend is unexpected and suggests homogeneity. The single T_g suggests a single-phase system or that if two phases are present they must be extremely small.

The two polymers comprising the polyblend are mutually soluble. The solubility parameter (δ) of PVC is 9.6. The solubility parameter of the copolymer was calculated from equation (7) to be also 9.6.

$$\delta_{mixture} = V_1\delta_1 + V_2\delta_2 \qquad (7)$$

V_1 and V_2 are volume fractions of the components of the copolymer. Since the solubility parameters of the two components are identical, they exhibit rarely observed polymer-polymer compatibility. The refractive index of the polyblend was found to be 1.546 at 23°C. The properties of a polyblend prepared by blending on a mill 50 weight-% PVC and 50 weight-% AMS-MAN copolymer are shown in Table IX. AMS-MAN azeotropic copolymer has tensile and flexural properties in the useful range with a DTUL of 129°C. Similarly, the polyblend has useful tensile and flexural properties with a DTUL of 95°C.

TABLE IX

Properties of Alpha - Methylstyrene-Methacrylonitrile Azeotropic Copolymer, Poly(vinyl chloride) and Polyblend

	PVC	AMS-MAN/PVC 50/50 wt.% Polyblend	AMS-MAN
Deflection temperature under load, °C	68	95	129
Izod impact, ft.lb./in. notch at 23°C	0.7	0.5	0.3
Elongation, %	110	5	3
Stress, psi	8,000	10,180	6,800
Tensile modulus, psi	180,000	214,000	265,000
Flexural strength, psi	15,000	17,375	14,600
Flexural modulus, psi	430,000	500,600	638,000

Summary

The data indicate that the properties of block copolymers and alternating copolymers are generally superior to the corresponding random copolymers, and graft copolymers are superior to the corresponding polyblends. By proper monomer unit arrangement, presently available monomers and polymers can be used to prepare unique polymeric materials containing desirable "built-in" properties to satisfy end-use performance requirements.

Acknowledgement

The author wishes to acknowledge Messrs. P. J. Patel and A. Birkmarer for their contributions to the AMS-MAN work and Professor H. James Harwood for the computer program.

References

1. Kenney, J. F., Polymer Engineering and Science 8, 216 (1968).
2. Matsuo, M. Polymer 9, 425 (1968).
3. Charch, W. H. and J. C. Shivers, Textile Research Journal, 29, 536 (1959).
4. British Patent 918,637 (1963), E. I. duPont de Nemours.
5. Wentworth, G., and J. R. Sechrist, J. Appl. Ply. Sci. 16, 1863 (1972).
6. Matsuo, M. A. Udea and Y. Kondo, Polymer Engineering and Science 10, 253 (1970).

7. Furukawa J. et. al. J. Poly. Sci, B 7, 561 (1969).
8. Yabumoto, S. et. al., J. Poly. Sci. A-1 7, 1683 (1969).
9. Powell, E. and B. Elgood, Chemistry and Industry 901 (1966).
10. Ragazzini, M. et. al., European Poly. J. 3 129 (1967).
11. Mukherjee, D. P. and C. Goldstein, Polymer Preprints, 14 #1, 36 May, 1973.
12. Harwood, H. J., Angew, Chem., 4 394 (1965).
13. Harwood H. J. and W. M. Ritchey, J. Poly. Sci., B2 601 (1964).
14. Goldfinger, G. and T. Kane, J. Poly. Sci., 3 462 (1948).
15. Johnston, N. W., Polymer Preprints, 10 #2, 608 September (1969).

BLOCK COPOLYMER THEORY. II. STATISTICAL THERMODYNAMICS OF THE MICROPHASES

Eugene Helfand

Bell Laboratories

Murray Hill, New Jersey 07974

SUMMARY

A theory of block copolymers has been developed based on an earlier mean-field approach to inhomogeneous macromolecular materials. Domains form in block copolymer systems. Then, as a result of the tendency of the block joints to stay in the interfacial regions, there is a loss of entropy in two ways. One is due to the confinement of the joints to a smaller region than when the system is homogeneous. The other has its origin in the vast number of polymer conformations which are suppressed because they create excessive density inhomogeneity. As domains grow, these two conformational free energy terms increase and must be balanced against the decrease of interfacial free energy to determine the equilibrium domain size and shape. The necessary statistics of the molecules are embodied in the solutions of modified diffusion equations. There are no adjustable parameters. Some numerical results for lamellar systems are presented. The paper concludes with a critique of prior theories, and a list of the type of experimental investigations which will shed light on block copolymer systems and other polymer blends.

I. INTRODUCTION

We look with awe upon the intricate formations nature has created by arranging sequences of monomers in macromolecules. Though not comparable, still admirable are man's achievements in tailoring nanometer structures.

Scientists and gadgeteers,
In a twink what they devise.
Nature takes a million years,
But, ah, she fashions butterflies.

During this conference we have heard frequent mention of polymer blends, graft copolymers, domains in ionomers, regions of crystallinity in PVC, islands of order in glasses, and so on. We come away with the impression that we must contend with heterogeneity as a factor in numerous polymer systems. The methods we shall discuss may be of value in analyzing a number of these nonuniform materials. However, we will concentrate here on block copolymers which separate into amorphous domains, as in the thermoplastic elastomers. Block copolymers are macromolecules in which there are long sequences, usually hundreds, of one type of repeat unit joined to long sequences of another.[1] The diblock copolymers we may denote by AB, where A and B each stand for a long sequence. The triblock copolymers, of type ABA, are better rubbers.

The physics controlling the morphology of the bulk block copolymers is quite simple to describe. In a material composed of units of types A and B, which have a positive heat of mixing, there is a tendency toward phase separation. The topology of the block copolymer molecules imposes restrictions on this segregation, and leads to the formation of a microdomain state. Domains which are essentially spherical, cylindrical, and lamellar have been observed. We schematize these domains in Figure 1.

From a thermodynamic point of view, associated with the interface between A and B domains is a positive surface free energy. This is a driving force toward growth of the domains in an effort to reduce surface-to-volume ratio. However, the system must contend with the consequences of the A-B joint being energetically confined to the interphase (interphase is a term used in the surface literature to denote an interfacial region of non-negligible thickness). One consequence of this confinement is that all the, let us say, A chains must have one end near the periphery of an A domain. In the absence of considerations other than the formation of a proper A-B interface, free random walk statistics would predict a density deficiency toward the center of the domain. Such an inhomogeneous density distribution is extremely unfavorable energetically. Hence the system will reject conformations which produce all but the slightest density nonuniformity; and accept, mostly, conformations which reach inward to produce virtually constant overall density. The consequent loss of conformational entropy, i.e., rejection of conformational states, becomes more and more severe as the domain grows.

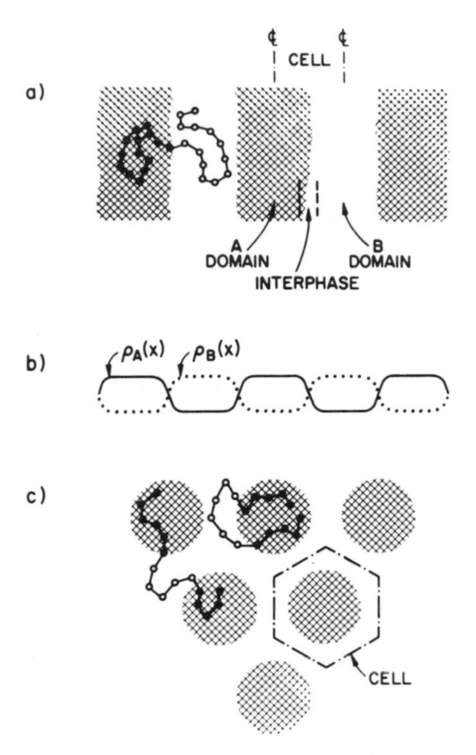

Figure 1. Schematic diagrams of domains in a block copolymer system. (a) A lamellar domain structure. (b) The polymer density pattern. (c) A cylindrical domain structure.

Another result of the confinement of the joint to the interphase is that as the domains grow, and the volume of interfacial region decreases, the density of joints in the interphase increases. The loss of entropy, which is associated with this diminution of volume available per joint degree of freedom, also opposes the tendency toward domain growth.

The equilibrium domain size and shape are a result of the balance of these three free energy terms.

A number of authors, notably Meier,[2-4] Leary and Williams,[5] and Krigbaum, Yazgan, and Tolbert[6] have formulated a view of block copolymers in many ways similar to this. Then they have attempted to develop quantitative expressions for the free energy. When the reader becomes familiar with the present work he should be able to appreciate their depth of insight, but also be aware of shortcomings of each effort. We delay to the end a critique of prior theories. I would like to point out one procedural sense in which our work is different. We have employed a general theory, developed earlier,[7] of inhomogeneous polymer systems, and applied it to block copolymers. Making a number of well-defined approximations, whose legitimacy we can assay, we see emerge the aspect of block copolymers portrayed above. Simultaneously we are delivered the correct calculational procedures. Furthermore, there is no problem, in principal, and probably also not in practice, in extending the theory to more complex situations: e.g., when there is added solvent or homopolymer; when the topology corresponds to graft copolymers, or multiblocks, ...ABABAB... ; or a variety of exotic systems mentioned in the literature, including this volume.[8]

II. GENERAL THEORY

We shall attempt to convey, in the simplest possible terms, the most important features of the general inhomogeneous polymer theory, with emphasis on application to block copolymers of the AB type. Those who wish to see more detail, in the context of the study of the interphase between two semi-infinite, immiscible polymer phases may look at the paper of the author and Tagami.[7] A more extensive report of the present work is in preparation.[9]

Consider two types of polymers, A and B. We need not specify, for the moment, how they are attached. For the sake of simplicity we will say that the two polymers, when pure, have the same properties; viz., each has

(i) bulk density, ρ_o, in segments per unit volume;
(ii) Kuhn statistical length of a segment, b; and
(iii) compressibility, κ.

The degree of polymerization of the two types of chains may differ, being Z_A and Z_B. The measure of incompatibility between A and B is the quantity χ, familiar in theories of mixing.[10,11] Definitions differ, so we will state ours (for a symmetric mixture only). The free energy of mixing at constant volume is assumed to be composed of the ideal entropy of mixing and a term $kT\chi\rho_A\rho_B/\rho_0$ per unit volume, where ρ_A and ρ_B are the densities of A and B in the mixture (χ is effectively equivalent to the μ_{12} of Scott[11]).

For the purpose of our statistical thermodynamic theory we focus attention on a function we call $Q(\underset{\sim}{r},t;\underset{\sim}{r}_0)$. What is it? One could say, "It is a constrained partition function (or an unnormalized distribution function) for a chain of t repeat units to start at $\underset{\sim}{r}_0$ and end at $\underset{\sim}{r}$." Let us try to use more tangible terms. Consider that to our system we add a "probe" molecule which has t segments. The probability density that one end of the probe will be found at $\underset{\sim}{r}_0$ and the other end at $\underset{\sim}{r}$ is equal to the function $Q(\underset{\sim}{r},t;\underset{\sim}{r}_0)$ times an appropriate normalization constant. We will ultimately be interested in chains, t long, which are part of larger macromolecules. Then the normalization and other factors will change, but all the required probabilities will be calculable in terms of Q.

Within the context of mean field theory, Q for a Gaussian molecule satisfies a modified diffusion equation. For the moment, consider an isolated A polymer chain (or, if you like, a chain which is part of a system of noninteracting molecules). Let it be in an external field, $U_A(\underset{\sim}{r})$, per repeat unit. For that molecule Q_A satisfies the equation

$$\frac{\partial Q_A(\underset{\sim}{r},t;\underset{\sim}{r}_0)}{\partial t} = \left[\frac{b^2}{6}\nabla^2 - \frac{U_A(\underset{\sim}{r})}{kT}\right] Q_A(\underset{\sim}{r},t;\underset{\sim}{r}_0), \tag{2.1}$$

with initial condition

$$Q_A(\underset{\sim}{r},0;\underset{\sim}{r}_0) = \delta(\underset{\sim}{r}-\underset{\sim}{r}_0). \tag{2.2}$$

In the field-free case the solution for Q is the familiar Gaussian distribution.

The type of approximation we shall use to go from non-interacting molecules to real bulk systems is mean field theory. There is no external field on the molecules in the system of interest. Rather the other molecules create a field on any piece of chain we single out as a probe. We take the mean value of this field.

In the type of system under consideration there are two sources of the field on a unit of A. First, it takes work to transfer a segment of A from pure A to a region with B content. At constant density this work is proportional to χkT and to the fractional composition of B, $\rho_B(\underset{\sim}{r})/\rho_0$. As a consequence of these repulsions between unlike units there is a tendency to develop deviations of the density from ρ_0. However, a counterfield then arises which tends to place more polymer in density poor regions and take it out of high density regions. The potential of this counterfield is

$$kT\zeta\,\frac{\rho_A(\underset{\sim}{r})+\rho_B(\underset{\sim}{r})-\rho_0}{\rho_0}, \tag{2.3}$$

$$\zeta \approx 1/(\rho_0 kT\kappa); \tag{2.4}$$

i.e., the field is proportional to the deviation of the density from ρ_0, with a constant of proportionality whose most important term is the inverse of the compressibility, κ. We should not be surprised to see the inverse compressibility entering as the measure of resistance to density deviations. (Polymers are highly incompressible so ζ is quite large. In fact, it is appropriate to take $\zeta \to \infty$ and the density deviation $\rho_A + \rho_B - \rho_0 \to 0$. Note that this creates an indeterminate form for the field, Eq. (2.3), which is finite and must be treated in a special manner, described previously.[12])

The equation for Q_A becomes

$$\frac{\partial Q_A}{\partial t} = \left[\frac{b^2}{6}\nabla^2 - w_A(\underset{\sim}{r})\right] Q_A, \tag{2.5}$$

$$w_A(\underset{\sim}{r}) = \chi \frac{\rho_B(\underset{\sim}{r})}{\rho_0} + \zeta\left[\frac{\rho_A(\underset{\sim}{r})}{\rho_0} + \frac{\rho_B(\underset{\sim}{r})}{\rho_0} - 1\right]. \tag{2.6}$$

This equation, and the similar one for B, do not form a closed set, because they contain the unknown density patterns, $\rho_A(\underset{\sim}{r})$ and $\rho_B(\underset{\sim}{r})$. However, the densities can be expressed in terms of the Q's to complete the set of equations. For example, in an AB block copolymer the density of A is

$$\rho_A(\underset{\sim}{r}) = \frac{N}{\mathfrak{Q}\Omega}\int_0^{Z_A} dt \int^{\Omega} d\underset{\sim}{r}_B d\underset{\sim}{r}_J d\underset{\sim}{r}_A$$

$$\times\, Q_A(\underset{\sim}{r}_A, Z_A - t; \underset{\sim}{r}) Q_A(\underset{\sim}{r}, t; \underset{\sim}{r}_J) Q_B(\underset{\sim}{r}_J, Z_B; \underset{\sim}{r}_B), \tag{2.7}$$

$$\mathfrak{Q} = \frac{1}{\Omega}\int^{\Omega} d\underset{\sim}{r}_A d\underset{\sim}{r}_J d\underset{\sim}{r}_B Q_A(\underset{\sim}{r}_A, Z_A; \underset{\sim}{r}_J) Q_B(\underset{\sim}{r}_J, Z_B; \underset{\sim}{r}_B), \tag{2.8}$$

where N is the number of molecules in the region Ω.

It is in the equations for densities ρ_A and ρ_B, and the partition-function-like quantity $\mathfrak{Q}$, that we first introduce the topology of the molecules. We see in $\mathfrak{Q}$ factors Q_A and Q_B with a common argument $\underset{\sim}{r}_J$, a consequence of the permanent joint. Equations (2.7) and (2.8) would be altered for other topologies; e.g., for an ABA copolymer there would be a partition function

$$\mathcal{Q} = (1/\Omega)\int^{\Omega} d\underset{\sim}{r}_{A1}d\underset{\sim}{r}_{J1}d\underset{\sim}{r}_{J2}d\underset{\sim}{r}_{A2}Q_A(\underset{\sim}{r}_{A1},Z_A;\underset{\sim}{r}_{J1})$$

$$Q_B(\underset{\sim}{r}_{J1},Z_B;\underset{\sim}{r}_{J2})Q_A(\underset{\sim}{r}_{J2},Z_A;\underset{\sim}{r}_{A2}), \tag{2.9}$$

and the form for other topologies should be clear.

To complete the statistical thermodynamic theory it is necessary to have an expression for the free energy. This has the form

$$F(N)/kT = \int^{\Omega} d\underset{\sim}{r}\{-(\chi/\rho_0)\rho_A(\underset{\sim}{r})\rho_B(\underset{\sim}{r})$$

$$+ (\zeta/2\rho_0)[\rho_0+\rho_A(\underset{\sim}{r})+\rho_B(\underset{\sim}{r})][\rho_0-\rho_A(\underset{\sim}{r})-\rho_B(\underset{\sim}{r})]\}$$

$$- N \log \mathcal{Q}. \tag{2.10}$$

(The derivation is based on a saddle function approximation to a functional integral.[9]) If we take Ω as a single cell in a periodic domain structure, we can determine the number of block copolymer molecules in this region, N, by minimizing F/N with respect to N:

$$\frac{\partial(F/N)}{\partial N} = 0. \tag{2.11}$$

Also the minimum free energy per molecule for various geometries determines the most favorable domain shape. In order to find the point at which the domain structure "melts," the free energy given by Eq. (2.10) must be compared with that of a homogeneous phase:

$$F_{homog}(N)/kT = \Omega\chi\rho_0\varphi_A\varphi_B + O(1/\zeta) \tag{2.12}$$

(the φ's are volume fractions). We have taken the zero of energy such that the free energy of pure A or B is zero.

III. NARROW INTERPHASE APPROXIMATION

It would be surprising if at this point the reader were not wondering about how the descriptive statements of the Introduction are related to the General Theory section. To clarify the link it will be necessary to consider an approximation; viz., that the interphase is much narrower than the domain size, so that there are domains of essentially pure A or B. This is an approximation implicit from the start in all previous theories. We make it here because it will help clarify the picture, and it appears to have some qualitative validity. It may, or may not, prove to be quantitatively useful.

Let us confine our attention to lamellar geometry, which we diagram in an idealized fashion in Figure 1. Our "cell" extends from the center of an A domain, through the interphase, to the center of the adjacent B domain. Thus it is really half an A and half a B domain. The system has reflection symmetry at the cell boundary. The cross sectional area we take as S.

We present a heuristic approach to the consequences of the approximation. A consistent solution can be found in which the density distribution across the interphase is essentially the same as that of an interphase between unattached A and B.[7] The interphase in the block copolymer is taken as large enough so that within it the transition from pure A to pure B is completed, but small enough so that we may regard all A molecules as having one end at a single plane. Do not think of this plane as the center of the interphase, but rather as the A-rich side. We are to join the A domain to A polymer in the interphase. A good way to do this is to assume that for every molecule which enters the domain from the interphase, a statistically equivalent one comes out of the domain and goes into the interphase. Mathematically, this is accomplished by using reflection boundary conditions, $\partial Q/\partial x = 0$ at the boundary. Meier[2-4] and others use absorptive boundary conditions, but that is because they are attempting at the same time to create the features of the interphase.

All this can be put into cleaner mathematical terms, and even made less severe.[9] Still we feel that for the parameters appropriate to many block copolymers the approximation is suspect. What we probably have to do is return to the fundamental equations of the previous section which couple the interphase and domain solutions more honestly. In other words, the drive to maintain uniform domain density may produce a deeper degree of interpenetration.

In the narrow interphase approximation the free energy reduces to a sum of terms, as described in the Introduction:

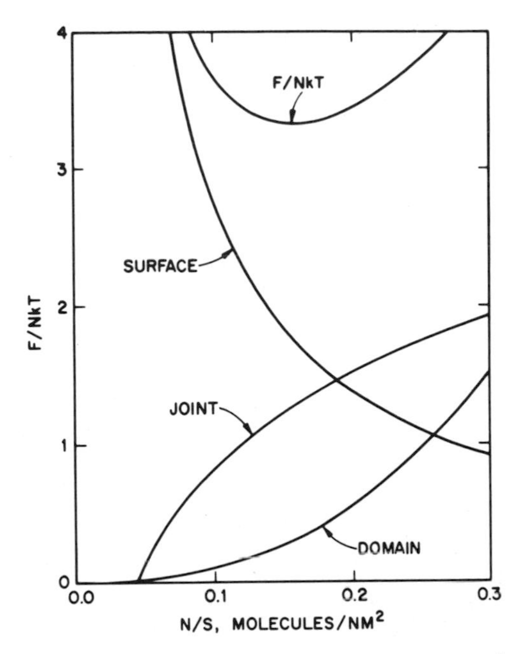

Figure 2. Theoretical free energy per molecule (divided by kT) plotted against number of block copolymer molecules per nm^2 of interface. The separate contributions from surface free energy, localization of joints, and maintaining uniform domain density are shown. The material is PS-PI described in the text.

$$F(N) = \left[\left(\frac{\chi}{6}\right)^{\frac{1}{2}} b\rho_0 kT\right]S + NkT \log\left(\frac{N}{a_J S}\right)\left(\frac{Z_A+Z_B}{\rho_0}\right)$$

$$- NkT\ (\log \mathcal{Q}_A + \log \mathcal{Q}_B), \tag{3.1}$$

$$a_J = \tfrac{1}{2}\ \pi b/(6\chi)^{\frac{1}{2}}. \tag{3.2}$$

The first term is surface free energy. The quantity $(\chi/6)^{\frac{1}{2}}\ b\rho_0 kT$ is the expression for the surface tension, found by the author and Tagami.[7] Note the $\chi^{\frac{1}{2}}$ dependence. (If available, a measured interfacial tension could be substituted.) S is the surface area.

The second term involves the logarithm of the ratio of the density of joints in the interphase (N joints divided by the volume Sa_J) to the density of joints in a homogeneous system.

Out of the theory has come the prediction that the effective transverse freedom for a joint in the interphase is a_J, a quantity inversely proportional to $\chi^{\frac{1}{2}}$. For b = .7 nm and a χ of .02, a_J is 3.2 nm, which is a rough measure of the predicted surface thickness.[7]

The final term arises from the loss of conformational entropy in maintaining uniform density in the A and B domains. The partition functions are those for chains of A or B which begin at the domain wall and reflect from that boundary:

$$\mathcal{Q}_A = \int^{A \text{ domain}} dx\ Q_A(x, Z_A; x_{wall}). \tag{3.3}$$

IV. CALCULATIONS

The theory, in the narrow interphase approximation, has been implemented with a computer program. However, we still regard the results of these calculations as tentative because of: (1) internal inconsistencies of the narrow interphase model for parameters of interest; (2) convergence difficulties in the calculation; and (3) uncertainties as to the values of the required parameters. Nevertheless, we feel the reader will want to see something of how the numbers are coming out and how they compare with experiments.

In Figure 2 we have plotted the free energy calculated for a PS-PI (polystyrene-polyisoprene) block copolymer studied by Douy, Mayer, Rossi, and Gallot.[13] It has a molecular weight for PS of 23,000 and for PI of 21,000. The parameter χ has been taken as equalling 0.019 from data of McIntyre and Rounds.[14] Curves are shown for the individual terms of the free energy per molecule vs. number of molecules of block copolymer per nm^2 of interface. The decreasing term is the surface free energy per molecule. Of the two increasing terms the less rapidly rising one is associated with the decreasing volume available to each joint. The more rapidly rising term, the free energy of maintaining uniform density, is fitted with surprising accuracy by an $(N/S)^{2.5}$ law.

The sum of these free energy terms has a minimum at $(N/S) = .16$ molecules/nm^2. The domain lamellar thickness, d_K for polymer K, is given by

$$d_K = 2(N/S)Z_K/\rho_{OK}, \quad K = \text{PS or PI}. \tag{4.1}$$

The predicted lamellar thicknesses are $d_{PS} = 11$ nm and $d_{PI} = 12$ nm. This is too far removed from the observed values (based on extrapolation to zero solvent content) of $d_{PS} = 15$ nm and $d_{PI} = 16$ nm to be explained by uncertainty in the parameters.

Here is an observation to ponder. Consider for the moment that the free energy associated with maintaining uniform density in the domains is small; so the free energy is given by

$$F/NkT = \left(\frac{\chi}{6}\right)^{\frac{1}{2}} b\rho_0 \frac{S}{N} + \log\left(\frac{N}{Sa_J} \frac{Z_A+Z_B}{\rho_0}\right). \qquad (4.2)$$

Differentiating with respect to N/S and equating the result to zero, we find that the equilibrium N/S is

$$(N/S) = (\chi/6)^{\frac{1}{2}} b\rho_0, \qquad (4.3)$$

a result that is actually independent of domain geometry. This formula predicts lamellar thicknesses of d_{PS} = 20 nm and d_{PI} = 21 nm (but parameter uncertainties could bring this to the experimental observation). The independence of N/S from molecular weight is roughly in accord with other data of Douy, et al.[13]

Next we consider data for an SBS (14,100-27,900-14,100 molecular weight) polystyrene-polybutadiene block copolymer found by Krigbaum, Yazgan, and Tolberg[6] to be lamellar. For the PS-PB pair we deduce χ = 0.21 based on Rounds and McIntyre's[14] results. Using the narrow interphase theory, modified for triblock copolymers, we predict N/S = .098 molecules/nm^2, d_{PS} = 9 nm, d_{PB} = 10 nm. The observed thicknesses are d_{PS} = 12-14 nm and d_{PB} = 15-16 nm (slightly dependent on orientation with respect to film face). Neglecting the free energy for maintaining uniform density, we find by Eq. (4.3) and (4.1) that d_{PS} = 14 nm and d_{PB} = 16 nm. However, when Krigbaum et al.[6] studied a 14,050-69,800-14,050 SBS sample, which assumes cylindrical geometry, they did not find the same value for N/S. Perhaps we should look forward to an improved theory decreasing the magnitude of the free energy associated with maintaining constant domain density, but not making the term negligible.

We will stop with these two cases and delay a more comprehensive comparison of experiment and theory until the next stage of our calculations is completed.

V. CRITIQUE OF PREVIOUS THEORIES

There have been a number of earlier efforts to analyze the statistical thermodynamics of the domain structure of block copolymers. The pioneer (all of six years ago) was Meier,[2] who exhibited a deep insight into the physics underlying these systems. In the light of the present work we can examine previous theories and reach judgments as to their merits and shortcomings. In particular we can make explicit their approximations. In this regard, we note that from the outset they all have begun with a picture of pure domains separated by either a sharp boundary or a narrow interphase.

Meier has produced a series of papers[2-4] in which his ideas evolve. In the first, he recognized the import of maintaining uniform density in the domains. He studied the density pattern of random walks confined to a domain and beginning near the domain boundary. Immediately he encountered the difficulty that, as a result of the way he handled confinement, the density had to fall to zero at the wall. He did not invoke our density deviation force, but circumvented the difficulty by arbitrarily declaring that the outer ten percent of the domain is an interphase. In addition to his spurious negative density deviation near the wall he observes the expected negative density deviation near the center. He balances the two negative deviations against a positive deviation he creates in the intermediate region by adding an excess of polymer. Finally he "swells" the polymer so that for any given amount of material the deviation is minimal. The concept is that for small swelling the density deviation at the center is large, whereas for large swelling the density deviation near the wall is great. One might worry here about the arbitrariness of the definition of interphase, but our objection is stronger. From the present work we expect that the structure of the interphase is a feature not greatly coupled to any swelling used to fill up the domain center. The near-wall distribution has the features of a normal polymer-polymer interphase. When we use the boundary condition $\partial Q/\partial x = 0$ at the wall, instead of Meier's $Q = 0$ at the wall, we completely decouple the interphase and domain. Thus the balance involved in Meier's minimization of density deviations would appear to have one defective arm.

We must examine another consequence of the arbitrary assumption that the interphase thickness is a fixed fraction of the domain size. The result of such an assumption on the entropy associated with joint localization is to render its contribution to the free energy per molecule independent of N (or, in Meier's terminology, independent of his expansion parameter, α). This qualitative error forecloses all hope of extracting meaningful results.

In a later paper[4] Meier attempts a more detailed theory of the interphase. However, he omits the loss of conformational entropy involved in joining two polymer phases with a given density distribution. Hence, he obtained results which he can only justify by a claim that the interfaces in block copolymers are drastically different from other interfaces.

Next, we should scrutinize Meier's "elastic entropy difference," which is the free energy term associated with swelling the polymer. The error here is really a carryover of an erroneous assumption of Flory,[10] although as employed by Flory it is not as serious. Flory says that the density distribution of a polymer is approximately Gaussian about one end (we agree). He writes this density distribution as

$$Z(3/\pi\langle r^2\rangle)^{3/2}\exp[-9r^2/\langle r^2\rangle].$$

It would follow that a change of $\langle r^2\rangle$ by a factor of α^2 would lead to a mere scaling of the density distribution. The conclusion is wrong. The degree of deviation from Gaussian shape increases.

Let us restate the problem in more general terms. Consider that we have a polymer molecule describable by a Gaussian random walk. The unconstrained mean square end-to-end distance is Zb^2, and the exact density pattern $\rho_0(\underset{\sim}{r})$ can be calculated. Let us impose the constraint that the mean square end-to-end distance be $\alpha^2 Zb^2$. Does it follow that the new density pattern is

$$\rho(\underset{\sim}{r}) = (1/\alpha^3)\rho_0(\underset{\sim}{r}/\alpha)?$$

No. Therefore the entropy loss $-(3/2)k(\alpha^2-1-\ell n\alpha)$, appropriate to end-to-end expansion[10] (as in affine rubber networks), does not represent the entropy loss associated with a uniform density swelling. We expect that one can get qualitative, but not quantitative, results from the Flory assumption.

Finally, Meier calculates the loss of entropy in confining the polymer to a finite domain. To the extent that Meier eliminates some A walks on the basis of them crossing into the B domain, he is determining part of the surface entropy. When he later adds a phenomenological surface free energy, proportional to the surface tension, he is counting some of the surface entropy twice.

Thus, though Meier's works must be studied for the deep insights they provide, his formulas contain defects.

Closely related to the theory of Meier is that of Inoue, Soen, Hashimoto, and Kawai.[15] These author's use a phenomenological surface free energy. They acknowledge the presence of an entropy associated with localizing the joint, but incorrectly assume that it is independent of domain size or shape. Finally, they equate the entropy loss associated with maintaining uniform density with the entropy loss involved in dictating that the mean square end-to-end distance be the domain radius. This is at best a qualitative estimate.

Leary and Williams[5] have given consideration to an interphase between domains. In many ways their work parallels the recent effort of Meier.[4] Their failure to include the conformational entropy of the interphase makes it impossible for them to find the correct interphase width by free energy minimization. Furthermore, they use Meier's ratios between chain expansion and domain size, which we have criticized above. Even within the context of this technique, it is difficult to see why this ratio is not made a function of interphase thickness.

Krigbaum, Yazgan, and Tolbert[6] do not take into account the interface thickness. Thus they do not obtain an accurate entropy associated with localizing the joints. However, in omitting surface thickness they avoid the error of others who have ended up making this thickness a function of domain size. When they differentiate the term as part of the procedure for calculating domain size [cf. Eq. (2.11)] they obtain the proper derivative. They include surface free energy phenomenologically via a surface tension. Their argument leading to an entropy for maintaining uniform density is hard to assess, though it appears to possess some of the defects of Meier's theory. But right or wrong, the fact that this term comes out small may be all that is needed for them to effectively obtain, to within a constant, our Eq. (4.2) [with (surface tension)/kT replacing our $(\chi/6)^{\frac{1}{2}} b\rho_0$].

We will not analyze in detail the theories of Krause,[16] Bianchi, Pedemonte, and Turturro,[17] Marker,[18] Krömer, Hoffman, and Kämpf,[19] or LaFlair,[20] because they possess previously discussed or other shortcomings.

VI. COMMENTS

We are still actively engaged in this research effort, so we hope to have much more in the way of calculations completed soon. Especially, we look forward to removing the narrow interphase approximation. Several improvements in the model are being considered, such as elimination of the requirement of symmetry in the properties of A and B.

I would like to close with a plea for the type of experiment which will help in the task of unravelling the mysteries of block copolymers and other inhomogeneous polymer systems.

1. Good measurements of χ should be made.
2. There should be a determination of the thickness of block copolymer and other interphases.[21]
3. Measurements are needed of the surface tension between immiscible polymers. Comparison with the theory of Helfand and

Tagami[7] should be made. I urge, also, a critical evaluation of the validity of the popular Fowkes theory.[22] His basic premise, that the density profile for an AB interface is a superposition of A against vacuum and B against vacuum, would not seem to be applicable to polymers, especially nearly compatible ones.[7]

4. Measurements should be made of the free energy of polymer mixtures using oligimers, copolymers, or solvents to achieve miscibility. This would enable us to experimentally determine mean fields in Eq. (2.6).

5. An effort should be made to separate equilibrium and non-equilibrium aspects of block copolymer domains and other blend morphologies.

VII. ACKNOWLEDGMENT

We thank Zelda Wasserman for her able programming assistance.

REFERENCES

1. The literature on block copolymers is now vast. Perhaps a good way to be led into this sea is through some of the published symposia: J. Polymer Sci. C, 26 (1969); Block Copolymers, ed. by S. L. Aggarwal (Plenum Press, New York, 1970); Block and Graft Copolymers, ed. by J. J. Burke and V. Weiss (Syracuse University Press, 1973). An excellent review of morphological aspects has been given by M. J. Folkes and A. Keller, The Physics of Glassy Polymers, ed. by R. N. Haward (Halsted Press, New York, 1973).

2. D. J. Meier, J. Polymer Sci. C, 26, 81 (1969).

3. D. J. Meier, Polymer Preprints, 11, 400 (1970).

4. D. J. Meier, Block and Graft Copolymers, ed. by J. J. Burke and V. Weiss (Syracuse University Press, 1973).

5. D. F. Leary and M. C. Williams, J. Polymer Sci. B, 8, 335 (1970); J. Polymer Sci. (Physics), 11, 345 (1973).

6. W. R. Krigbaum, S. Yazgan, and W. R. Tolbert, J. Polymer Sci. (Physics), 11, 511 (1973).

7. E. Helfand and Y. Tagami, J. Polymer Sci. B, 9, 741 (1971); J. Chem. Phys., 56, 3592 (1971); J. Chem. Phys., 57, 1812 (1972).

8. C. H. Bamford and G. E. Eastmond, in this volume, present a novel example; while L. H. Sperling attempts to classify usual and unusual materials, with an aim toward suggesting novel topologies.

9. E. Helfand, in preparation.

10. P. J. Flory, Principles of Polymer Chemistry (Cornell University Press, Ithaca, 1953).

11. R. L. Scott, J. Chem. Phys., 17, 279 (1949).

12. E. Helfand, Polymer Preprints, 14, 970 (1973).

13. A. Douy, R. Mayer, J. Rossi, and B. Gallot, Mol. Cryst. Liq. Cryst., 7, 103 (1969).

14. D. McIntyre and N. A. Rounds, private communication; N. A. Rounds, Doctoral Dissertation, University of Akron (1970).

15. T. Inoue, T. Soen, T. Hashimoto, H. Kawai, J. Polymer Sci. A-2, 7, 1283 (1969).

16. S. Krause, J. Polymer Sci. A-2, 7, 249 (1969); Macromolecules, 3, 84 (1970).

17. U. Bianchi, E. Pedemonte, and A. Turturro, Polymer, 11, 268 (1970).

18. L. Marker, Polymer Preprints, 10, 524 (1969).

19. H. Krömer, M. Hoffman, and G. Kampf, Ber. Bunsenges, physik. Chem., 74, 859 (1970).

20. R. T. LaFlair, IUPAC Symposium, Boston, July 1971; Supplement to Pure and Appl. Chem., 8, 195 (1971).

21. Interesting X-ray experiments have been done by D. G. LeGrand, J. Polymer Sci. B, 8, 195 (1970) and by H. Kim, Macromolecules, 5, 594 (1972).

22. F. M. Fowkes, Ind. Eng. Chem., 56 (12), 40 (1964).

CATIONIC GRAFTING: THE SYNTHESIS AND CHARACTERIZATION OF BUTYL RUBBER-*g*-POLYSTYRENE AND PVC-*g*-POLYISOBUTYLENE*

J. P. Kennedy, J. J. Charles, and D. L. Davidson

Institute of Polymer Science, The University of Akron, Akron, Ohio 44325

The discovery that certain alkylaluminum compounds (e.g. Et_2AlCl, Et_3Al, Me_3Al, etc.) in conjunction with suitable alkyl halides (e.g. tBuCl, $PhCH_2Cl$, etc.) are efficient initiating systems for the polymerization of isobutylene, styrene, etc. provides the basis for novel efficient cationic graft synthesis. By an extension of our mechanism developed to rationalize initiation in homopolymerization systems, it was postulated and demonstrated that graft copolymerizations could be initiated by suitable macromolecular halides in the presence of suitable alkylaluminum compounds:

$$P\text{-}Cl + Me_3Al \rightleftharpoons P^{\oplus}MeAlCl^{\ominus} \xrightarrow{M} PM^{\oplus} \longrightarrow \text{graft}$$

where PCl is a macromolecular halide initiator such as chlorobutyl rubber or polyvinyl chloride and M is a cationically polymerizable monomer.

We present some research in which the polymerization of styrene and isobutylene have been initiated by halogenated butyl rubber and PVC, respectively, in the presence of Et_2AlCl and Me_3Al.

*<u>Rubber Age</u>, in press.

Experimental - 1. Materials: The halogenated butyl rubbers were commercial products obtained from Enjay Chemical Co. (chlorobutyl rubber, Enjay HT1066 containing ~ 1.2 wt % Cl) and Polysar X-2 containing ~ 2 wt % Br. Polyvinyl chloride was obtained from B. F. Goodrich Chemical Co. (Geon 103 EP) and Pechiney Saint-Gobain (RB 9010). These materials were reprecipitated and characterized by intrinsic viscosity and osmotic pressure. Alkylaluminums (Texas Alkyls Co.) were freshly distilled under reduced pressure.

The overall composition of the grafts (weight % grafted branches) was determined by gravimetry, I.R. analysis, and in some cases chlorine analysis. Grafting efficiency, G.E., is the ratio of grafted polystyrene to total polystyrene formed. The % polystyrene, % Pst, is the ratio of grafted polystyrene in the graft to the total graft copolymer formed.

1. Halobutyl-g-polystyrene

Chlorobutyl rubber is a random copolymer of isobutylene and isoprene containing 1-2 mole % chlorine. It is suitable to initiate the graft polymerization of cationic monomers in conjunction with Et_2AlCl, etc. since this rubber contains labile allylic chlorine atoms which when abstracted result in the formation of a macromolecular carbenium ion.

We have examined the stability of chlorobutyl rubber under simulated grafting conditions in the presence of Et_2AlCl coinitiator. M_n data shown in Table I indicate that under our conditions backbone degradation does not occur.

a) Graft Synthesis and Characterization

Grafting can be effected both in polar and nonpolar media. The use of polar solvents, i.e., CH_2Cl_2, results in higher G.E. and % Pst in the graft for comparable styrene conversions as compared to nonpolar solvent, i.e. n-pentane. In n-pentane, the G.E. and % Pst in the graft remain constant whereas in mixed solvents,

TABLE I

Degradation studies with Et_2AlCl

$[Et_2AlCl]$ x 10^5 moles/l	$\overline{M}_n \times 10^{-3}$
-	156 ± 5
15.5	161 ± 5
30.5	155 ± 7

Conditions: 60 min; -50^0C; $n\text{-}C_5/CH_2Cl_2$ (70/30); [Cl-Butyl]: 20 gm/l.

e.g. $n\text{-}C_5/CH_2Cl_2$, the G.E. decreases with conversion while the % Pst increases. Preliminary experiments indicate that G.E. is reduced at high conversions and that it is significantly improved by the addition of polar solvent. Representative data are shown in Table II.

The fact that the calculated and experimental $\overline{M}_n$ values agree within what is considered to be experimental variation demonstrate successful grafting. According to solubility studies 20% Pst in the graft is needed to solubilize the graft copolymer in a good solvent for polystyrene such as MEK, i.e. the MEK-soluble polystyrene branches pull the MEK-insoluble poly(isobutylene-co-isoprene) backbones into solution. These solubility studies along with I.R. analysis are additional proof for successful grafting.

b) Grafting from Bromobutyl

We have studied the effect of the nature of the halogen on G.E. and % Pst in the graft. It was expected that bromobutyl rubber would be more effective in producing graft copolymers than chlorobutyl. However, in all cases G.E. and % Pst have been found to be lower for bromobutyl than for chlorobutyl (Table III). The data indicate that bromobutyl is slower to initiate grafting than chlorobutyl. Conceivably the larger, less basic bromine reacts slower with the alkyl-

TABLE II

Grafting Styrene from Chlorobutyl Rubber

Time min.	Solvent (ml)	Conv. %	G.E. %	Pst %	$\bar{M}_n \times 10^{-3}$ Theor.	$\bar{M}_n \times 10^{-3}$ Exp.
5	$n\text{-}C_5/CH_2Cl_2$ (70/30)	15	65	43	270	257 ± 8
5	$n\text{-}C_5$ (100)	18	45	29	200	185 ± 5

Conditions: -50^0C; [Cl-Butyl]: 20 g/l; [Et_2AlCl]: 0.024 M [Styrene]: 1.0 M

TABLE III

Grafting from Bromobutyl and Chlorobutyl

	Time min.	Conv. %	G.E. %	Pst %	$\bar{M}_n \times 10^{-3}$ Theor.	$\bar{M}_n \times 10^{-3}$ Exp.
Bromobutyl:						
	0	-	-	-	-	126
	1.0	20	30	4	134	131
Chlorobutyl:						
	0	-	-	-	-	146
	1.0	22	71	30	189	190

Conditions: $n\text{-}C_5$/MeCl (70/30); -55^0C; [Et_2AlCl]: 0.024 M [Styrene]: 1.0 M; [X-Bu] where X = Cl or Br: 20 g/l

aluminum to produce the active initiating species. Slower initiation allows chain transfer to compete to a greater extent with polymerization resulting in lower G.E.'s, with bromobutyl as compared to chlorobutyl under the same conditions.

2. PVC-g-Polyisobutylene

The chemical stability of PVC toward Et_2AlCl and Me_3Al coinitiators has been examined. Data in Table IV show that under conditions identical to grafting but in the absence of monomer, the intrinsic viscosity of PVC remains unchanged in the presence of Et_2AlCl or Me_3Al. Thus alkylaluminum coinitiation should provide a clean copolymer synthesis without backbone degradation.

TABLE IV

Attempted PVC Degradation with Et_2AlCl or Me_3Al

Rx.	$[\eta]^{27}_{THF}$, dl/g	
Time, min.	Et_2AlCl	Me_3Al
0	0.93	0.93
60	0.93	0.93

$[Et_2AlCl] = 3 \times 10^{-2}$ M; $[Me_3Al] = 4 \times 10^{-2}$ M; PVC: Geon 103, Solvent = 1,2-DCE; -30^0C

a. Graft Synthesis and Characterization

Data assembled in Table V show the results of some representative grafting experiments in which we used PVC initiator in conjunction with Et_2AlCl or Me_3Al coinitiators.

The final n-pentane insoluble products contained polyisobutylene, indicating successful graft copolymerization. Copolymers containing the largest amount of polyisobutylene have been made in homogeneous 1,2-dichloroethane solution. The highest grafting efficiencies have been obtained by using CH_2Cl_2. Use of

TABLE V

Representative Products of PVC-g-PIB

PVC	Coinitiator	Solvent	Temp., ^0C	Pentane Insol. PIB wt.%	Pentane Insol. G.E. %
GEON 103	Et_2AlCl	1,2-DCE	-30	60	17
"	"	CH_2Cl_2	"	11	62
"	Me_3Al	"	"	6	50
"	"	1,2-DCE	"	12	59
RB 9010	"	CH_3Cl	-50	3	53

[Coinitiator] = 3-5 x 10^{-2} M; [Isobutylene] = 1-2 M

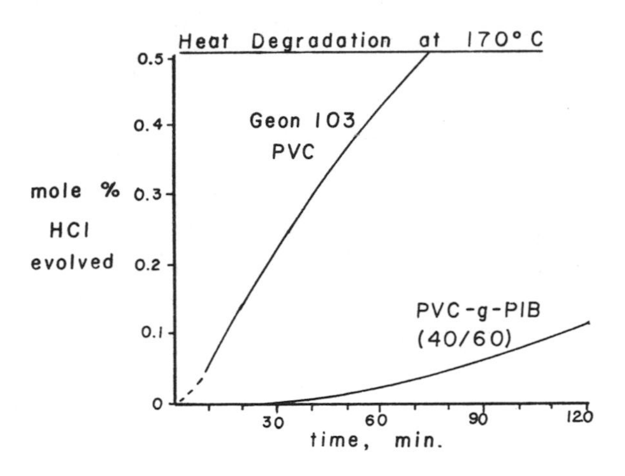

Fig. 1

Me_3Al in either 1,2-DCE or CH_2Cl_2 also gives rise to high G.E., although low PIB contents result. Heterogeneous grafting can be obtained by using Me_3Al co-initiator and CH_3Cl, a nonsolvent for PVC.

Graft copolymers with high PIB content are tough, leathery materials, and initial solubility studies reveal that they are insoluble in good solvents for PVC. For example, a PVC-g-PIB (40/60) graft copolymer is totally insoluble in DMF, an excellent solvent for PVC and a nonsolvent for PIB. This indicates that the PVC backbone is prevented from being solubilized by the PIB branches.

b. Heat Degradation of PVC-g-PIB

Using labile chlorine sites for grafting-from PVC provides enhanced thermal stability, since these sites are thought to be initiating sites for dehydrochlorination. Fig. 1 shows a dramatic improvement in the heat stability of PVC-g-PIB (40/60), as there is no HCl evaluation by the graft copolymer in 20 min. at $170^{0}C$.

SYNTHESIS, MORPHOLOGY AND PROPERTIES OF AB CROSSLINKED COPOLYMERS

C.H. Bamford and G.C. Eastmond

Department of Inorganic, Physical and Industrial Chemistry

University of Liverpool, Liverpool L69 3BX, England

INTRODUCTION

The past decade has witnessed a tremendous upsurge of interest in the whole field of multicomponent polymers which has led to both scientific and technological advances. Commercial developments include new impact-resistant plastics and new types of elastomers; in both cases the advantageous properties depend on microphase separation in the bulk polymers resulting from the incompatibility of the components. Simultaneous scientific advances have produced a partial understanding of the natures and properties of the new materials[1]. It has been realised that the character of the interactions between phases is important[2]; some guide lines have been laid down which allow morphologies of the materials to be reconciled with the structures of the constituent molecules[3] and in some cases morphologies can be predicted from a knowledge of molecular architecture. Limited progress has also been made in understanding physical properties in terms of morphology and molecular structure. The possibility of rapid commercial success in this new area of polymer research has resulted in a large proportion of the total effort being devoted to the search for novel materials; so for a full scientific understanding of the structures and properties of the many multicomponent polymers which have been developed has lagged behind technological developments.

Multicomponent polymers which have been subject to detailed investigation include graft copolymers, such as ABS (acrylonitrile-styrene-butadiene copolymer), whose structural parameters are not accurately known, and linear block copolymers, such as SBS (styrene-b-butadiene-b-styrene), which can be prepared with well-defined structures[4]. Naturally, the latter polymers have been subjected to a very detailed scientific examination in an attempt to establish their structure-property relationships. In view of the paucity of materials with different but known structures it has not been possible to study the general applicability of trends in behaviour which have been established for a limited number of linear block copolymers. However, it became apparent that use of a new series of free-radical initiating systems, the initiation mechanisms of which we have studied in depth, permitted the synthesis of multicomponent crosslinked polymers of known structure[5-8], thus extending the range of materials available for detailed investigation.

The multicomponent crosslinked polymers, which may be designated ABCPs, consist of chains of polymer A crosslinked with chains of a second polymer B. They are prepared by generating free radicals at specific sites X on chains of a preformed polymer A (known as the prepolymer) and using the resulting macroradicals to initiate the graft polymerization of a monomer M, as shown schematically in reaction (1). Combination termination of the growing chains then

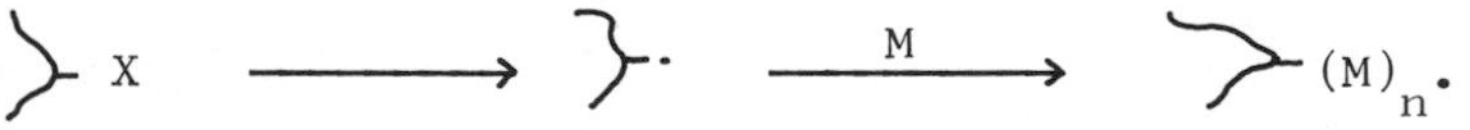

$$\succ X \longrightarrow \succ\cdot \xrightarrow{M} \succ(M)_n\cdot \qquad (1)$$

leads to the formation of a crosslink of polymer B, derived from M, between the A chains, as depicted in reaction 2(a). Structures such as (I) bear obvious analogies to linear ABA block copolymers,

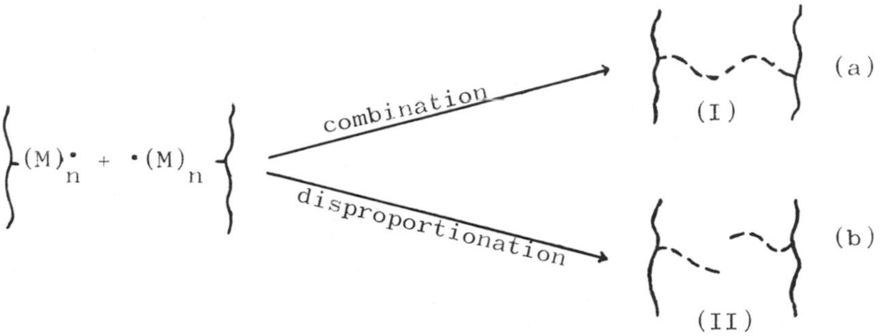

$$\}(M)_n^{\cdot} + {\cdot}(M)_n\{ \quad \begin{cases} \xrightarrow{\text{combination}} \text{(I)} & \text{(a)} \\ \xrightarrow{\text{disproportionation}} \text{(II)} & \text{(b)} \end{cases} \qquad (2)$$

except that instead of the chains being linked end-to-end the junction points may be situated anywhere along the length of the prepolymer chain. As reaction is extended to additional X sites on the same A chains, more complex structures such as (III) are built up and eventually the reaction mixture gels with the formation of infinite network structures, such as (IV).

In principle, it is possible to prepare ABCPs incorporating any polymer, into which suitable reaction sites can be introduced, as the A component and any polymer derived from a free-radically polymerizable monomer as the B component. Thus a wide variety of ABCPs may be synthesised and we have prepared many of these materials incorporating chemically different component polymers. This paper reports the techniques employed in the preparation of ABCPs of

(III)

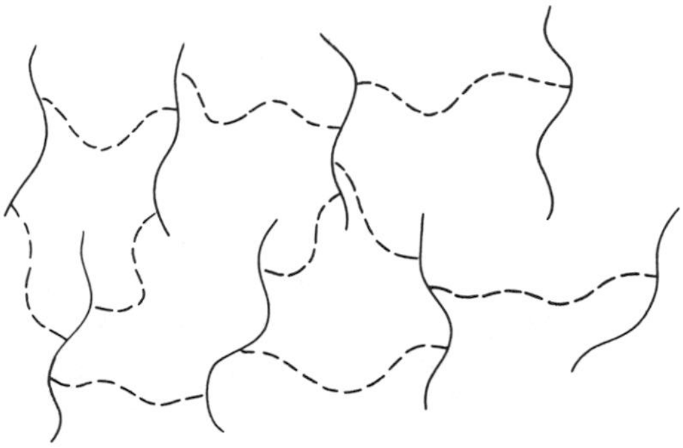

(IV)

predetermined structure and reviews the conclusions of a preliminary survey of their morphologies and properties.

SYNTHESIS OF ABCPs

Chemical Aspects

Many derivatives of transition metals in low oxidation states, in association with suitable organic halides and in the presence of suitable solvents, act as effective initiators of free-radical polymerization[9]. The kinetics and mechanisms of the radical-forming processes have been studied in depth for a variety of metal derivatives and, while the reaction mechanisms differ in detail according to the metal derivative employed, they have many features in common. The overall reaction is an electron-transfer from transition metal $\underline{M}$ to halide in which the oxidation state of the former increases while the latter generally dissociates into a halide ion and a radical fragment, e.g.

$$\underline{M}^{0} + CCl_4 \longrightarrow \underline{M}^{I}Cl^{\ominus} + \dot{C}Cl_3 \qquad (2)$$

However, under the conditions normally holding in polymerization processes, electron-transfer does not occur directly between halide and organometallic derivative; instead, the latter is first converted into a more reactive species, generally, in thermal reactions, by ligand exchange with an electron-donor in the system. For example, with molybdenum carbonyl the first, and rate-determining, step is the S_N2 process

$$Mo(CO)_6 + S \rightleftharpoons S\cdot\cdot Mo(CO)_5 + CO \qquad (a) \qquad (3)$$

in which the donor S is either a polar solvent or the vinyl monomer; this is followed either by (3b) or the redox reaction (3c).

$$S\cdot\cdot Mo(CO)_5 \longrightarrow \text{inactive products} \qquad (b)$$
$$S\cdot\cdot Mo(CO)_5 + CCl_4 \longrightarrow Mo^{I}Cl^{\ominus} + \dot{C}Cl_3 \qquad (c) \qquad (3)$$

It has been amply demonstrated that the radicals generated from CCl_4 which are responsible for initiating polymerization are $\dot{C}Cl_3$; the most direct evidence has been obtained by determination of the incorporation of labelled atoms in the polymer when isotopically labelled CCl_4 is used as the halide component of the initiating system[7,10]. The Mo^{I} derivative formed in (3c) is subsequently oxidized by the halide to a Mo^{V} species, with generation of additional $\dot{C}Cl_3$ radicals.[11] In other systems, the initial ligand-exchange occurs through an S_N1 reaction, for example

$$Ni\{P(OPh)_3\}_4 \rightleftharpoons Ni\{P(OPh)_3\}_3 + P(OPh)_3$$
$$Ni\{P(OPh)_3\}_3 + S \rightleftharpoons S\cdot\cdot Ni\{P(OPh)_3\}_3 \qquad (4)$$

The absolute rate coefficients of the reactions in (3) and (4) have been deduced from detailed studies, a description of which is given in references (12,13). This work has elucidated the nature of the structural features which determine the activity of the halide component. Bromo-derivatives are much more active than the corresponding chloro-derivatives and multiple substitution at a single C atom greatly increases reactivity. Fluoro-compounds which do not contain additional halide atoms are inactive in the type of reaction we are considering, no doubt an account of the high strength of the C-F bond, and iodo-derivatives yield elemental iodine, which functions as a retarder. Introduction of an electronegative group such as carboalkoxy into the molecule greatly increases reactivity.

Many initiating systems containing organometallic derivatives and halides are photochemically active.[14] For practical purposes those based on manganese carbonyl $Mn_2(CO)_{10}$ are particularly convenient since they initiate in visible light.[14b] The reaction involves scission of the Mn-Mn bond and the overall process is represented by (5)

$$Mn_2(CO)_{10} + CCl_4 \xrightarrow{h\nu} \dot{C}Cl_3 + \frac{1}{2} Mn_2(CO)_{10} + Mn(CO)_5Cl \qquad (5)$$

The quantum yield for radical production at $\lambda = 435$nm is close to unity for "high" halide concentration ($> 10^{-3}$ mol l^{-1}, approximately).

Most of our work on the synthesis of ABCPs has been carried out with thermal initiation by $Mo(CO)_6$ at $80^{\circ}C$ or with photoinitiation by $Mn_2(CO)_{10}$ at $25^{\circ}C$. From the discussion above, it will be seen that the initiation process is essentially the scission of a halogen atom from the halide group. Confirmatory evidence for this conclusion has been deduced from additional radiotracer experiments and studies of gelation kinetics with other halides with $Mo(CO)_6$, $Mn_2(CO)_{10}$ and $Pt\{PPh_3\}_4$ as organometallic derivatives; nickel derivatives react differently with ethyl trichloracetate[7].

In the synthesis of multicomponent polymers with these initiating systems, simple halides are replaced by polymers carrying reactive halogen-containing groups. Such groups are incorporated into the side-chains of the polymers and are the reaction sites X depicted in the schematic reaction (1). Typical polymers used in the preparation of ABCPs contain of the order of 10^3 such groups per polymer molecule. Polymers which have been used in this connection include polyvinyltrichloracetate (PVTCA), polyvinylbromoacetate, cellulose acetate containing a proportion of trichloracetate groups, the polycarbonate

$$\left[O - C_6H_4 - \underset{CCl_3}{\overset{H}{C}} - C_6H_4 - O - CO \right]_n$$

and polystyrene functionalised by conversion of some units to (V) below

$$-CH_2-\underset{\substack{| \\ C_6H_4 \\ | \\ CH_2O\cdot OCCl_3}}{CH}- \qquad (V)$$

Virtually the only restriction on the monomer which is used to form the B component of the ABCPs is that it should be polymerizable by free radicals in solution. In most of the synthetic work carried out to date methyl methacrylate[15] (MMA), styrene[15] (St) and chloroprene[16] (Cp) have been employed, but several other monomers have been used in kinetic studies.

Control of Structural Parameters of ABCPs

One of the most important structural parameters of the ABCPs which can be varied in a controlled manner is the crosslink density. We quantify this parameter in terms of the crosslinking index, defined[17] as the number of crosslinked units per primary molecule in the system. For a primary polymer of given number-average degree of polymerization, the critical value of the crosslinking index corresponding to the gel point depends on the molecular weight distribution of the primary polymer (which is the prepolymer or A component in the ABCPs) but is determined uniquely by the weight-average degree of polymerization. For convenience, we define the the relative crosslinking index γ_r as the ratio of the actual crosslinking index in the sample relative to the critical crosslinking index at the gel point. On this basis $\gamma_r = 1$ at the gel point and under other conditions γ_r is equal to the reaction time expressed as multiples or fractions of the gel time, assuming a constant rate of crosslink formation. If necessary values of γ_r calculated from the simple ratio of reaction times can be corrected to allow for decreasing rates of radical and crosslink formation during reaction due to consumption of initiator. The crosslinking index is obviously related to the average length of A chains between crosslinking points.

The average length of the A chains is determined prior to the formation of the ABCP. The mean crosslink length is controlled by the polymerization kinetics during the crosslinking process. In the simpler cases, up to the gel point, the average chain length of the crosslinks can be calculated from the rate of initiation, the monomer concentration and the kinetic parameter $k_p k_t^{-\frac{1}{2}}$ in the same way as chain lengths are calculated for a simple homopolymerization; k_p, k_t are the rate coefficients for propagation and second-order termination of the growing radicals, respectively. Very often, following the onset of gelation, the polymerization reactions show a marked autoacceleration arising from a decrease in k_t [16]; under such conditions it is not possible to calculate crosslink lengths simply from kinetics but they may be determined from a knowledge of monomer conversion and the total number of initiating radicals generated during the reaction period.

As we have already pointed out, crosslinked (I) or branched (II) species may be formed in these reactions, according to the nature of the termination reaction (2). In the absence of transfer processes the branch:crosslink ratio is determined by the relative rates of the two termination reactions. Applying Flory's gelation theory[18] to the systems under consideration, we see that the gel point corresponds to one crosslinked unit per weight-average prepolymer chain, and the reaction time required to reach the gel point (i.e. the gel time t_g) is given by[5] equation (6), in which c is the concentration of reactive sites on the prepolymer

$$t_g = \frac{c}{\bar{P}_w \mathcal{I}} \; \frac{k_{tc} + k_{td}}{k_{tc}} \qquad (6)$$

chains (in base mol l^{-1}), $\bar{P}_w$ is the weight average degree of polymerization of the prepolymer (in terms of reactive sites), $\mathcal{I}$ is the rate of radical formation and k_{tc}, k_{td} are the rate coefficients for combination and disproportionation termination, respectively. We have previously established [5] reaction conditions under which equation (6) can be applied quantitatively to gelation kinetics in the present systems. From a study of gelation kinetics we have thus been able to determine the relative rates of combination and disproportionation termination for several monomers[8]. It appears that for those monomers which do not have an α-methyl group, e.g. styrene, acrylates, acrylonitrile, chloroprene, the propagating radicals undergo termination exclusively by combination. Consequently in such cases the multicomponent polymers contain no branches of B chains. By contrast, propagating radicals of monomers which carry an α-methyl group, e.g. methacrylates and methacrylonitrile, undergo termination by both mechanisms simultaneously and the multicomponent polymers contain both crosslinks and branches of B chains. The incidence of combination in the termination reaction and the resulting branch:crosslink ratios in the ABCPs are summarised in table 1 for a series of methacrylate esters.

Table 1. Influence of the Termination Reaction on Structures of ABCPs.

Methacrylate ester	$k_{tc}/(k_{tc}+k_{td})$	T^oC	Branch:Crosslink Ratio
Methyl	0.34	25^oC	3.85
	0.20	80^oC	7.70
Ethyl	0.32	25^oC	4.16
n-Butyl	0.25	25^oC	5.88

Equation (6) can be readily modified to allow for transfer to monomer and for initiator consumption during the formation of the ABCPs[6], and such corrections were applied in calculating the relative proportions of the termination reactions. As can be seen from the data for methyl methacrylate, when both termination processes are operative the branch:crosslink rates may be controlled to some extent by taking advantage of the temperature dependence of $k_{tc}/(k_{tc}+k_{td})$.

In all cases branch:crosslink ratios can be modified by deliberately introducing branches of the B polymer in a controlled manner. Two convenient methods exist for producing these structural modifications, and both rely on the generation, during the crosslinking process, of active radicals which are unattached to the prepolymer. First, an active low molecular weight halide such as ethyl trichloracetate, may be added to the reaction mixture. Radicals are then produced simultaneously from the small molecule and polymeric halides during the crosslinking process. Initiating radicals are thus produced which are unattached to the A polymer and lead to propagating chains which cannot be incorporated into crosslinks but may take part in branch formation through combination termination with a growing grafted B chain. Naturally, disproportionation termination between attached and unattached growing chains, or termination between two unattached chains, yields linear homopolymer B which is not incorporated into the crosslinked structures. The relative proportions of polymer B as crosslinks, branches and soluble homopolymer can be calculated simply from a knowledge of the relative rates of formation of attached and unattached radicals. It is interesting to note that branches formed as a result of combination termination between attached and unattached growing chains are of the same average length as the crosslinks in the absence of complications associated with the gel effect. By contrast, branches formed as a result of disproportionation termination of attached growing chains are, on average, half as long as the crosslinks.

The second method of increasing the branch:crosslink ratio involves the use of a chain-transfer agent during crosslinking. Transfer from a propagating grafted chain to a low molecular weight transfer agent obviously leads to unattached growing B chains, the subsequent fate of which is then the same as in the preceeding case, i.e. they may be incorporated into branches or form linear homopolymer. Thus by the uses of a transfer agent branches of B polymer are generated in two ways, directly during the transfer process and also as a consequence of bimolecular termination of attached and unattached chains. Again, relative rates of formation of crosslinks, p^x, branches, p^b, and soluble homopolymer, p^u, may be calculated from a knowledge of the polymerization kinetics, and are given in equations (7).

$$
\begin{aligned}
p^x &= \oint \frac{2}{(n+1)^2} \\
p^b &= \oint \frac{(n-1)(n+3)}{(n+1)^2} \\
p^u &= \oint \frac{(n-1)^2(n+2)}{2(n+1)^2} \\
n &= \bar{P}_o/\bar{P} = 1 + 2\, C_s[S]k_p/(\oint k_t)^{\frac{1}{2}}
\end{aligned}
\qquad (7)
$$

$\bar{P}_o$, $\bar{P}$ are degrees of polymerization of the whole crosslinking polymer in the absence and presence of transfer, respectively; C_s is the transfer constant for transfer agent S; k_p, k_t are the rate coefficients for propagation and termination in the polymerization of the crosslinking monomer. It can also be shown, from a detailed kinetic analysis that, at a constant rate of initiation, as transfer becomes more extensive the average degree of polymerization of crosslinks, P_n^x, branches, P_n^b, and linear polymer P_n^u, decrease; these parameters are given by equations (8). This situation contrasts with the use of low molecular weight initiators to introduce branches when, at a constant rate of initiation, the crosslink and branch lengths remain constant as their proportions

$$\bar{P}_n^x = \bar{P}_o \left\{ \frac{2}{1 + n} \right\}$$

$$\bar{P}_n^b = \bar{P}_o \left\{ \frac{5 + n}{(1 + n)(3 + n)} \right\} \qquad (8)$$

$$\bar{P}_n^u = \bar{P}_o \left\{ \frac{3 + n}{(1 + n)(2 + n)} \right\}$$

are varied by adjusting the relative rates of formation of attached and unattached polymers.

The synthetic route to ABCPs we have described is extremely versatile in allowing the preparation of materials based on a wide range of component polymers. In the absence of kinetic complications such as gel effects, it allows a high degree of control of the detailed structures of the polymers to be obtained. However, it should be clear that these structures are only calculable in a statistical manner since they embody the randomness normally associated with free-radical polymerization. Thus, although the synthesis is much more versatile than could be obtained with anionic polymerization, the latter yields products with more precisely-defined structures.

MORPHOLOGY AND PROPERTIES OF ABCPs

It is now well established that virtually all chemically different polymers are mutually incompatible and that an homogenised mixture of two such polymers will tend to undergo phase separation. Consequently there is an inherent tendency for microphase separation in ABCPs. Under conditions of low crosslink density it is easy to visualise the manner in which molecular reorganisation and phase separation can occur. For example, an ⟩—⟨ -shaped structure such as (I) is analogous to a linear ABA block copolymer; the crosslink corresponds to the central B block and the prepolymer chains to the A blocks. The only real distinction between structures such as (I) and the linear ABA polymers is that the ABCPs have two end blocks attached to each end of the central block so that (I) might be designated an A_2BA_2 block copolymer; the end blocks generally have different lengths. Thus, it might be anticipated that ABCPs of low crosslink density will undergo microphase separation in much the same way as ABA block copolymers. It should be remembered that the whole polymers prepared at low crosslink density are blends of crosslinked polymer and unreacted prepolymer.

At high crosslink densities, where the polymers contain infinite network structures such as (IV), the same polymer-polymer interactions are present and thermodynamic parameters still favour microphase separation, but it is by no means clear whether the geometrical constraints present in such structures will allow sufficient molecular reorganisation to permit phase separation to an extent consistent with the observation of multiphase phenomena. Our initial investigations into the morphologies and properties of ABCPs, therefore, were aimed at discovering whether or not microphase separation does occur in highly crosslinked ABCPs, establishing basic morphological features of ABCPs and exploring the possibility of observing the influence of one component on the properties of the other.

Dilatometric Observations

Bamford, Eastmond and Whittle[15] prepared a series of networks based on PVTCA crosslinked with PSt and with PMMA, designated PVTCA/PSt and PVTCA/PMMA, respectively, and investigated the thermal expansion behaviour of these polymers with the aid of a dilatometric technique. The PVTCA was prepared from polyvinyl alcohol by esterification with trichloracetyl chloride to about 60% conversion and had a glass-transition temperature of 59°C. PVTCA is incompatible with PSt and with PMMA, as evidenced by phase separation in solutions of the polymers in a common solvent and from the observations of well-defined glass-transitions, in blends, at temperatures corresponding to the T_gs of the homopolymers. Structural parameters of some PVTCA/PSt and PVTCA/PMMA networks are presented in table 2.

Samples I-IX (Table 2) were prepared in the absence of transfer agents or low molecular weight initiator so that branch:crosslink ratios were determined solely by the radical termination mechanisms. Samples I-III and VII-IX were prepared by crosslinking at 25°C using photoinitiation by $Mn_2(CO)_{10}$ and samples IV-VI were prepared at 80°C using $Mo(CO)_6$ as thermal initiator. In all cases the solid ABCPs were isolated by drying the polymers from gels swollen by the monomers used to form the crosslinks, i.e. from methyl methacrylate (samples I-VI) and from styrene (samples VII-IX).

Dilatometric data for the PVTCA/PSt samples (VII-IX) showed the existence of two well-defined glass transitions at 55-60°C(T_{g1}) and 97-100°C (T_{g2}); corresponding closely to those of the component homopolymers. In addition, the coefficients of expansion of the ABCPs throughout the temperature range investigated (25-115°C) corresponded closely with those expected, on the basis of simple volume additivity, for heterogeneous mixtures of the component homopolymers of corresponding compositions. These data were interpreted as indicating virtually complete microphase separation of the components, even for highly crosslinked materials (at least up to γ_r = 6).

It was also reported that two glass transitions could be identified in dilatometric plots (fig.1) for PVTCA/PMMA ABCPs (except for

Table 2. Structures and Properties of PVTCA/PMMA and PVTCA/PSt ABCPs

Sample No.	Crosslinking Polymer	% PVTCA (w/w)	γ_r	$\bar{P}_n$ crosslinks	Branches: crosslinks	W	T_{g1} obs.	T_{g2} obs.	T_{g1} calc.
I	PMMA	21.5	3	4560-4740	4	1.2	74	97.5	82
II	PMMA	8	9	5060-5440	4	1.7	71	102	84
III	PMMA	3.5	16-18	6900-8000	4	2.1	69	101	88
IV	PMMA	21	1	11820	8	2.8	74	99	88.5
V	PMMA	7.5	3.5	11700	8	4.4	71	102	92
VI	PMMA	3	7	16760	8	-	-	103	-
VII	PSt	77	1	2880	0	0.17	55	97	-
VIII	PSt	40.5	3	4750	0	0.12	60	97	-
IX	PSt	21.5	6	5800	0	0.22	60	100	-
	PVTCA						56		
	PMMA							~100	
	PSt							~100	

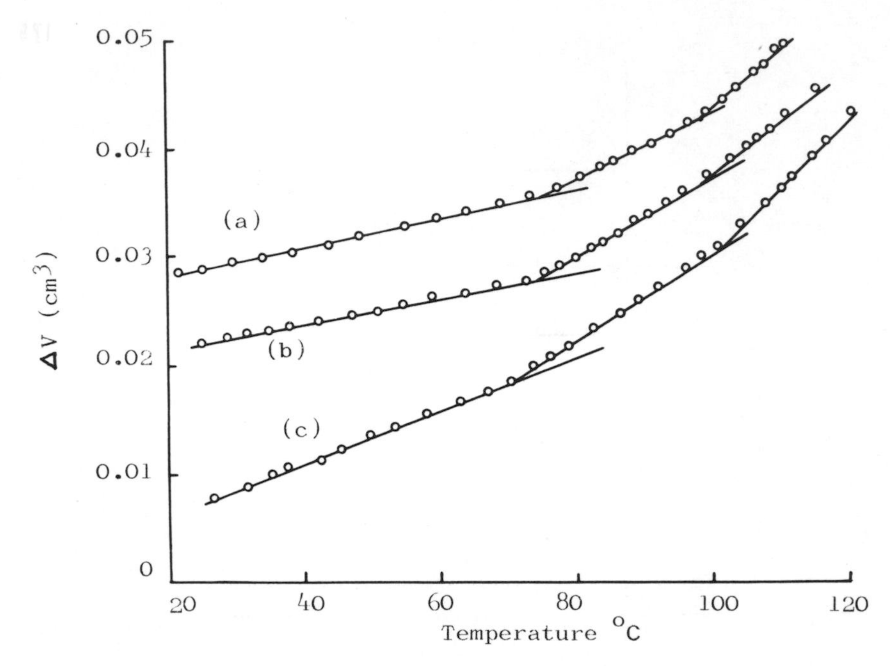

Fig.1 Dilatometric data for PVTCA/PMMA networks VII,IV,V, (curves a,b,c, respectively), showing the presence of two glass transition temperatures T_{g1} (lower) and T_{g2}. Curves below T_{g1} and above T_{g2} calculated from the composition of the polymer and coefficients of expansion of the components.

sample VI which contains only 3% w/w PVTCA), but it was apparent that their thermal behaviour differed significantly from that of the PVTCA/PSt networks. While the upper glass transition T_{g2}, appeared to correspond closely with that of pure PMMA, the lower glass transition T_{g1} was located at temperatures higher than T_g for PVTCA by 10-15 °C. Further, although the slopes of the dilatometric plots at temperatures below T_{g1} and above T_{g2} corresponded closely to those expected for mixtures of the two glassy and two liquid polymers of appropriate composition, respectively, the slopes of dilatometer plots in the temperature range between T_{g1} and T_{g2} were much higher than expected for a mixture of liquid PVTCA and glassy PMMA. These data are compatible with the view that microphase separation in the PVTCA/PMMA ABCPs is incomplete. Since there appeared to be a glass transition at about 100°C it was thought that one phase in the solid polymer was essentially pure PMMA and that the second phase consisted of a mixture of PVTCA and PMMA. It was visualised that the mixed phase existed as a solid solution of the two normally incompatible polymers and that this mixed phase exhibited a single T_g, corresponding to T_{g1}. The inclusion of some PMMA into the PVTCA phase would be expected to raise T_{g1} above the T_g of PVTCA, as observed. Consequently, at temperatures just above T_{g1} the PMMA in the mixed phase would be undergoing segmental motion cooperatively with the PVTCA and would therefore partake in a glass transition at temperatures lower than in pure PMMA. Coefficients of expansion at temperatures between T_{g1} and T_{g2} would reflect the total volume of polymer involved in the lower glass transition and would thus be higher than if PVTCA alone were involved.

On the basis of the above model for the PVTCA/PMMA ABCPs, compositions of the mixed phases were calculated[15] from the coefficients of expansion between T_{g1} and T_{g2} derived from the dilatometric data. Values of w quoted in table 2 represent these compositions expressed as the weight of PMMA associated with unit weight of PVTCA in the mixed phase. Values of w calculated in this way show that in all cases the mixed phase contains a large proportion of PMMA.

A more detailed examination of the dilatometric data has revealed that the initial model proposed requires some modification. For homogeneous mixtures of amorphous materials in which the free volumes of the components are additive, the glass transition temperature of the mixture can be calculated from the glass transitions, volume fractions and coefficients of expansion of the components with the aid of the Kelley-Bueche equation.[19] The Kelley-Bueche equation is often used to calculate T_gs of amorphous polymers plasticised with low-molecular-weight diluents but Noland et al.[20] have shown that it describes the variation in T_g with composition for compatible mixtures of PMMA and polyethyl methacrylate with polyvinylidene fluoride. Assuming that the mixed phase is essentially homogeneous and that simple additivity of volumes applies we should therefore be able to calculate the glass transition temperature of the mixed phase from the magnitudes of w given in table 2. It turns out that values of T_{g1} calculated in this way are considerably higher than those observed (table 2). We believe that this discrepancy arises from a non-uniform composition of the mixed phase. If we make the assumption that the mixed phase forms the domains, non-uniformity of composition could be the result of a radial distribution of composition within the domains, or of the presence of domains of different composition, or a combination of both. In agreement with this a smooth curve can be drawn through the volume-temperature data above T_{g1}, the slope gradually increasing through the apparent T_{g2}; fig.2 contains the same experimental data as fig.1 but with a smooth curve drawn through the data points above T_{g1}. On this view, the smooth curve results from the superposition of a series of T_gs arising from regions of different composition, giving an apparently continuous glass transition from T_{g1} to T_{g2}. T_{g1} corresponds to the glass transition temperature of the regions which are richest in PVTCA while the values of w given in table 2 are average compositions of

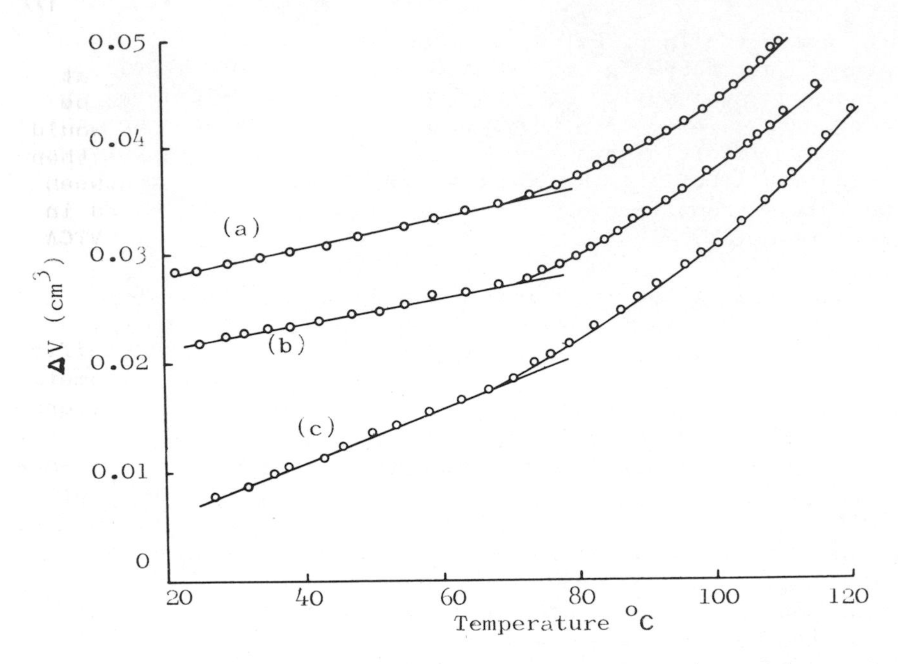

Fig.2 Dilatometric data, as in fig.1, for PVTCA/PMMA networks VII,IV,V, (curves a,b,c, respectively), showing a 'continuous' glass transition above T_{g1}.

the mixed regions in the various samples.

From the dilatometric data described above it therefore appears that the PVTCA/PSt and PVTCA/PMMA ABCPs, exhibit quite different morphologies and thermal properties. In the absence of other data it would be easy to attribute these to different degrees of incompatibility arising from variations in polymer-polymer interaction parameters in the two types of network. However, there are major structural differences between the two types of ABCP examined: PVTCA/PSt networks contain no PSt branches while in PVTCA/PMMA a large proportion of the PMMA is in the form of branches. Further, the data in table 2 indicate that in the PVTCA/PMMA polymers the PMMA content of the mixed phase increases with increasing branch: crosslink ratio. It seemed of interest, therefore, to consider the possibility that the presence of branches is at least partially responsible for the formation of mixed phases in PVTCA/PMMA ABCPs. Consequently, we synthesised PVTCA/PSt networks containing PSt branches using the techniques outlined in the preceding section. The relevant data[21] for three such samples prepared with the aid of the low molecular weight initiator technique are presented in table 3; the solid polymers were obtained by drying gels swollen with styrene monomer.

Table 3. Structures and Properties of PVTCA/PSt Networks Containing PSt Branches

Sample	% PVTCA	γ_r	$\bar{P}_n$ branch[a]	$\bar{P}_n$ crosslinks[a]	branches: crosslinks	W	T_{g1}	T_{g2}
X	38.3	1.3	1150	1150	1.0	0.45	74	89
XI	37.3	1.3	926	926	2.0	0.74	70	91
XII	30.3	1.5	1044	1044	2.0	0.90	72	92
XIII	55	3.5	-	500	0	0.05	65	90

(a) Estimated gravimetrically, as described in text.

The data in table 3 show that in samples X-XII, which contain PSt branches, T_{g1} is higher than the corresponding value for a similar sample (XIII) without branches or T_g for PVTCA (table 2); fig.3 shows a typical dilatometer plot for PVTCA/PSt network containing branches. In addition, the coefficients of expansion show anomolies which are consistent with the incorporation of some PSt

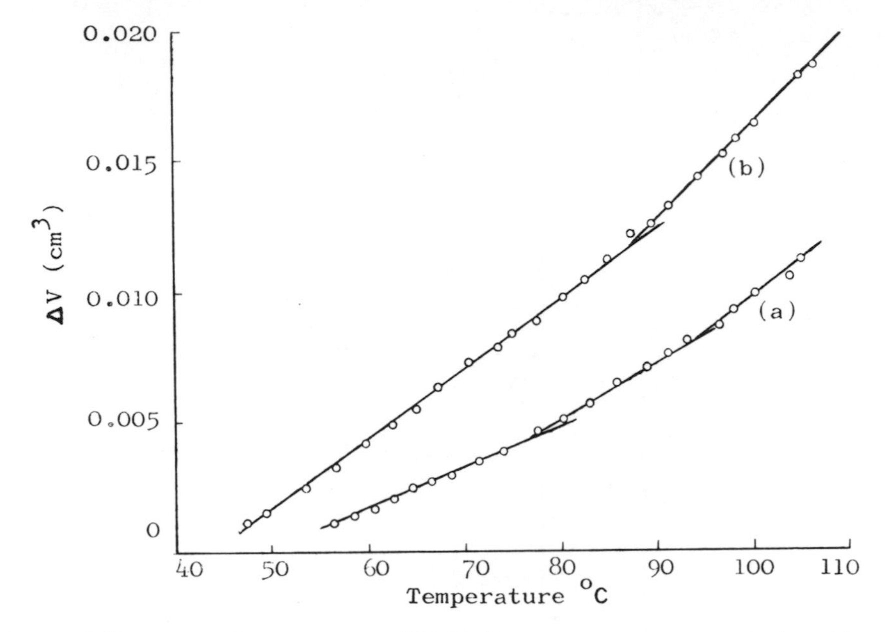

Fig.3 Dilatometric data for PVTCA/PSt networks containing PSt branches. (a) 38.3% PVTCA (w/w), $\gamma_r = 2$, $\bar{P}_n$ crosslinks = 1.1×10^3, cast from styrene, showing two glass transition temperatures at 75°C and 95°C; (b) 56.2% PVTCA (w/w), $\gamma_r = 2$, $\bar{P}_n$ crosslinks = 1.7×10^3, cast from styrene + ethyl acetate (9:1 v/v), showing one glass transition temperature at 87.5°C.

along with the PVTCA in a mixed phase; values of w calculated from these data are also given in table 3. During the preparation of PVTCA/PSt ABCPs containing PSt branches it is inevitable that considerable quantities of linear PSt homopolymer are produced and no attempt was made to remove this material from the networks. In separate experiments[21] it was shown that addition of considerable proportions of polystyrene homopolymer to PVTCA/PSt ABCPs did not enhance the formation of mixed phases but instead the added PSt entered the pure PSt phase.

In addition, PVTCA/PSt networks containing PSt branches were prepared by means of the transfer-agent technique and these materials also provided evidence for mixed-phase formation; experimental details for these samples are not given here since the simultaneous occurrence of chain transfer and the autoacceleration which follows gelation introduces considerable complications in the calculation of structural parameters.

Comparison of the various dilatometric data for PVTCA/PSt and PVTCA/PMMA ABCPs shows quite clearly that microphase separation can occur in these multicomponent polymers, even up to high crosslink densities, and that the presence of branches of the crosslinking polymer encourages the formation of mixed phases. This phenomenon may be of general occurrence in ABCPs having branches of the B component and it is interesting to reflect on the mechanistic implications. Consider a small portion of an ABCP shown in fig.4(a), which represents, schematically, two A chains in separate domains joined by a crosslinking B chain in the matrix; the junction points are located on the domain-matrix interface. A consequence of the necessity of uniform space-filling within the domains is that the A-chains are unable to adopt all their normal conformations (as in pure homopolymer), especially in the vicinity of the domain-matrix interface. Any tendency for the junction points to move into the domains to allow the A chains to take up more favourable conformations will be opposed by a similar tendency for the junction point to move into the matrix to allow the B chain to adopt more favourable conformations and retractive forces

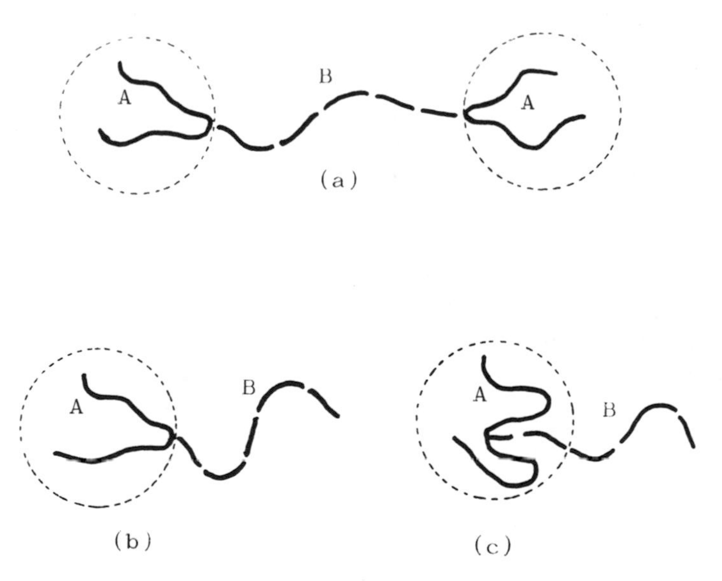

Fig.4 Schematic representation of microphase separation in ABCPs where the prepolymer A enters the domains, showing complete microphase separation in (a) H-shaped structures, and (b) structures having branches of the B component, (c) represents the incorporation of B branches into A domains.

in the B chain, so stabilising the junction points in the vicinity of the interface.

Fig.4(b) shows an equivalent situation for an A chain in a domain attached to a branch of B in the matrix, again with the junction point at the interface. In this case, displacement of the junction points into the domain, to allow more favourable conformations of A chains, will not be opposed by retractive forces in the B chain because the other end of that chain is not covalently linked to the remainder of the network. The only forces which oppose the movement of the junction points into the domain arise from the various polymer-polymer interactions and if these effects can be overcome at least some portion of the B branch will be drawn into the A-domain and contribute to the formation of a mixed phase, as shown schematically in fig.4(c).

During our investigations of PVTCA/PSt networks an interesting solvent-effect was observed.[21] All the dilatometric data in the preceding discussion refer to networks obtained by drying gels swollen by the monomer of the crosslinking chains; i.e. styrene for the PVTCA/PSt networks. It was observed that if the PVTCA/PSt networks containing PSt branches (samples X-XII, table 3) were dried from styrene + ethyl acetate (9:1 v/v) the solid polymers exhibited only one well-defined T_g, indicating a complete absence of microphase separation; a dilatometer plot for such a sample is shown in fig.3. The observed T_gs agreed, to within 3^oC, with those calculated from the Kelley-Bueche equation for mixtures of the homopolymers, assuming simple volume additivity. Thus, it appears that changing the swelling agent from styrene to styrene + ethyl acetate results in the formation of a completely homogeneous mixture of the two incompatible components. The effect is reversible. If a sample cast from the mixed solvent and exhibiting only one T_g is reswollen in pure styrene and dried the sample shows two T_gs; reswelling in styrene + ethyl acetate and drying gives a polymer with one T_g, and so forth. PVTCA/PSt networks containing no PSt branches (e.g. sample XII) did not show this effect under the conditions investigated.

Confirmation of the morphological features of PVTCA/PSt networks described above has recently been obtained from investigations using differential scanning calorimeter and torsion pendulum techniques.

Electron Microscope Studies of ABCPs

Morphological investigations of multicomponent polymers in the electron microscope are, in general, hindered by lack of contrast between the phases. If one of the components is unsaturated it is possible to increase the inherent contrast by incorporating heavy atoms by staining with osmium tetroxide. Many workers have used this technique in investigations of the morphologies of multi-component polymers containing polybutadiene or polyisoprene components. We have applied the technique to ABCPs containing polychloroprene[16]; in these systems the polychloroprene component can be stained and the morphologies of the polymers revealed.

In a preliminary study, samples of PVTCA/PCp ABCPs of low and high crosslink densities were examined and direct evidence of microphase separation was obtained.[16] Subsequently investigations were extended to PSt'/PCp ABCPs which were also shown to exhibit phase separation[22] (here PSt' denotes polystyrene containing functionalised units IV, used as prepolymer). Fig.5(a) presents a typical electron micrograph of a PVTCA/PCp polymer of low crosslink density (γ_r = 1.2) in which the polychloroprene crosslinks form the domains; fig.5(b) is a corresponding micrograph of a PVTCA/PCp network at high crosslink density (γ_r = 9).

In very lightly crosslinked specimens the copolymer species are mainly in the form of simple H-shaped molecules such as (I) (equation 2), but the major proportion of the whole polymer is unreacted prepolymer. When such a material undergoes microphase separation, with the crosslinks entering the domains, the morphology of the

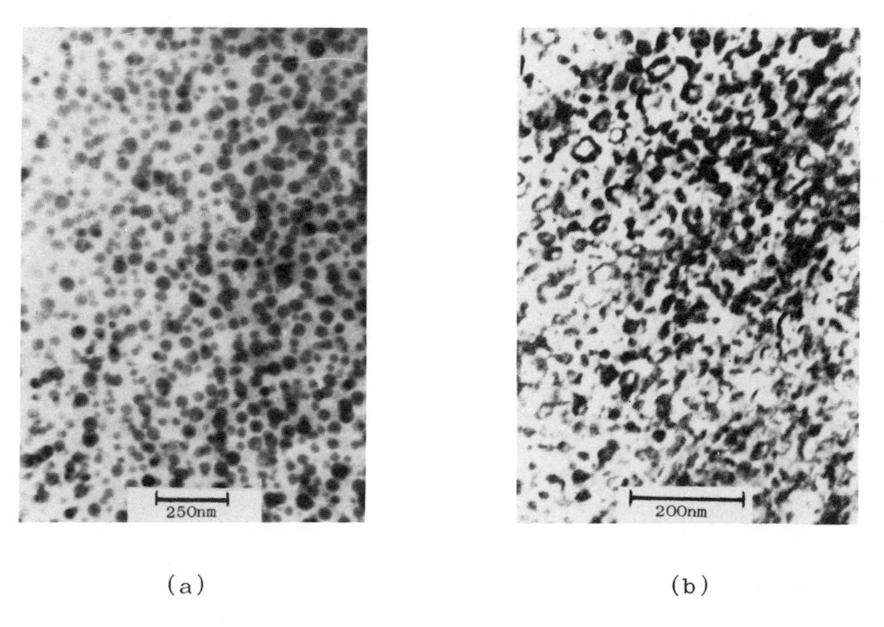

Fig.5 Electron micrographs of stained sections of PVTCA/PCp ABCPs of (a) low crosslink density, γ_r = 1.2 (PCp 8% w/w, $\bar{P}_n$ = 900), (b) high crosslink density, γ_r = 9 (PCp 75% w/w, $\bar{P}_n$ = 3720).

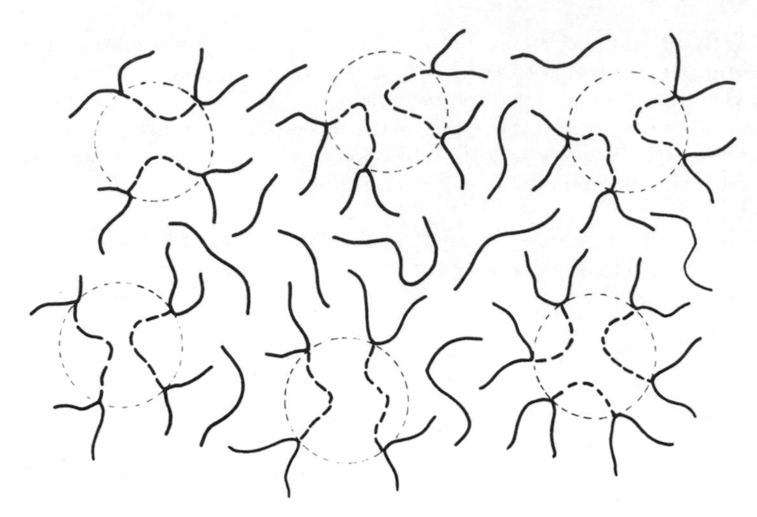

Fig.6 Schematic representation of an ABCP, of low crosslink density, in which the B component enters the domains and the matrix contains both reacted and unreacted A chains. ——— A component, ------- B component.

system can be schematically represented as in fig.6. If phase separation is virtually complete the junction points between A and B chains lie at the domain-matrix interface. By analogy with linear block copolymers, it might be expected that some correlation will exist between the morphology and the chain dimensions in the ABCPs. When the crosslinks of an ABCP constitute the domains, as is the case with the systems under discussion, the domain sizes must be limited by the lengths of the PCp chains. On the other hand, in lightly crosslinked materials prepared by our techniques, it is not to be expected that the interdomain separation will be determined by the lengths of the prepolymer chains. The matrix of such a polymer contains the unreacted prepolymer, which naturally increases the interdomain separation over and above that which would be observed in a simple assembly of H-shaped molecules. Thus, the major factor determining the interdomain separation under these conditions is the overall composition; this situation is represented in fig.6, which illustrates that the interdomain distance is greater than the length of the prepolymer chain.

It has been observed that in these polymers containing PCp the domain diameters remain constant with increasing crosslink density, for constant PCp chain-length.[22] As the crosslink density increases and more prepolymer is incorporated into copolymer species the number of domains per unit volume also increases and the interdomain separation decreases. Eventually interdomain separations become comparable with the chain lengths of the prepolymer molecules and it then becomes possible for a prepolymer chain carrying two or more crosslinks to link two or more domains. The crosslinks

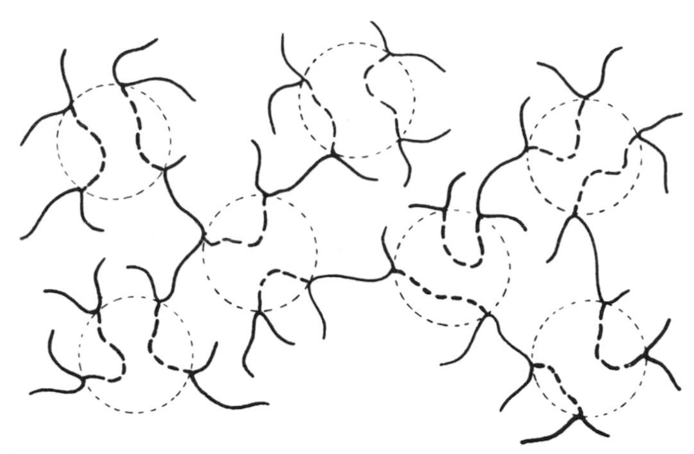

Fig.7 Schematic representation of an ABCP, of higher crosslink density than fig.6, showing A chains which have reacted more than once providing interconnections between domains composed of B chains. ——— A component, ------ B component.

attached to one prepolymer chain may be in different domains, as represented schematically in fig.7. With further increase in crosslink density more interdomain connections must be produced.

At high crosslink densities (e.g. $\gamma_r > 5$) the morphology is not so clearly defined and while electron microscope studies show that microphase separation does occur under these conditions the morphology presents a more confused picture (fig.5(b)). Examination of micrographs of networks with high crosslink density suggest that the domain sizes are smaller than in corresponding samples with small values of γ_r. Apparently, at the highest values of γ_r the geometrical constraints imposed on molecular reorganisation play an important part in determining the morphologies of the materials, preventing aggregation of the crosslinking chains into domains of sizes expected from studies of ABCPs with crosslink density.

Domains in ABCPs containing PCp at relatively low crosslink densities are small;[22] typically domain diameters being in the range 20-100nm. Meier[3] has carried out theoretical calculations for linear AB block copolymers which provide a relationship between domain diameter and degree of polymerization of the domain-forming segment. The calculations were based essentially on space-filling considerations aimed at achieving uniform density within the domains. The result was the prediction that domain diameters should be proportional to the square root of the degree of polymerization of the domain forming chains, the proportionality factor depending on the interfacial tension between the component polymers. More recent calculations by Helfand[23] lead to similar conclusions.

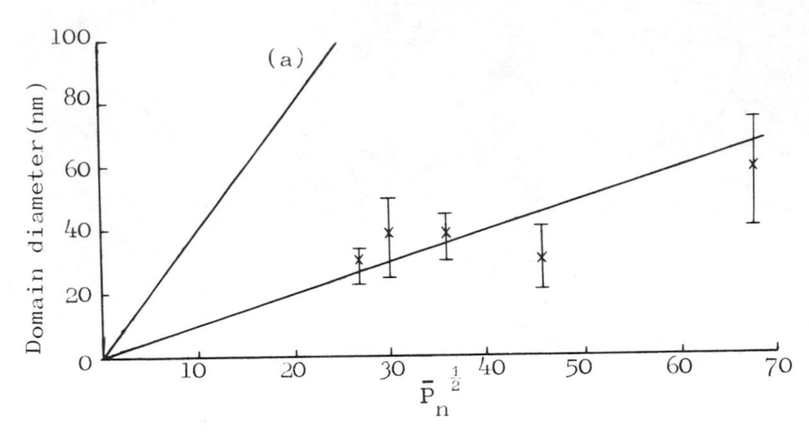

Fig.8 Variation of the diameter of PCp domains in PSt'/PCp ABCPs (γ_r<2) with the square root of the degree of polymerization of PCp chains. Curve (a) is the theoretically predicted variation of domain size for AB block copolymers, according to the theory of Meier.

It is to be expected that similar factors will be involved in domain formation in ABCPs and in AB block copolymers. Fig.8 shows the variation in domain diameter with degree of polymerization ($\bar{P}_n$) of the polychloroprene chains for various PSt'/PCp ABCPs; the bars represent the distribution of domain sizes and the crosses their average diameter. From these preliminary data it appears that the domain diameters are approximately proportional to the square root of the degree of polymerization of the PCp chains, in conformity with the proposed model. No information is at present available on the interaction parameters of the polymers used in this work, but, if we assume that they are similar to those for PSt and polybutadiene, Meier's figures[3] can be used to estimate domain diameters in block copolymers having the same chain dimensions as the ABCPs. Meier's theory predicts domain diameters shown by the line (a) in fig.8. Evidently the domains in the ABCPs are much smaller than would be expected for AB block copolymers composed of the same chains. This apparent discrepancy is partly resolved if we consider the differences in molecular architecture in the two systems. Fig.9(a) is a schematic representation of an AB block copolymer molecule with one group in the domain and the junction point between the blocks located at the domain-matrix interface. In this case the other end of the domain-forming block may reside anywhere within that domain. The corresponding situation for a system containing H-shaped molecules in which the crosslinks form the domains is shown schematically in fig.9(b). For good phase separation in the ABCP, the junction points must again be at the interface, so that if one end of a crosslinking chain is first placed at the interface then the chain must adopt a conformation such that the other end of the chain is also located at the

Fig.9 Schematic representations of microphase separation in (a) an AB block copolymers, (b) an H-shaped ABCP molecule in which the B component enters the domain.

interface. Transformation of the situation in fig.9(a) to that in fig.9(b) by bringing the free end of the domain-forming chain to the interface, is tantamount to removing material from the centre of the domain and so reducing the density in this region and increasing the density near the interface. If uniform density is to be maintained it is readily apparent that, for similar chain lengths, domains in the ABCPs must be smaller and contain fewer chains than those in the corresponding AB block copolymers.

When ABCPs undergo microphase separation with the A component (the prepolymer) forming the domains, similar considerations will apply. However, in such cases it is to be expected that it will be interdomain separation which is controlled by the chain lengths of the prepolymer since the domains will be expanded by unreacted prepolymer.

Nmr Observations on ABCPs

To date there have been no reports of studies of molecular motions in multicomponent polymer systems made with the aid of nmr techniques. We believe that this is a potentially valuable field for investigation. In materials which exhibit microphase separation the chains within the domains or the matrix, or both, will be forced to adopt conformations not normally found in bulk polymers, especially near the interfaces. Such differences will result in changes in chain packing which might influence molecular motions in those chains. Further, since chains in the separate phases are chemically linked at the interface, it would seem to be of interest to ascertain whether motions in one polymer are influenced by those in the other. Special effects may also be associated with the interface itself and may be detectable if the interfacial region occupies a sufficiently large volume fraction of the total polymer. We have therefore embarked on a programme to study these effects in ABCPs.

The PVTCA/PMMA networks I-VI (table 2), whose morphologies have already been discussed in terms of dilatometric data, were examined by broad-line nmr spectroscopy[24] over the temperature range -200 to 150°C. This range covers both the onset of α-methyl group rotation in the PMMA and the onset of segmental motions in the component polymers. Since the PVTCA/PMMA networks under consideration are largely composed of PMMA, and since each MMA unit contains eight protons whereas the vinyl trichloracetate unit contains only three protons, over 90% of the protons in these samples are contained in the PMMA chains. Virtually the whole of the nmr signal, therefore, originates from the PMMA chains and, to a first approximation, changes in the nmr signal can be considered to reflect changes in the motion of the PMMA chains.

The variation in the second moment (ΔH_2^2) of the nmr absorption line for PMMA homopolymer is shown by the dotted line in fig.10. The decrease in ΔH_2^2 at low temperatures results from the onset of α-methyl group rotation about the triad axis in the polymer; the onset of ester methyl group rotation occurs at much lower temperatures. The onset of segmental motion at the glass transition is responsible for the decrease in ΔH_2^2 above 100°C.

The transition region corresponding to the onset of α-methyl group rotation in the PMMA homopolymer covers the temperature range -140°C to 80°C, approximately. For PVTCA/PMMA samples I,IV,V the

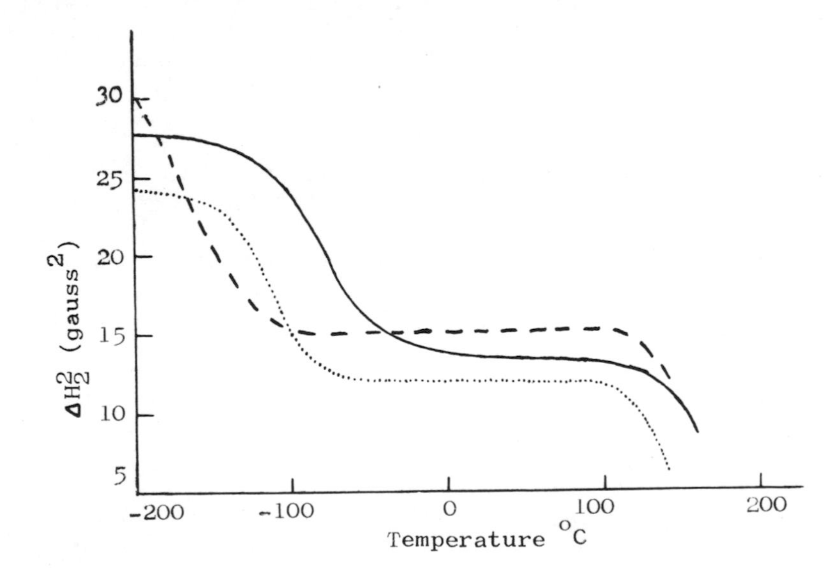

Fig.10 Variation of second moments (ΔH_2^2) of nmr absorption lines with temperature for PVTCA/PMMA networks II (-) and IV (---) and for PMMA homopolymer (......).

temperature range for this transition is shifted to lower temperatures; the effect is greatest for sample IV and decreases through V to I. Data for IV are shown in fig.10 (dashed line) for which it is seen that all α-methyl groups are rotating at $-100^{o}C$; motional narrowing of the nmr line starts at temperatures around $-200^{o}C$.

In complete contrast to the above, networks II, III show some restriction to the rotation of the α-methyl groups. Data for sample II presented in fig.10 (solid line) show that α-methyl group rotation appears to commence in the ABCP at the same temperature as in PMMA homopolymer, but that only a fraction of the methyl groups are rotating at sufficient frequencies to give motional narrowing at $-80^{o}C$. Only at temperatures above about $0^{o}C$ are all methyl groups rotating rapidly. The transition in these networks therefore covers a very broad temperature range. Experimental data for sample VI, which contains only a very small proportion of PVTCA, are almost identical with those for pure PMMA, and blends of PVTCA and PMMA also provide data indistinguishable from those of PMMA homopolymer.

Nmr data, therefore, provide the surprising result that the α-methyl group rotations in PMMA may be enhanced or retarded in ABCPs which contain the same components and have only relatively minor differences in structure. We believe these phenomena can be attributed to variations in the morphologies of the polymers. From the data in table 2 we have determined the distribution of PMMA between the mixed phase and the pure PMMA phase, with results presented in table 4.

Table 4. Distribution of PMMA in PVTCA/PMMA networks.

Sample	% of total polymer w/w			Weight fraction of total copolymer in mixed phase
	PVTCA	PMMA in mixed phase	'free' PMMA	
I	21.5	25.8	52.7	0.473
II	8.0	13.6	78.4	0.216
III	3.5	7.4	89.1	0.109
IV	21.0	58.9	20.1	0.799
V	7.5	33.0	59.5	0.405

Examination of table 4 indicates that enhancement of methyl group rotation occurs in systems containing a large proportion of the PMMA in the mixed phase (samples I,IV,V). Retardation of methyl group rotation is apparently only observed when almost all of the PMMA is in the pure PMMA phase.

In a mixed phase there is presumably an intimate mixture of PVTCA and PMMA and, as already described, the phase must adopt conformations consistent with good space filling. The PMMA chains will therefore be in an unusual environment and we believe that in these regions the methyl groups are located in regions in which there is more free volume than in PMMA homopolymer, thus allowing rotation of those groups at lower temperatures than normal.

In view of the distribution of the PMMA between the phases, the retardation of methyl group rotation must originate in the PMMA phase, i.e. in the continuous matrix. Consideration of the molecular packing again leads to a possible explanation of the observed effect. For simplicity let us assume that a well-defined boundary exists between a domain and the matrix. To obtain uniform space-filling in the matrix in the vicinity of that boundary the PMMA chains involved must again adopt conformations different from those in the bulk homopolymer. It therefore seems reasonable to suggest that as a result of the unusual conformations some α-methyl groups in the PMMA chains will be in closer proximity to other groups than in the homopolymer so that their rotation may be hindered. The morphology of the PVTCA/PMMA ABCPs has not yet been observed by electron microscopy but, by drawing a parallel with known morphologies of other ABCPs, we may assume that the domain diameters will probably not exceed about 50nm. A simple calculation based on this radius and the composition data in table 4 indicates that the thickness of the matrix between domains is also of the order of 50nm. If perturbations in packing extended only 20nm into the matrix from the domain surfaces a considerable fraction of the PMMA in the matrix could be affected. Additional evidence for restrictions to rotations of α-methyl groups in PMMA chains in ABCPs containing a large proportion of PMMA in the matrix has been adduced from pulsed nmr studies; it was observed that the increase in spin-spin relaxation time associated with the onset of α-methyl group rotation occurred more gradually in such samples than in PMMA homopolymer.[22]

So far it has not been possible to explain quantitatively the high absolute values of ΔH_2^2 of the PVTCA/PMMA ABCPs (from 0-100°C) compared with PMMA homopolymer. Theoretical calculations indicate that the high observed values are not a simple consequence of broadening by chlorine atoms in the PVTCA. However, if the assumptions we have made about the conformation of PMMA chains in the ABCPs are correct, it is possible that these effects can be explained by changes in both the intra- and intermolecular contributions to ΔH_2^2 since these depend critically on the average proton-proton distances in the polymers.

REFERENCES

1. See, for example, various papers in J. Polymer Sci. Pt.(C), 26, 1969; 'Colloidal and Morphological Behaviour of Block and Graft Copolymers', ed. G.E. Molau, Plenum Press, N.Y., 1971;

'Multicomponent Polymer Systems', ed. N.A.J. Platzer, Advances in Chemistry Series 99, Amer. Chem. Soc., Washington D.C. 1971; other papers in this volume.

2. S.G. Turley, J. Polymer Sci., Pt.(C), 1, 101 (1963).

3. D.J. Meier, J. Polymer Sci., Pt.(C), 26, 81 (1969).

4. M. Morton, J.E. McGrath, P.C. Juliano, J. Polymer Sci., Pt.(C), 26, 99 (1969).

5. C.H. Bamford, R.W. Dyson and G.C. Eastmond, J. Polymer Sci. Pt.(C), 16, 2425 (1967).

6. C.H. Bamford, R.W. Dyson, G.C. Eastmond and D. Whittle, Polymer, 10, 759 (1969).

7. C.H. Bamford, G.C. Eastmond and D. Whittle, Polymer, 10, 771 (1969).

8. C.H. Bamford, R.W. Dyson and G.C. Eastmond, Polymer, 10, 885 (1969).

9. For a summary see C.H. Bamford in 'Reactivity, Mechanism and Structure in Polymer Chemistry' ed. A.D. Jenkins and A. Ledwith, John Wiley and Sons, 1974, Chapter 3.

10. C.H. Bamford, G.C. Eastmond and V.J. Robinson, Trans. Faraday Soc., 60, 751 (1964).

11. C.H. Bamford, G.C. Eastmond and F.J.T. Fildes, Proc. Roy. Soc., A326, 431, 453 (1972).

12. C.H. Bamford and I. Sakamoto, J. Chem. Soc., Faraday Trans. I, in course of publication.

13. C.H. Bamford and E.O. Hughes, Proc. Roy. Soc., A326, 469, 489 (1972).

14. For summaries see (a) E. Koerner von Gustorf and F.W. Grevels, Fortschritte der Chemische Forschung, 13, 366 (1969); C.H. Bamford, Pure and Applied Chem., 34, 173 (1973).

15. C.H. Bamford, G.C. Eastmond and D.Whittle, Polymer, 12, 247 (1971).

16. J. Ashworth, C.H. Bamford and E.G. Smith, Polymer, 13, 57 (1972).

17. P.J. Flory, J. Amer. Chem. Soc., 63, 3083 (1941).

18. P.J. Flory, 'Principles of Polymer Chemistry', Cornell Univ. Press, Ithaca, N.Y., 1953.

19. F.N. Kelley and F. Bueche, J. Polymer Sci., 50, 549 (1961).

20. J.S. Noland, N.N.-C. Hsu, R. Saxon and J.M. Schmitt, in 'Multicomponent Polymer Systems' ed. N.A.J. Platzer, Advances in Chemistry Series 99, Amer. Chem. Soc., Washington, D.C.1971.

21. D. McGuire, unpublished results.

22. E.G. Smith, Ph.D Thesis, University of Liverpool, 1973.

23. E. Helfand, this volume.

24. D. Whittle, Ph.D Thesis, University of Liverpool, 1971.

Section II. Introduction

The papers in this section cover a tremendous range: from the details of polymer-polymer solubility in the styrene-α-methyl styrene system to the mechanism of orientation of domains in polyurethanes and on to the commercial uses of different types of formulations containing block copolymers. I suspect that the greater understanding of block copolymers provided by the more fundamental investigations will soon lead to better and possibly different products, while the novel products will lead to new sorts of fundamental investigations.

Robeson, Matzner, Fetters, and McGrath have obtained some provocative data on styrene-α-methyl styrene diblock copolymers and on mixtures of the corresponding homopolymers. Apparently, poly-α-methylstyrene is more soluble in polystyrene than polystyrene is in poly-α-methylstyrene. Furthermore, the homopolymers appear to be more soluble in each other at lower temperatures than they are at higher temperatures. Both of these results imply that something beyond Flory-Huggins theory is needed to explain all aspects of polymer-polymer solubility. As discussed by the authors, this "something" is probably the new "equation-of-state" theory of solutions. An additional result of this work, suitable for pondering, is the glass transition temperature of the block copolymer which appears to be lower than that of a random copolymer of similar composition. This has not yet been explained.

Work, in his paper on segmented polyurethanes, has indicated that the block lengths necessary to achieve domain formation vary with the chemical nature of the block. This result, of course, is compatible with the various theoretical treatments of domain formation in block copolymers. Seymour, Estes, and Cooper used differential infrared dichroism and birefringence measurements to measure the orientation behavior and mechanism of the domains in some polyurethanes. Differences were found between the orientation mechanism in a sample with independent hard domains and the orientation mechanism in some samples containing higher percentages of hard segments, presumably interlocking.

Kraus and Railsback studied styrene-butadiene diblock graded copolymers and found no evidence for a separate interlayer phase in addition to the polystyrene and polybutadiene phases in their samples using dynamic mechanical measurements. In blends with other rubbers, evidence for three phases was shown. The authors also present a discussion of commercial uses and applications of styrene-butadiene block copolymers. Holden, in his paper, presents a more extensive discussion of block copolymer applications which includes tabulations of specific commercial products which include block copolymers.

McGrath, Robeson, and Matzner present mechanical properties and water adsorption data for sulfone-nylon 6 block copolymers in order to show that such tough, semi-crystalline block copolymers may exhibit good solvent and environmental stress crack resistance.

Sonja Krause

POLYSULFONE-NYLON 6 BLOCK COPOLYMERS AND ALLOYS

J. E. McGrath, L. M. Robeson and M. Matzner

Union Carbide Corporation

Bound Brook, New Jersey 08805

INTRODUCTION

We have been interested in the field of block copolymers and alloys for over a decade. Our work published to date[1-10] has described some of our studies in the area of thermoplastic elastomers. Several interesting materials were uncovered and characterized. Moreover, some important rheological phenomena were also described[10].

The ability of block copolymers to display thermoplastic elastomeric behavior is, of course, well established[11-13]. The literature now abounds with examples of such macromolecules[14-16]. The case where two rigid blocks are chemically combined has been overlooked. Current knowledge of such systems and their potential utility is extremely sparse although some reports are appearing in the literature[17-21]. It was clearly of interest to investigate this area to establish structure-property relationships and the potential usefulness of such materials. One very intriguing aspect, namely the effect of the differential solubility parameter (Δ)[10] on processability was of great interest in these studies as well.

An attractive area that we would like to discuss in this paper is that of semi-crystalline block copolymers and the effect of this superstructure on properties such as chemical and environmental stress crack resistance. As is well known[22-24], all amorphous rigid thermoplastics (both ductile and brittle) display limited solvent resistance and undergo catastrophic failure when exposed to certain environments while under stress. Typically, the behavior is illustrated in Figure 1. Examples of plastics which behave in this manner include ABS, impact polystyrene, polycarbonates, polyphenylene oxide, and aromatic polyethers.

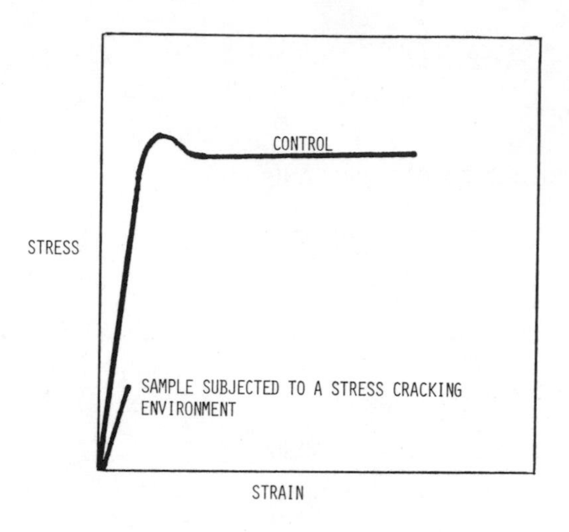

FIGURE I

Effect of Solvent Induced Stress Cracking on Amorphous Engineering Thermoplastics

It was reasoned that block copolymers prepared from such rigid amorphous engineering thermoplastics with a second crystalline segment would lead to vastly improved material performance. Such a system possessing an A-B-A structure in which the respective blocks are a polyarylether (polysulfone (PSF)) and crystalline nylon 6 (N6) is schematically depicted in Equation I.

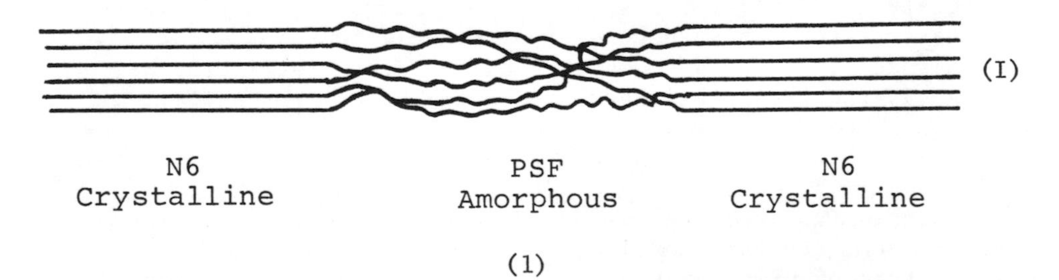

N6 Crystalline PSF Amorphous N6 Crystalline

(1)

The presence of the crystalline nylon 6 should result in an overall enhanced solvent and stress crack resistance. The preparation and characterization of the block copolymer (1) is described in this paper.

RESULTS AND DISCUSSION

A. Polymer Synthesis

It was shown[25] that activated aromatic halides such as 4,4'-dichlorodiphenylsulfone (2) are excellent initiators for

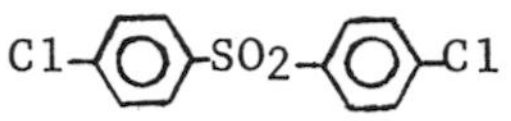

$$Cl-C_6H_4-SO_2-C_6H_4-Cl \qquad (II)$$

(2)

the anionic polymerization of ε-caprolactam. The mechanism of the anionic polymerization of ε-caprolactam[26] indicates that the initiator becomes permanently attached to the polymeric nylon 6 chain. Therefore, the use of a polymeric analog of (2) such as structure (3) should yield the desired segmented copolymer.

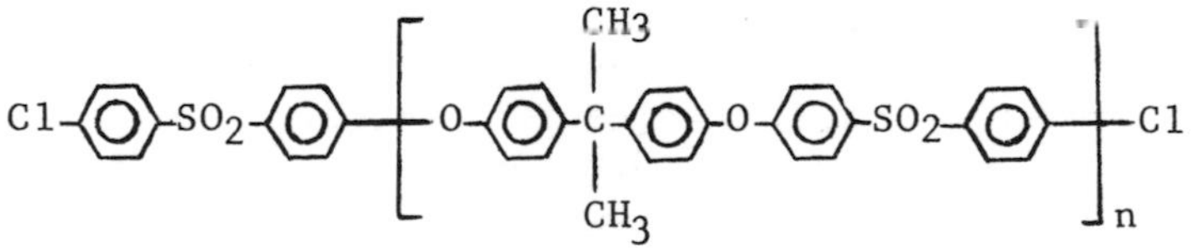

$$Cl-C_6H_4-SO_2-C_6H_4-\left[O-C_6H_4-C(CH_3)_2-C_6H_4-O-C_6H_4-SO_2-C_6H_4\right]_n-Cl \qquad (III)$$

(3)

The chlorine terminated polysulfone (3) is easily prepared[27] by conducting the preparation of the poly(aryl ether) with an excess of the activated halide as shown in Equation IV.

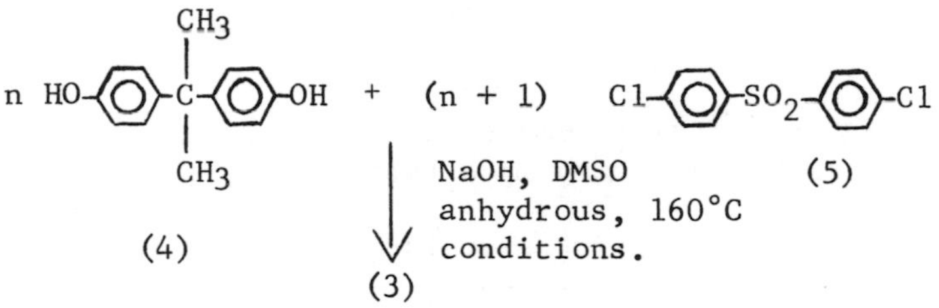

$$n\ HO-C_6H_4-C(CH_3)_2-C_6H_4-OH + (n+1)\ Cl-C_6H_4-SO_2-C_6H_4-Cl \xrightarrow[\text{conditions.}]{\text{NaOH, DMSO anhydrous, 160°C}} (3) \qquad (IV)$$

(4) (5)

It was found that (3) is soluble in molten ε-caprolactam. The addition of two to four mole percent (based on ε-caprolactam) of a base catalyst (sodium hydride) to this solution followed by reaction at elevated temperature (160°C or higher) resulted in rapid polymerization of the lactam to its equilibrium value. The reaction leads to a block copolymer based on two facts:

(1) Extraction with a typical polysulfone solvent (e.g., chloroform) indicates that a significant part of (3) is chemically attached to nylon 6.

(2) The mechanical properties of the block copolymer are attractive while those of the corresponding physical blends are poor.

In addition to the previously described method, a second approach was also developed. It is based upon the cleavage of the polysulfone chain by strong base at high temperature. The latter is visualized, as shown, for the case of the lactamate anion:

$$\sim\!\bigcirc\!-SO_2-\bigcirc-O-\bigcirc\!\!+\!\!\sim + \bar{N}\underset{CO}{\overset{(CH_2)_5}{\Big|}} \rightarrow \bigcirc\!-SO_2-\bigcirc-N\underset{CO}{\overset{(CH_2)_5}{\Big|}} + \bar{O}-\bigcirc\!\!+\!\!\sim \quad (V)$$

(6) (7) (8) (9)

It is readily seen that species (8) formed during the process is equivalent to that which one would obtain from a chlorine-terminated polymer. Therefore, (8) should behave as a polymeric initiator and give the block copolymer when heated with ε-caprolactam monomer and base. In practice, this process is very easy to conduct: commercial polysulfone homopolymer (P-1700) is dissolved in molten ε-caprolactam containing 2-4 mole % base (sodium hydride). The solution is heated to 160°C or higher and the copolymer is obtained. The polymerization can be carried out in a flask or in a continuous manner by metering the solution into an extruder. Obviously, in both methods one must strip or extract the unreacted lactam.

The crude polymers prepared via either method were submitted to a series of extractions. Methanol was used to determine the amount of unreacted lactam. The product was then extracted with chloroform in order to eliminate the free polysulfone. This latter procedure established the following:

(a) the presence of significant amounts of free PSF possessing lower molecular weight than the starting material;

(b) that the maximum amount of PSF ever incorporated into the copolymer was approximately 25 weight percent, and

(c) the chloroform extractables became negligible as the PSF weight fraction charged was progressively decreased to the level of ∿25 percent.

The rationale for (a) is of course in line with the degradation mechanism shown in Equation V. As far as (b) and (c) are concerned, they are explained by the much faster rate of nylon 6 propagation as compared to its initiation. This is in agreement with previous findings[25].

The remaining problem, namely the preparation of compositions containing higher amounts of PSF, was solved by applying the well-known alloying behavior of block copolymers with their respective homopolymers[28-30]. Our experimentation has shown that the extracted block copolymer was easily blendable and mechanically compatible with either PSF or nylon 6. Mechanical compatibility was rather limited for the ternary blends of the copolymer with both homopolymers. Thus, all compositions of interest were readily obtained via proper alloying procedures using either a two-roll mill or an extruder.

The foregoing description of the block copolymer synthesis clearly indicates that the desired compositions were obtained. However, their architecture (AB or ABA) is not known and it is assumed that both types of chains are present in the product.

Some typical polymers prepared in the course of this work are shown in Table I. The compositions were ascertained via sulfur and nitrogen analyses.

B. Polymer Properties

Mechanical properties of the block copolymers and their alloys are summarized in Table II. For comparison purposes, the corresponding data on homopolymers and their blends are also included. As the data indicate, tough rigid materials are obtained over the entire compositional range.

The melt flow results shown in Table II deserve some comment. Note that the block copolymer could be processed at lower temperatures than polysulfone. The data show that the melt viscosity of the material is within the range expected if the viscosity values of both homopolymers were averaged. This is further substantiated by the ease of melt fabrication of these resins. This might be interpreted as being in contradiction with reports in the literature[10-11]. However, the phenomena can be rationalized in the following manner. First of all the material probably contains both A-B and A-B-A structures. Therefore, the physical network characteristic of ABA and $(AB)_n$ block copolymers is present only to a limited extent in our material. The value of Δ is uncertain. Due to the large hydrogen bonding contribution in nylon 6, widely different values can be calculated for its solubility parameter. The values of Δ could thus vary appreciably and no definite correlation of Δ with melt processibility is feasible for this system at this time. Modulus-temperature behavior is graphically illustrated in Figure 2. The curve shows that both the copolymer and alloys are two phase systems displaying the characteristic glass transitions and crystalline melting points of the constituents. Generally, the compositions were translucent due to the polyamide crystallinity.

TABLE I

Charge: Wt. % ε-Caprolactam	Charge: Wt. % Polysulfone	R.V. of Polysulfone Oligomer[3]	NaH Mole %	Polymerization Time	Temp., °C	R.V.[5]	After Extractions: % N6	After Extractions: R.V.[5]
99	-[1]	--	2.0	3 hr.[2]	160	2.67	--	--
50	50	0.32	2.0	1.5 hr.	170	--	--	--
50	50	0.32	3.2	20 min.	170	1.37	--	--
50	50	0.32	3.5	20 min.	170	2.50	79.5	--
50	50	0.32	2.0	27 min.	170-180	1.20	78.5	1.30
80	20	0.32	0.6	30 min.	175	2.53	78	3.51
50	50	0.5[4]	2.0	30 min.	175	1.53	80	--
64	36	0.55	1.2	30 min.	180	1.77	82	--
80	20	0.26	1.1	1 hr.[6]	170	4.3	82.5	--
76	24	0.5[4]	1.4	5 min.	190	1.85	--	--
80	20	0.26	1.1	3 min.	210	2.62	--	--
80	20	0.26	1.0	30 min.	180	--	--	--
45	55	0.38	0.8	35 min.	170	1.45	--	--

[1] Polymerization initiated with 4,4' dichlorodiphenyl sulfone. The product was thus pure nylon 6.
[2] Solid after 20 minutes but total heating time was three hours.
[3] 0.2 gm. polymer per 100 ml. of $CHCl_3$ at 25°C.
[4] Commercial polysulfone P-1700.
[5] 0.1 gm. polymer per 100 ml. of m-cresol at 25°C.
[6] Solid after 15 minutes, but total heating time was one hour.

TABLE II

Mechanical Properties of a Polysulfone-Nylon 6 Block Copolymer, a PSF/N6-Polysulfone Blend and the Two Respective Homopolymers[1]

Sample	Wt. % Nylon 6	Melt Flow dg/min.[8]	R.V.[2] dl/gm	1% Secant Modulus psi	Tensile Strength psi	% Yield Elon-gation	% Elon. @ Break	Pen-dulum Impact ft.lb/in[3]	Glass Tran-sition Temp,°C	Crystal-line Melting Point, °C	Heat Distortion Temperature °C(264 psi)[7]
Polysulfone[3]	0	6	0.51	268,000	10,500	7.0	93	136	190	--	175
Polysulfone blends with the poly-sulfone-Nylon 6 Block Copolymer[4]	40	4	0.8	217,000	8,000	10.0	51	97	50,175	205	140
Polysulfone-Nylon 6 Block Copolymer[5]	78	38	2.6	148,000	7,500	-	150	200	50,175	205	55
Nylon 6[6]	100	30	1.5	150,000	10,000	-	345	360	50	210-225	50

[1] All data on compression molded films equilibrated at 50% R.H.

[2] Reduced viscosity values (η^{sp}/c) measured at 0.2 gm/dl in chloroform at 25°C for polysulfone. Data on the blends, block copolymer and nylon 6 determined in m-cresol at 0.1 gm/dl at 25°C.

[3] UCC (P-1700).

[4] Blended on a two roll mill at 240°C.

[5] See Experimental for preparation.

[6] Plaskon Molding Resin (Allied Chemical), after 28 days at 70°F, 45% R.H.

[7] ASTM test D648-56.

[8] Polysulfone melt flow measurements were at 350°C, all others at 250°C. (IP)

Of particular interest are the chemical and stress-crack resistance data shown in Tables III and IV for these novel compositions. Significant improvement in both properties are noted.

Improvement in the environmental stress aging (ESA) of polysulfone becomes significant at nylon 6 contents of 30 wt. %. This is shown in Figure 3 in which a one minute rupture time vs. applied stress in an acetone environment is plotted. It is believed that at this particular point the contribution of the copolymer to the continuous matrix becomes appreciable.

FIGURE 2

E-T Curves for Polysulfone Nylon-6 Compositions

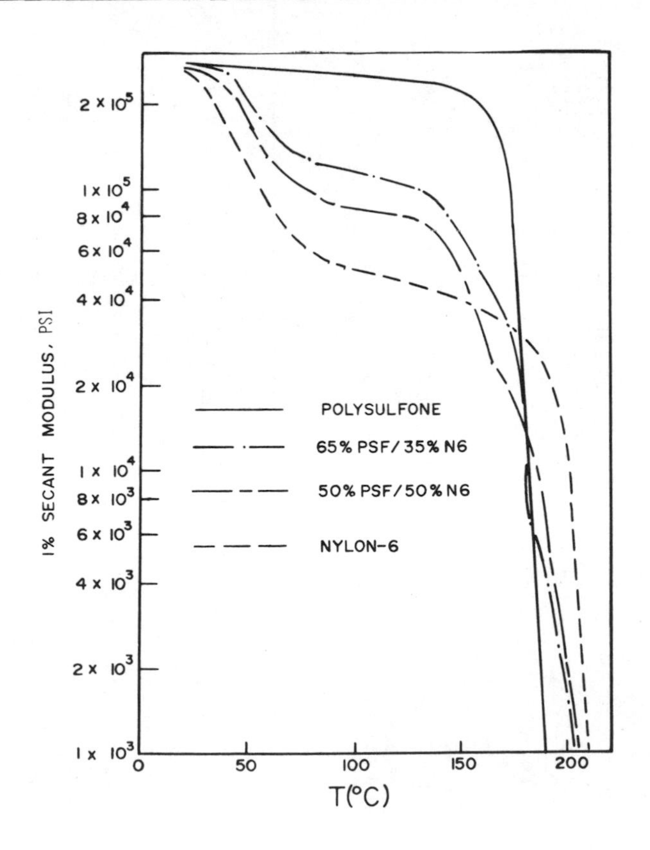

TABLE III

Chemical Resistance of Polysulfone of PSF/N6 (35 Wt. % N6) and of Nylon-6

Solvent	Polysulfone	% Weight Gain for PSF/N6 (35 Wt.% N6)	Nylon-6
Trichloroethylene	Swells and forms gel	11.6%	0.43%
Chlorobenzene	Dissolves	9.3%	0.11%
Acetone	Swells and forms gel	1.8%	0.33%
Ethanol	0.20%	2.0%	8.5%
Benzene	Swells and forms gel	1.1%	0.30%
Heptane	0.12%	.16%	0.18%

Determined after 18 days immersion in the above solvents of a 1/8" x 1/2" x 5" injection molded sample.

One disadvantage of Nylon 6 is its hydrophilicity and consequently the effect of water adsorption on its mechanical and electrical properties and dimensional stability.

Water adsorption data for the PSF/N6 compositions and the equilibrium values are given in Table V. Diffusion coefficients (50 and 100% RH) are listed in Table VI. Plots of q_t/q_∞(q_t =quantity of water adsorbed at time t,q_∞ = equilibrium water adsorption) versus time are shown in Figures 4 and 5. It is clearly seen that as the Nylon 6 concentration increases, so does also the sensitivity of the copolymer to water.

TABLE IV

Effect of Time and Stress on the ESA Characteristics of PSF/N6 Copolymers and Alloys[1]

Solvent	Compositions			
	35 Wt.% N6	50 Wt.% N6	60 Wt.% N6	77 Wt.% N6
Trichloroethylene	2,000 psi - no change 3,000 psi - brittle	3,000 psi - no change 5,000 psi - brittle	-- 5,000 psi - brittle	5,000 psi - no change 6,000 psi - brittle
Acetone	2,000 psi - no change 4,000 psi[2] - rupture	3,000 psi - no change 6,500 psi[3] - rupture	-- 5,000 psi[4] - rupture	5,000 psi - no change 7,500 psi[5] - rupture
Ethyl Acetate	1,000 psi - no change 2,000 psi - brittle	3,000 psi - no change 5,000 psi[6] - rupture	-- 5,000 psi[7] - rupture	5,000 psi - no change --
Xylene	2,000 psi - no change 3,700 psi[7] - rupture	3,000 psi - no change 5,000 psi[7] - rupture	5,000 psi - no change --	5,000 psi - no change 7,000 psi - brittle

[1] All experiments refer to a duration of 10 minutes unless otherwise stated.
[2] 15 sec. duration.
[3] Conducted for 40 seconds.
[4] For 5 minutes.
[5] 3 Minutes duration.
[6] Within 1 minute.
[7] In 2 minutes.

FIGURE 3

Stress Level Required for One Minute Rupture in Acetone Versus Nylon-6 Composition

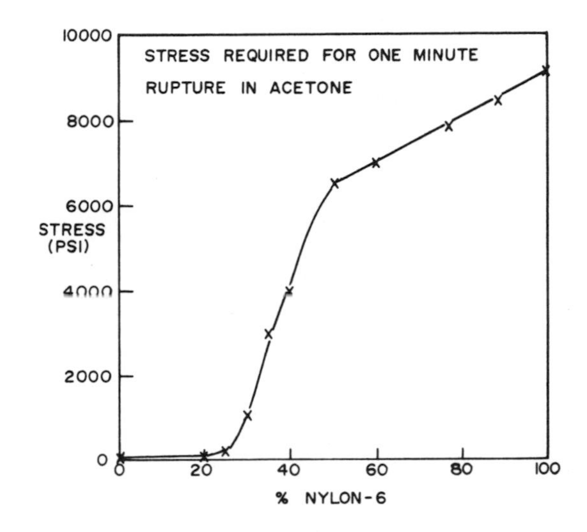

TABLE V

% Nylon 6	Water Adsorption, % ASTM-D-570	Equilibrium Data	
		50% R.H.	100% R.H.
100 (Plaskon 8201)	1.83	2.67	8.72
88.5	1.53	2.19	8.04
77	1.51	1.99	6.86
60	-	1.72	6.00
50	1.00	1.48	4.99
35	.71	1.16	3.57
35	-	1.17	3.65

TABLE VI

Water Diffusion Coefficients for Polysulfone-Nylon 6 Composites

% Nylon	D* cm^2/second 50% R.H.	D* cm^2/second 100% R.H.
100 (Plaskon 8201)	0.722×10^{9}	1.85×10^{-9}
77	1.04×10^{-9}	2.06×10^{-9}
50	1.94×10^{-9}	2.41×10^{-9}
35	2.95×10^{-9}	3.34×10^{-9}

* Calculated from: $q_t/q_\infty = .049/(t/l^2)$ at the time where $q_t/q_\infty = 0.5$.
See: Crank, J., "The Mathematics of Diffusion", Oxford University Press, London, 1956.

The diffusion coefficient of water in Nylon 6 is concentration dependent. In the case of polysulfone it is not. Interestingly, the diffusion coefficient of water is lower for Nylon 6 than for polysulfone. Thus, the PSF/N6 materials will adjust to equilibrium properties faster than the N6 homopolymer.

In order to determine the effect of water adsorption on dimensional stability, two experiments were conducted. The first was to determine the linear dimensional change in 1/8" x 1/2" x 5" injection molded bars by measuring the distance between scribed marks and measuring the distance change with water adsorption with a measuring microscope. The results for Nylon 6 and the 65% polysulfone/35% Nylon 6 composition are given below (one month water immersion).

Sample	% Linear Change
Nylon 6	2.1
65% Polysulfone/35% Nylon 6	.765

The second method consisted in measuring the densities of dry and wet samples in a density gradient column. The analysis of these results yields a volumetric dimensional change. The results are given in Table VII for 50% R.H. and 100% R.H., respectively. These data point out that increasing polysulfone content improves both the linear and volumetric dimensional stability in water.

FIGURE 4

Water Adsorption Data (50% R.H.)
for Various Polysulfone/Nylon 6 Compositions

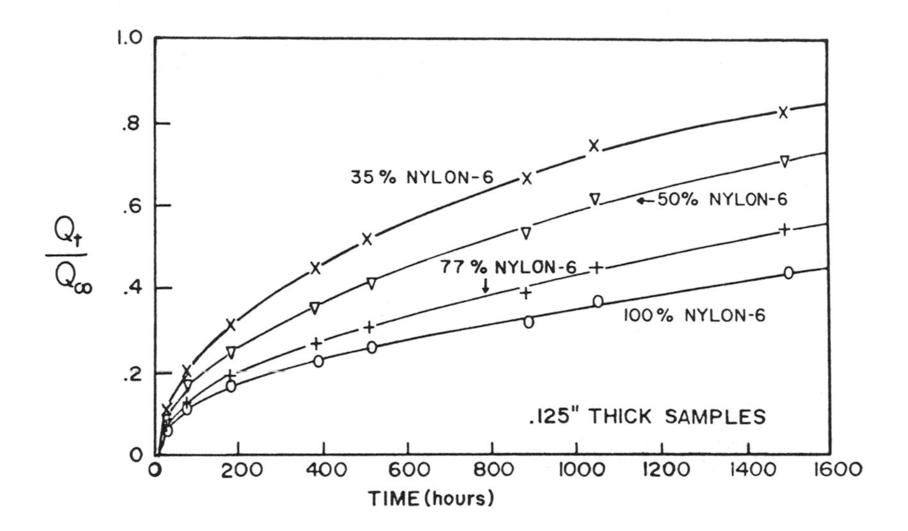

FIGURE 5

Water Adsorption Data (100% R.H.)
for Various Polysulfone/Nylon 6 Compositions

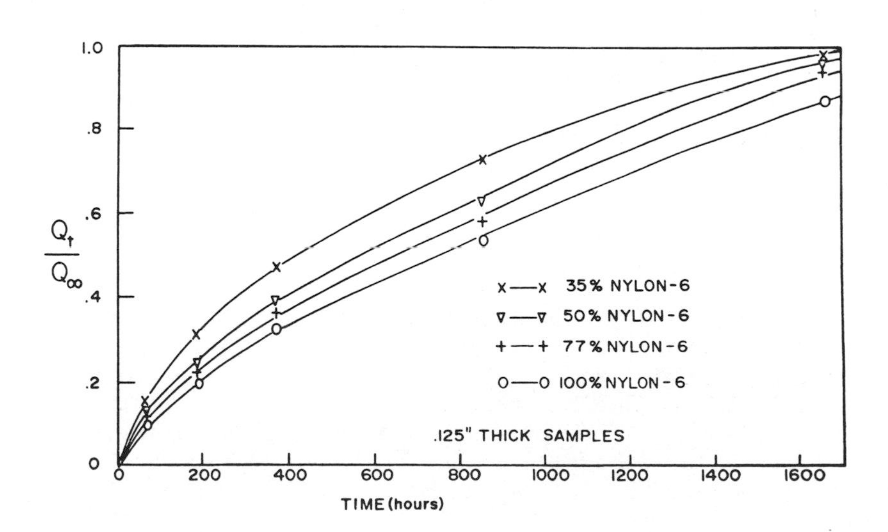

TABLE VII

Dimensional Stability Data of
Polysulfone/Nylon 6 Compositions in Water Environment

% Nylon 6	Volumetric Change cm^3 (dimensional change)/cm^3 (original dimensions)	
	50% R.H.	100% R.H.
100 (Plaskon 8201)	1.45×10^{-2}	8.43×10^{-2}
88.5	1.20×10^{-2}	8.17×10^{-2}
77	1.14×10^{-2}	6.93×10^{-2}
60	$.99 \times 10^{-2}$	6.05×10^{-2}
50	$.835 \times 10^{-2}$	4.94×10^{-2}
35	$.66 \times 10^{-2}$	3.55×10^{-2}

Work with other amorphous rigid-rigid block copolymers is in progress and will be reported in the near future.

CONCLUSIONS

In conclusion, a rigid-rigid block copolymer containing polysulfone and nylon 6 segments was prepared. The material is a two phase system displaying the characteristic glass transition and crystalline melting point of the constituents. It is a tough, high modulus thermoplastic which is easily processible via standard melt fabrication techniques. Due to its semi-crystalline nature, its solvent and environmental stress crack resistance are dramatically improved over those of the polyether homopolymer.

EXPERIMENTAL

Synthesis of Chlorine-Terminated Polysulfone

To a five-liter, four-neck Morton flask fitted with a dropping funnel, thermometer, argon inlet, mechanical stirrer, water separator and condenser were added 343 gms. (1.5 mole) of Bis A, 2,000 ml. chlorobenzene (MCB) and 1,000 ml. of dimethylsulfoxide (DMSO). After the solution became homogeneous and was well purged with argon, 122.8 grams of 97.9% sodium hydroxide (3.0 moles) was added as a 50% freshly made solution in boiled, distilled water. An anhydrous solution of the Bis A disodium salt in a predominately DMSO media was obtained by azeotropic distillation of the MCB and water. After the pot temperature had reached 153°C, a clear, amber solution was obtained. At this point, the 4,4'-dichlorodiphenyl-sulfone (DCDPS) 448.28 grams (1.5625 mole) (a 4 mole % excess) was added as a hot, 50% solution in dry MCB. The reaction temperature was adjusted to 160°C and the mixture was kept under those conditions for 2.3 hours. The cooled solution was removed and was diluted with 1,000 ml. of MCB, filtered through celite, then coagulated in methanol. The precipitated polymer was washed twice with methanol and once with water in a Waring blender. The white polymer was vacuum dried at 100°C for twenty-four hours and @ 125°C for four hours. The yield was 620 grams (94%) and the R.V. (0.2% in $CHCl_3$ at 25°C) was 0.318 dl/gm. The latter value suggested a molecular weight ($\bar{M}n$) of around 12,000 gm./mole. The chlorine analysis (Galbraith Labs) on polymer redried four hours at 130°C under vacuum was 0.53 ± .01% (duplicate determinations) which agrees well with an $\bar{M}n$ of 12,000 (0.59%).

Preparation of a Block Copolymer Containing ∿78 Wt. % of Poly(ε-caprolactam)

To a dry, 500 ml., 3-neck flask fitted with a mechanical stirrer, argon inlet-tube, a 3-way teflon stopcock attached to a condenser, and a graduated receiver, were added 56.6 grams of the chlorine-terminated oligomer described above, 100 ml. of MCB, and 220 ml. of freshly distilled ε-caprolactam (1.95 moles = 220 grams). Stirring and circulation of dry argon were started. The solution was heated (oil-bath) and >95% of the chlorobenzene was removed by distillation.

In a separate, dry 100 ml. flask fitted with a magnetic stirrer, thermometer, an argon-inlet tube, and a condenser, were placed 30 ml. of distilled ε-caprolactam. The temperature of the molten lactam was kept at 130°C, dry argon was circulated over its surface, and 0.4 grams (0.8 grams of the dispersion in the mineral oil) of NaH (1.67×10^{-2} moles) was added to it. Rapid evolution of hydrogen took place and yielded a clear solution of the catalyst.

Twenty-nine and one-half milliliters of this catalyst solution (1.64×10^{-2} moles) were now transferred to the oligomer solution in ε-caprolactam. The oil-bath temperature was kept at 175°C. Rapid increase in viscosity was observed, and at the end of a twenty-three minute period, the mixture was solid. Heating was continued for another nine minutes, at which time the reaction was considered completed. The product was cooled under argon. The cold product recovered from the flask after breaking the latter was highly crystalline and tough. A saw and a mechanical grinder were used in order to break the material into small particles. Yield of this crude product was quantitative. Reduced specific viscosity, $\eta_{sp./c}$, RV, (0.1 grams/100 ml.; m-cresol, 25°C) was 2.53 dl/gram.

Extraction of the crude material with methanol was followed by a soxhlet extraction with chloroform (24 hours). Approximately 9 wt. percent were removed by the methanol; the chloroform insoluble fraction amounted to about 90 wt. percent indicating a very high degree of PSF incorporation into the copolymer. Reduced viscosity was 3.5 dl/g (0.1 g/100 ml.; m-cresol, 25°C). Elemental analysis (Galbraith Laboratories) gave 9.7% N and 1.45% S, which corresponds to 78% wt. of nylon 6 in the copolymer. Block copolymers utilizing the second route were similarly prepared.

The ESA tests were conducted as follows. A 20 mil thick 1/8" wide strip was subjected to a constant stress. A cotton swab saturated with the organic solvent was wrapped around the test specimen. Various time and stress levels were investigated. The effects were followed visually.

REFERENCES

1. M. Matzner, A. Noshay and C. N. Merriam, U. S. 3,539,657 (to Union Carbide Corp.) Nov. 10, 1970.
2. M. Matzner and A. Noshay, U. S. 3,579,607 (to Union Carbide Corp.) May 18, 1971.
3. M. Matzner, A. Noshay and R. Barclay, U. S. 3,701,815 (to Union Carbide Corp.) Oct. 31, 1972.
4. A. Noshay, M. Matzner and C. N. Merriam, Polymer Preprints 12, 247, (1971).
5. A. Noshay, M. Matzner and C. N. Merriam, J. Poly. Sci., A-1 9, 3147, (1971).
6. A. Noshay, M. Matzner, G. Karoly and G. B. Stampa, Polymer Preprints, 13, 292, (1972); J. Appl. Poly. Sci.
7. M. Matzner, A. Noshay, L. M. Robeson, C. N. Merriam, R. Barclay and J. E. McGrath, NASA Conf. on Polymers for Unusual Service Conditions, Nov. 30, 1972, to be published in Appl. Polymer Symposia.
8. A. Noshay, M. Matzner and T. C. Williams, I & E Chem. Prod. Dev. in Press (1973).
9. L. M. Robeson, A. Noshay, M. Matzner and C. N. Merriam, Die Angewante Makromolekulare Chemie, 29/30, 47 (1973).
10. M. Matzner, A. Noshay, and J. E. McGrath, Polymer Preprints, 14, No. 1, 68 (1973).
11. G. Holden, E. T. Bishop and N. R. Legge, J. Poly. Sci., Part C, No. 26, 99 (1969).
12. M. Morton, J. E. McGrath and P. C. Juliano, J. Poly. Sci., Part C, No. 26, 99 (1969).
13. R. P. Zelinski and C. W. Childers, Rubber Chem. & Tech., 41, 161 (1968).
14. W. K. Witsiepe, Polymer Preprints, 13, No. 1, 588 (1972).
15. Rubber World, 167, No. 5, p. 49, Feb. (1973).

16. "Advances in Polyurethane Technology", edit. by J. Burst and H. Gudeon, Interscience (1968).
17. E. Lanza, H. Berghmans and G. Smets, J. Polymer Sci. Physics, 11, 96 (1973).
18. M. Baer, J. Poly. Sci., A, 2, 417, (1964).
19. M. Matzner, L. M. Robeson, L. G. Fetters and J. E. McGrath, Polymer Preprints, 14, No. 2 1063 (1973).
20. J. J. Robertson, U. S. 3,378,602 (to Firestone) 1968.
21. Belg. 706, 153 (to PPG).
22. O. K. Spurr and W. D. Niegisch, J. Appl. Poly. Sci., 6, 585 (1962).
23. R. P. Kambour, Polymer, 5, 107 (1964); ibid, Polymer Eng. and Sci., 8, 281 (1968).
24. I. Norisawa, J. Poly. Sci., A-2, 10, 1789 (1972).
25. M. Matzner, L. M. Robeson, R. J. Greff and J. E. McGrath, Die Angew. Makromol. Chemie, 26, 137 (1972).
26. M. Matzner, D. L. Schober and J. E. McGrath, Polymer Preprints, 13, No. 2, 754, 1972; Europ. Poly. J., 9, 469 (1973).
27. M. Matzner and J. E. McGrath, U. S. 3,657,385 (to Union Carbide Corp.) April 18, 1972.
28. J. E. McGrath and M. Matzner, U. S. 3,655,822 (to Union Carbide Corp.) April 11, 1972.
29. M. Morton, p. 1 in Block Polymers, edited by S. L. Aggarwal, Plenum Press, New York (1970).
30. G. Riess, J. Periard and A. Banderet, p. 173 in "Colloidal and Morphological Behavior of Block and Graft Copolymers", edited by G. E. Molau, Plenum Press, New York (1971).
31. H. Kawai, T. Soen, T. Inoue, T. Ono and T. Uchida, Progress in Polymer Science, Japan, Vol. 4, K. Imahori, editor, Kodansha Ltd. (1972).
32. A. Noshay, M. Matzner, B. P. Barth, R. K. Walton, U. S. 3,536,657 (to Union Carbide Corp.) Oct. 27, 1970.

DYNAMIC MECHANICAL PROPERTIES OF UREA-URETHANE BLOCK POLYMERS

J. L. Work

Armstrong Cork Company, Research and Development Center

Lancaster, Pennsylvania 17604

I. INTRODUCTION

Segmented polyurethanes containing different molecular weight glycols and urea-urethane block polymers have been the subject of numerous investigations in the past decade (1-16). The results of these investigations indicate clearly that these polymers form highly aggregated structures. If the aggregation occurs preferentially between certain segments in the macromolecule, concentration gradients arise in the polymer which should eventually lead to the formation of domains of the type observed in ABA-type block polymers (17). Only recently has a domain structure been directly observed in these systems (10,19,20). The domains are much smaller than those observed in the styrene-butadiene block polymers and, therefore, are more poorly defined. This probably occurs because the block lengths are shorter and more dispersed in size in the urethane system than they are in the styrene-butadiene system.

It is difficult to assess from these studies the influence of the block lengths on polymer properties because the size and distribution of the blocks in the macromolecules are ill defined in most cases. It is the purpose of this report to describe methods for preparing urea-urethane block polymers with more carefully controlled molecular structures. Attempts will be made to deduce the state of aggregation of the polymers in their solid state from the results of dynamic mechanical property measurements. The control of the molecular structures in these preparations is not the ultimate which would be to have each kind of segment be exactly the same size and to have the same number of segments in each molecule. L. L. Harrell (21) has reported on a method for preparing segmented urethanes with carefully controlled structures. He likewise was unable to achieve the ultimate in control. To date, the only systems that have come close to achieving this kind of control are the ionically prepared block polymers.

II. EXPERIMENTAL

All of the polymers used in this study are based on poly (oxypropylene)glycol, 4,4'-methylenebis(cyclohexyl isocyanate) ($H_{12}MDI$), and 4,4'-methylenebis(cyclohexyl amine) ($H_{12}MDA$). The glycols were obtained from Wyandotte Corp. The diisocyanate was obtained from E. I. duPont de Nemours Co. under the trade name of Hylene W. The diamine was obtained from the Aldrich Chemical Co. All of these materials were used without further purification.

The general formula for these block polymers is

$$\left\{ \left[\bigcirc \overset{O}{\underset{H}{\overset{\|}{NC}}}O(CH_2\overset{CH_3}{C}HO)_q\overset{O}{\underset{H}{\overset{\|}{CN}}} \bigcirc CH_2 \right]_m \left[\bigcirc \overset{O}{\underset{H\ H}{\overset{\|}{NCN}}} \bigcirc CH_2 \right]_n \right\}_p \qquad (1)$$

where q is the number of polyoxypropylene repeat units in the glycol, m is the number of repeat units in the urethane block, n is the number of repeat units in the urea block, and p is the number of times this sequence is repeated in the polymer chain. The general procedure for the preparation of the block polymers is to prepare and characterize prepolymers containing either the urea or urethane units. These blocks are terminated with either isocyanate or primary amine groups and then reacted to produce the desired block polymer.

III. PREPARATION OF BLOCK POLYMERS

A 30 w/w% benzene solution of glycol and $H_{12}MDI$ and 0.4% dibutyltinbis(octyl thioglycolate) (T-31) are reacted at 80°C for 6 hours under anhydrous conditions. The value of m is varied by using different excesses of diisocyanate. This isocyanate-terminated prepolymer is precipitated with hexane and extracted with petroleum ether in a liquid-liquid extractor for 48 hours. The solvent is removed in a vacuum dessicator at 20°C and 0.11 mm Hg. The equivalent weight of this prepolymer is determined using the procedure of Sorenson and Campbell (18). A portion of the prepolymer is reacted with methanol, dissolved in THF, and passed through a GPC column to ascertain that there is less than 1% monomer present in the prepolymer. Infrared spectroscopy was used to ascertain the absence of urea bonds which could arise from the reaction of the diisocyanate with water. The urea bond has a characteristic band at 1640 cm^{-1} and the urethane bond has one at 1710 cm^{-1}.

This prepolymer is chain extended in toluene solution with excess water at 80°C using 0.4% T-31 catalyst to produce a block polymer having n = 1. If equal molar amounts of prepolymer and $H_{12}MDA$ in benzene are reacted, a block polymer having n = 2 is produced. The degree of polymerization (p) of this polymer can be lowered by reacting excess quantities of $H_{12}MDA$ with the prepolymer. To increase n further, it is necessary to prepare an amine terminated urea prepolymer. This is accomplished by slowly adding a 10% DMF solution of $H_{12}MDI$ to a 25% DMF solution of $H_{12}MDA$. The $H_{12}MDA$ is always maintained in excess. The product precipitated during the reaction and was filtered and dried under vacuum. The equivalent weight of the prepolymer is determined by titration of a cresol solution with ethanolic HCl. To a cresol solution of one mole of amine terminated urea is added one mole of isocyanate terminated urethane in benzene to produce a block polymer having n > 2. The urethane block in all of these cases is not monodispersed.

Block polymers having m = 1 are prepared by reacting equal molar quantities of an isocyanate terminated urea prepolymer with a glycol at 75°C in DMF using 0.4% T-31 catalyst. The isocyanate terminated urea is prepared by reacting 0.1 moles of water with 0.5 moles of $H_{12}MDI$ at 75-80°C using 0.4% T-31 catalyst. The reaction mixture is filtered hot to remove the high molecular weight oligomers. The filtered solution is extracted with hexane in a liquid-liquid extractor maintained at 64°C for 24 hours. A gummy solid results, which when dried under vacuum produces a crystalline solid. A portion of this prepolymer is reacted with methanol and passed through an automatic liquid chromatography column to ascertain that the concentration of $H_{12}MDI$ is less than 1%. The equivalent weight of the prepolymer was determined by the procedure of Sorensen and Campbell (18) with the exception that cresol was used as the solvent and 0.1N ethanolic HCl was used as the titrant.

Poly(oxypropylene)glycols having molecular weights of 774(P710), 440(P410), and 192(tripropylene glycol) were used in the preparation of these block polymers. Only the tripropylene glycol is monodisperse in molecular weight. The dynamic mechanical properties of these polymers were determined using a Rheovibron. Test specimens were prepared by compression molding at 200°C and annealed at 80°C for 24 hours under 0.1 mm Hg.

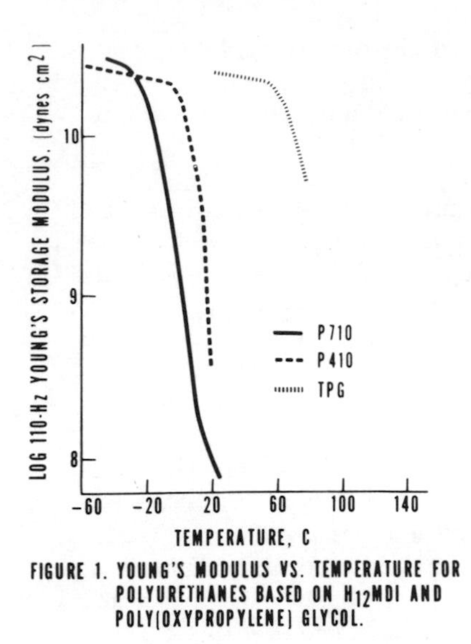

FIGURE 1. YOUNG'S MODULUS VS. TEMPERATURE FOR POLYURETHANES BASED ON H_{12}MDI AND POLY(OXYPROPYLENE) GLYCOL.

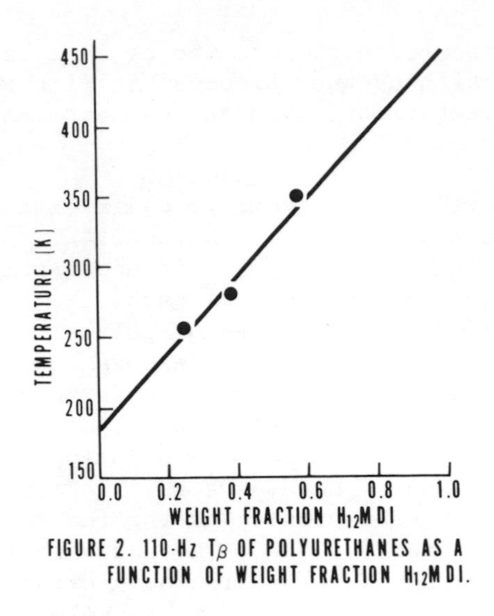

FIGURE 2. 110-Hz T_β OF POLYURETHANES AS A FUNCTION OF WEIGHT FRACTION H_{12}MDI.

IV. RESULTS AND DISCUSSION

A. Variation of Properties of Polyurethanes with Glycol Molecular Weight

The 110 Hz dynamic Young's storage modulus as shown in Figure 1 as a function of temperature for polyurethanes based on P710, P410, and TPG indicates that these materials exist as single phase amorphous materials. The change in modulus with increasing H_{12}MDI concentration is very similar to that observed for random copolymers (22). Further, the transition temperature associated with the glass transition temperature of these polymers increases linearly with increasing concentration of H_{12}MDI (Figure 2) as would be expected for a random copolymer with varying composition based on free volume considerations (23).

Thus it can be concluded that there is no segregation of different parts of the polyurethane molecules into crystalline or noncrystalline domains. Therefore, any segregation observed in the block polymers must be due to the presence of the blocks.

TABLE I

Characteristics of P710-Based Urea-Urethane Block Polymers

Urea conc. (w/w %)	Urethane conc. (w/w %)	m	n	110 Hz T_β (C)
7.0	93.0	3.03	1.00	-16
12.2	87.8	1.64	1.00	-14
13.1	86.9	3.03	2.00	-18
21.8	78.2	1.64	2.00	-16
22.9	77.1	1.00	1.31	-20
34.7	65.3	2.90	6.76	-10

B. Variation of Properties with Block Lengths for P710-Based Urea-Urethane Block Polymers

The composition of these block polymers is given in Table I. Within experimental error, the T_β does not change with block length or concentration with the possible exception of the material containing 34.7% urea and is the same as T_β for the polyurethane based on P710. The 110 Hz Young's storage modulus at temperatures above T_β increases with increasing urea concentration at constant m (Figure 3). With m ≈ 3.0 and n = 1, the block polymer has the

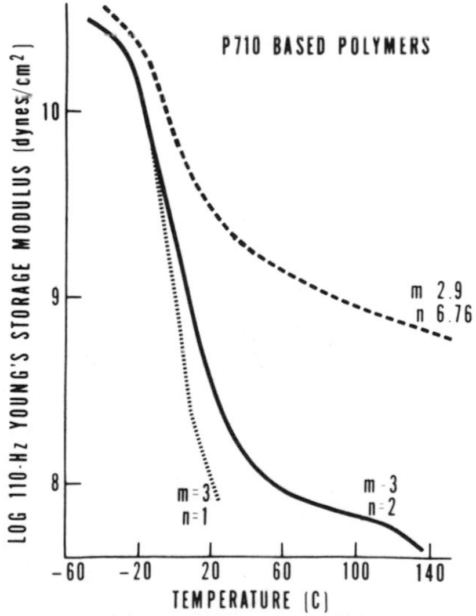

FIGURE 3. YOUNG'S MODULUS VS. TEMPERATURE FOR UREA-URETHANE BLOCK POLYMERS BASED ON $H_{12}MDI$ AND P 710.

TABLE II

Characteristics of P410 Based Urea-Urethane Block Polymers

Urea conc. (w/w %)	Urethane conc. (w/w %)	m	n	110 Hz T_β (K)
3.3	96.7	11.0	1.10	279
6.1	93.9	11.0	2.10	281
12.6	87.4	2.39	1.00	299
16.4	83.6	1.74	1.00	308
17.3	82.7	11.0	6.86	286
22.3	77.7	2.39	2.00	318
31.3	68.7	1.00	1.39	317

modulus-temperature profile of a homogeneous material. When n is increased to 2 or greater, a long plateau occurs which is typical for block polymers having a domain structure (3,14). Since T_β does not increase with increasing urea concentration, it can be concluded that one of the domains is polyurethane while the second domain may be polyurea or a mixture of polyurethane and polyurea. As m is decreased, the long plateau region becomes less distinct indicating that the structure or composition of the second domain becomes more heterogeneous with decreasing m (Figure 4). The

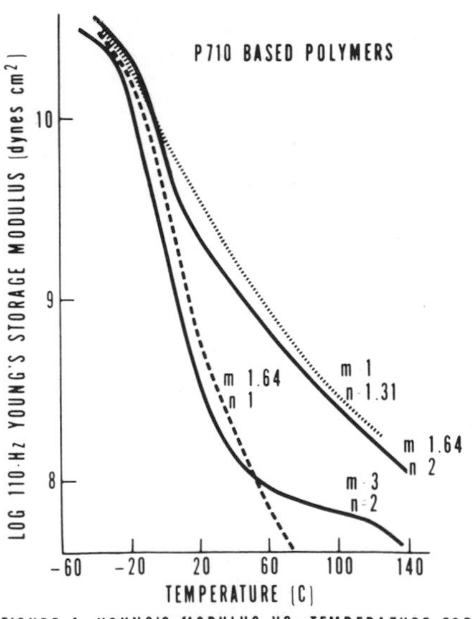

FIGURE 4. YOUNG'S MODULUS VS. TEMPERATURE FOR UREA-URETHANE BLOCK POLYMERS BASED ON $H_{12}MDI$ AND P 710.

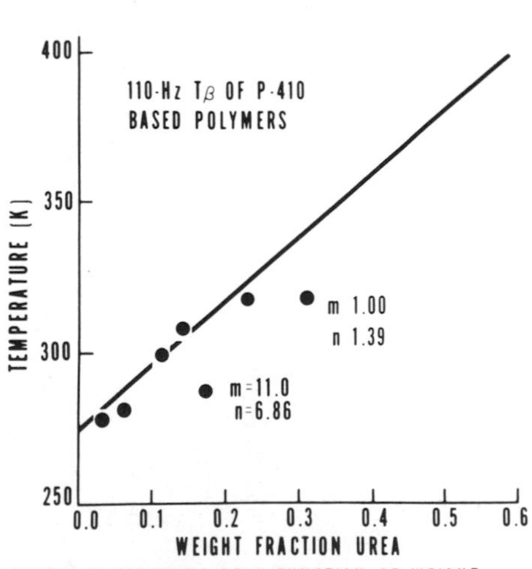

FIGURE 5. 110-Hz T_β AS A FUNCTION OF WEIGHT FRACTION UREA FOR UREA-URETHANE BLOCK POLYMERS BASED ON $H_{12}MDI$ AND P 410.

effects observed are not due to differences in urea concentration but rather to differences in block lengths as can be ascertained by comparing the polymers having m = 3, n = 2, and m = 1.64, n = 1 in Figure 4. Thus, a well developed domain structure can be expected only when $m \geq 3$ and $n \geq 2$. The molecular weight of these polymers were not determined because of aggregation in their solutions. However, the molecular weight of m = 3.03 and n = 2 was decreased by reacting one mole of isocyanate terminated urethane prepolymer with 1.3 moles of $H_{12}MDI$. No effect on properties was observed with this decrease in molecular weight.

C. Variation of Properties with Block Lengths for P410-Based Urea-Urethane Block Polymers

The composition of these block polymers is given in Table II. The T_β of these polymers increases linearly with increasing urea concentration with two exceptions (m = 11.0, n = 6.86 and m = 1.00, n = 1.39)(Figure 5). The modulus temperature profiles (Figures 6 and 7) show that only the polymer with m = 11.0 and n = 6.86 exhibits a well-defined plateau region and therefore is the only polymer with a well developed domain structure. Thus, longer blocks are required for domain formation to occur in polymers using P410 than in polymers using P710. The broad transition region for the polymer having m = 1.00 and n = 1.39 shows that

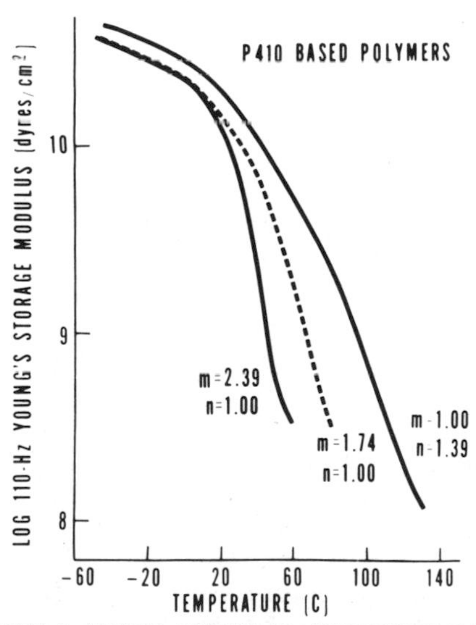

FIGURE 6. YOUNG'S MODULUS VS. TEMPERATURE FOR UREA-URETHANE BLOCK POLYMERS BASED ON $H_{12}MDI$ AND P 410.

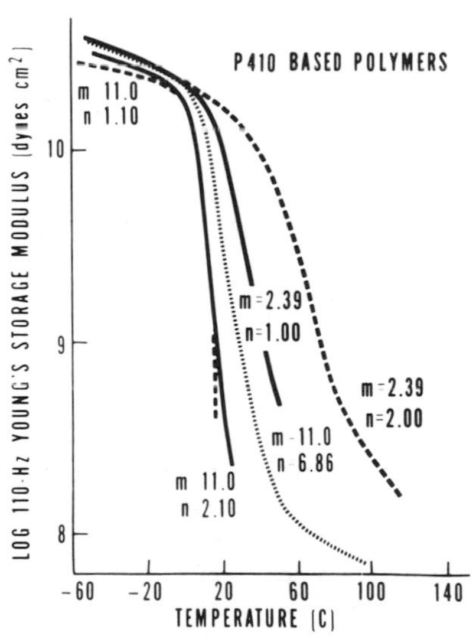

FIGURE 7. YOUNG'S MODULUS VS. TEMPERATURE FOR UREA-URETHANE BLOCK POLYMERS BASED ON $H_{12}MDI$ AND P 410.

TABLE III

Characteristics of TPG-Based Urea-Urethane Block Polymers

Urea conc. (w/w %)	Urethane conc. (w/w %)	m	n	110 Hz T_β (K)
0	1	"∞"	0	349
0.091	0.909	11.0	2.10	361
0.181	0.819	2.36	1.00	373
0.245	0.755	11.0	6.86	361
0.307	0.693	2.36	2.00	405

this polymer is neither well segregated into domains nor homogeneous. This is unexpected since polymers having longer block lengths are homogeneous. At this time, there is no explanation for this behavior. If the linear relationship in Figure 5 is extrapolated to a weight fraction of urea of 1, the extrapolated T_β of the polyurea is 485K.

D. Variation of Properties with Block Lengths for TPG-Based Urea-Urethane Block Polymers

The composition of these block polymers is given in Table III. The T_β of these polymers increases linearly with increasing urea concentration with the exception of m = 11.0 and n = 6.86 (Figure 8). The modulus-temperature profiles (Figure 9) show that only this latter polymer exhibits any type of plateau region. These observations show that this is the only polymer of the series that is segregated into domains. Thus, longer blocks are required for domain formation to occur in polymers using TPG than in polymers using P710. More data at intermediate block lengths would be required to determine if there is a difference in the block length requirements for domain formation between TPG and P410-based polymers.

If the linear relationship in Figure 8 is extrapolated to a weight fraction of urea of 1, the extrapolated T_β of the polyurea is 498K. This value of T_β is in good agreement with the value obtained for the P410 series considering the long extrapolations and the experimental errors involved in the T_β measurements. This result lends support to the argument that the block polymers having T_β's which fall on these linear relationships are homogeneous.

V. CONCLUSIONS

The block lengths required to produce domain formation in urea-urethane block polymers based on H_{12}MDI, H_{12}MDA, and poly (oxypropylene) glycol increases with decreasing glycol molecular weight. The T_β of the homogeneous block polymers is a linear function of weight fraction urea for a given glycol molecular weight. The T_β for the polyurethanes is a linear function of the weight fraction H_{12}MDI in the polyurethane. There is no detectable crystallinity or domain formation in the polyurethanes themselves. Therefore, the domain formation observed in the block polymers is due to the presence of the blocks. Further work must be done to more clearly define the nature of the domains formed in these polymers.

VI. ACKNOWLEDGMENTS

Thanks are due to Mrs. Margaret Brooks for obtaining the dynamic mechanical spectra.

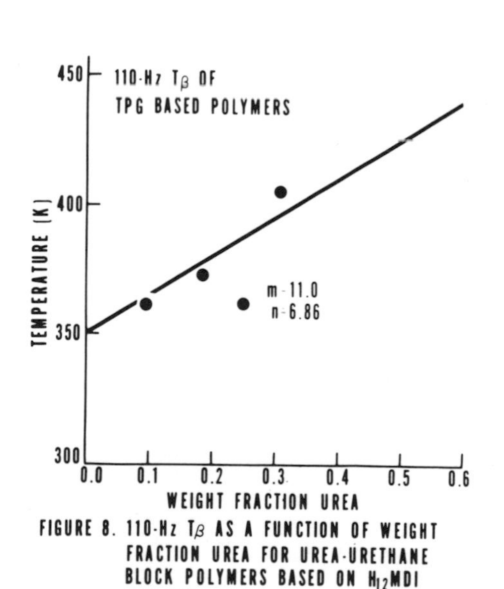

FIGURE 8. 110-Hz T_β AS A FUNCTION OF WEIGHT FRACTION UREA FOR UREA-URETHANE BLOCK POLYMERS BASED ON H_{12}MDI AND TPG.

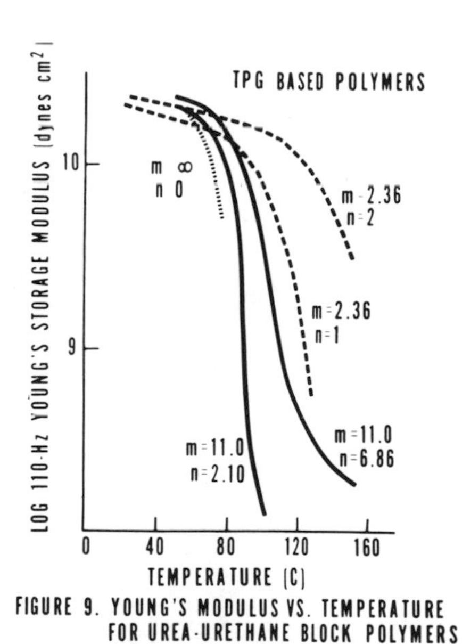

FIGURE 9. YOUNG'S MODULUS VS. TEMPERATURE FOR UREA-URETHANE BLOCK POLYMERS BASED ON H_{12}MDI AND TPG.

VII. REFERENCES

1. R. W. Seymour, G. M. Estes, and S. L. Cooper, Macromolecules, 3, 579 (1970).

2. G. M. Estes, R. W. Seymour, and S. L. Cooper, Macromolecules, 4, 452 (1971).

3. D. S. Huh and S. L. Cooper, Polym. Eng. Sci., 11, 369 (1971).

4. R. Bonart, L. Morbitzer, and H. Rinke, Kolloid-Z. u. Z. Polymere, 240, 807 (1970).

5. J. L. Illinger, N. S. Schneider, and F. E. Karasz, Polym. Eng. Sci., 12, 25 (1972).

6. T. Kajiyama and W. J. MacKnight, Trans. Soc. of Rheo., 13, 527 (1969).

7. A. J. Havlik and T. L. Smith, J. Polymer Sci., Part A, 2, 539 (1964).

8. A. M. North, J. C. Reid, and J. B. Shortall, Eur. Polymer J., 5, 565 (1969).

9. R. W. Seymour, G. M. Estes, D. S. Huh, and S. L. Cooper, J. Polymer Sci., Part A-2, 10, 1521 (1972).

10. S. L. Samuels and G. L. Wilkes, J. Polymer Sci., Part B, 9, 761 (1971).

11. G. M. Estes, S. L. Cooper, and A. V. Tobolsky, J. Macromol. Sci., C4, 313 (1970).

12. R. Bonart, L. Morbitzer, and G. Hentze, J. Macromol. Sci., 133, 337 (1969).

13. T. Kajiyama and W. J. MacKnight, Macromol., 2, 245 (1969).

14. J. Stoelting, F. E. Karasz, and W. J. MacKnight, Polym. Eng. Sci., 10, 133 (1970).

15. S. B. Clough and N. S. Schneider, J. Macromol. Sci., B2, 553 (1968).

16. S. B. Clough, N. S. Schneider, and A. O. King, J. Macromol. Sci., B2, 641 (1968).

17. T. Soen, T. Inoue, M. Katsumori, and H. Kawai, J. Polym. Sci. Part A2, 10, 1757 (1972).

18. W. R. Sorensen and T. W. Campbell, Preparative Methods of Polymer Chemistry, pp. 138, Interscience Publishers, Inc., New York, 1961.

19. J. A. Koutsky, N. V. Hien, and S. L. Cooper, J. Polym. Sci., Part B, 8, 353 (1970).

20. C. E. Wilkes and C. S. Yusek, J. Macromol. Sci.-Phys., B7, 157 (1973).

21. L. L. Harrell, Macromolecules, 2, 607 (1969).

22. L. E. Nielsen, Mechanical Properties of Polymers, Van Nostrand and Reinhold, New York, 1962.

23. F. Bueche, Physical Properties of Polymers, Chap. 5, Interscience, New York (1962).

ORIENTATION STUDIES OF POLYURETHANE BLOCK POLYMERS

Robert W. Seymour[a], Gerald M. Estes[b], and S. L. Cooper

DEPARTMENT OF CHEMICAL ENGINEERING

UNIVERSITY OF WISCONSIN — Madison, Wisconsin 53706

ABSTRACT

Recent results from orientation studies on polyurethane block polymers are presented. Birefringence measurements are of limited value in the study of multiphase systems because of the importance of non-orientational sources of birefringence. However, both conventional and differential infrared dichroism are very useful methods. The application of these techniques in the study of orientation mechanisms, orientation hysteresis, and time dependent orientation processes are discussed and illustrated with pertinent results.

INTRODUCTION

Thermoplastic elastomers, of which the polyurethanes are a well known class, are block polymers composed of alternating "hard" and "soft" segments. Physical crosslinking and reinforcement are provided by the hard segments, which in polyurethanes are generally formed from the extension of an aromatic diisocyanate with a low molecular weight diol. Elastomeric properties are imparted by the soft segments, commonly aliphatic polyethers or polyesters 1000-5000 in molecular weight.

---

[a]Present address: Tennessee Eastman Co., Kingsport, Tenn.
[b]Present address: Elastomer Chemicals Dept., E. I. duPont de Nemours and Company, Wilmington, Del.

The existence of microphase separation, caused by clustering of at least some of the hard and soft segments into separate domains, has been well established (1-3). The mechanical properties of these materials reflect their heterophase nature and are characterized by an enhanced rubbery modulus and high, nearly reversible extensibility (4,5). Block polymer systems exhibit the major glass or melting transitions of the two components as well as various secondary loss mechanisms (6).

The typical polyurethane is extensively hydrogen bonded (7), the donor being the NH group of the urethane linkage. The hydrogen bond acceptor may be in either the hard urethane segment (the carbonyl of the urethane group) or the soft segment (an ester carbonyl or ether oxygen). The relative amounts of the two types of hydrogen bonds are determined by the degree of microphase separation, with increased phase separation favoring interurethane hydrogen bonds.

Hydrogen bonding considerations have been shown to influence morphological features such as chain ordering in partially crystalline polyurethanes (8,9). The importance of hydrogen bonds in determining mechanical properties is less clear, however. The hetero-bonded solid state concept (10) predicts that such secondary bonding will play a major role in dynamic mechanical transition behavior. Studies of hydrogen bonded polymers have generally concluded, though, that the presence of such bonds does not necessarily enhance mechanical properties (11,12,13). Other authors are of the opposite opinion, however (14).

Much attention has been directed towards an elucidation of the domain morphology in polyurethane block polymers, as their physical properties may be directly ascribed to the presence of this two phase microstructure. The behavior of the elastomer under deformation is a function of both the domain size and degree of order within the domains (8,9,15). The original morphology may be altered through stretching to high elongations (8,9,15), annealing (16), and annealing under strain (heat setting)(8,17). The importance of such studies is apparent, as an understanding of how the morphological features may be altered is fundamental to the development of structure-property relationships.

Bonart studied polymers with "paracrystalline" and/or crystalline hard segments by x-ray techniques. On elongation, the "paracrystalline" hard segments oriented initially transverse to the stretch direction. At higher strains, the original ordered regions were disrupted and segmental orientation into the stretch direction occurred. Truly crystalline hard segments oriented transversely at all elongations, and were described as acting more as an inert filler. The soft segments were found to orient into the stretch direction at all elongations.

Kimura also studied materials with well ordered hard segments. Low angle light scattering indicated the presence of a supramolecular texture whose orientation could be observed in a manner similar to crystalline homopolymers. X-ray and infrared results suggested a model similar to that of Bonart.

This paper will discuss the application of two other methods of orientation measurement, birefringence and infrared dichroism. Data from both conventional and differential dichroism studies will be included.

BIREFRINGENCE METHODS

Initial orientation studies on polyurethane block polymers employed the measurement of strain induced birefringence (19,20). A part of this work was an attempt to understand the morphological changes accompanying stress softening. Stress softening is a phenomenon common to block polymers and filled rubbers, whereby the modulus at a given strain level decreases with cyclic strain application. Substantial hysteresis is seen in stress-birefringence curves; while the strain-birefringence behavior is less dependent on strain history. Examples of this are shown in Figures 1 and 2 for a polyether-urethane containing 38 wt. % aromatic urethane segments.

Interpretation of these results is extremely difficult, however, because of the two phase structure present in nearly all block copolymer systems. It had generally been assumed that the birefringence of a heterophase system could be effectively expressed as a linear combination of contributions from each phase. The existence of an environmental birefringence (including the so-called form birefringence) was recognized, but its importance was discounted. Later analysis demonstrated its significance, however (21-24).

A preliminary comparison of the total birefringence response mentioned above (19,20) and the molecular orientation in the separate phases of polyurethane elastomers (22) has indicated that orientational birefringence alone cannot account for the observed birefringence response. Due to the small domain sizes and the high levels of orientation which are achieved during extension, the environmental birefringence contribution can be expected to make a significant contribution to the overall birefringence response, and appears to be considerably more im-

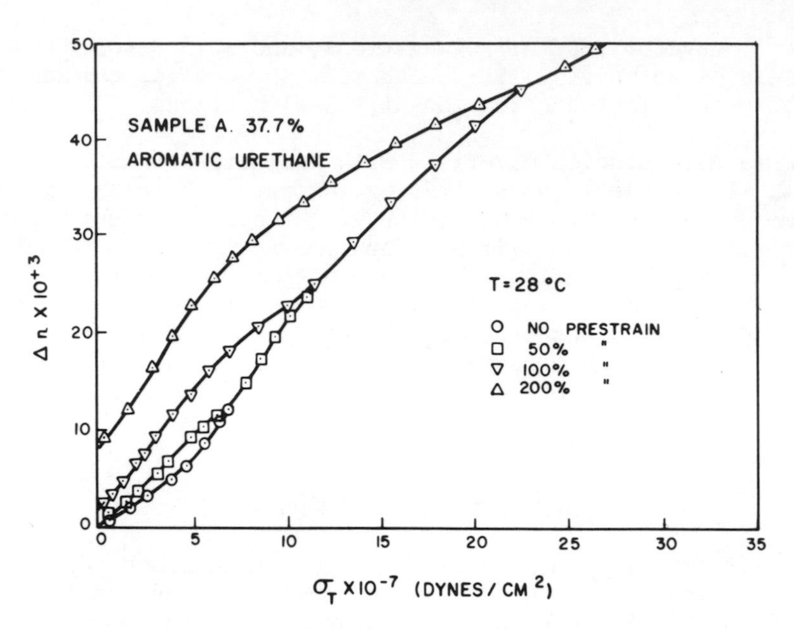

Figure 1 Stress-Birefringence Hysteresis in a Segmented Polyurethane Elastomer, ET-38-1.

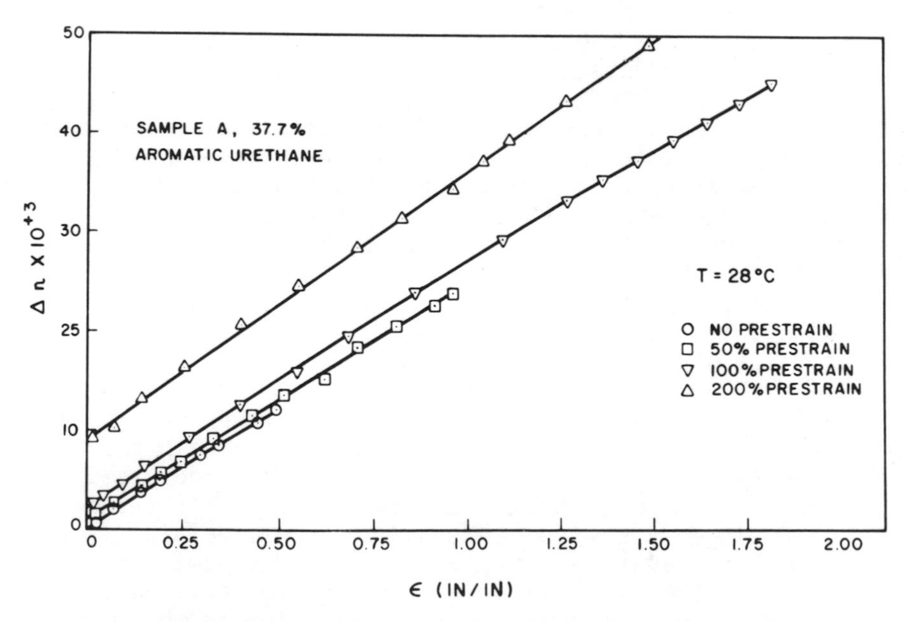

Figure 2 Strain-Birefringence Hysteresis in a Segmented Polyurethane Elastomer, ET-38-1.

portant in the polyurethane elastomers and perhaps other block polymer systems than has been previously thought.

INFRARED DICHROISM

Because of the difficulties in interpretation of birefringence results, the technique of infrared dichroism has been used in studying orientation. It is of particular value in polyurethane block polymer systems where the dichroic behavior of bands characteristic of the individual blocks can be studied. This is possible because there are infrared active groups peculiar to each segment which absorb in regions free from other bands. A quantitative description of segmental orientation also requires a reasonably accurate knowledge of the transition moment directions for the vibrations of interest. In polyurethanes, the NH group is characteristic solely of the hard segment and may be conveniently used to study hard block orientation. The CH_2 group can generally be used for soft segment orientation, although methylene groups are also generally present at small concentrations in the hard segment. The transition moment direction for both of these vibrations can be taken as 90° (21).

The two techniques available for the measurement of infrared dichroism are compared in Figure 3. The most common and straightforward technique used in measuring dichroic effects is to record two spectra of the oriented polymer, one with the beam polarized parallel to the stretch direction and one perpendicular. The ratio of the absorbances obtained for a given band from these two experiments defines the dichroic ratio from which the uniaxial orientation function, f, can be calculated (21).

A schematic of the experimental set up used to measure dichroism in this way is shown in Figure 3a. A conventional double beam spectrophotometer is employed, with a single polarizer in the common beam. In practice the polarizer is set at 45° to the slit direction to minimize machine polarization effects. Two spectra are run with the sample at ±45°.

An alternative double beam double polarizer arrangement has been described by Gotoh (25) and is shown schematically in Figure 3b. Two polarizers (at ±45° to the slits) are used, so that the chopped common beam going to the monochromator contains both $A_{\parallel}$ and $A_{\perp}$ information. The quantity recorded is the dichroic difference, $A_{\parallel} - A_{\perp}$. Thus the output for an unoriented sample will be a straight line, while orientation of any vibration will result in an upwards or downwards peak depending on the relative magnitude of $A_{\parallel}$ and $A_{\perp}$. The dichroic difference may also be related to the orientation

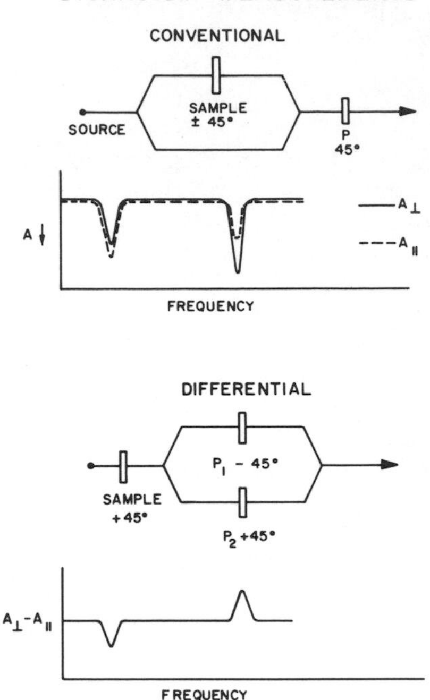

Figure 3 Techniques for Measurement of Dichroism. a) Conventional b) Differential.

function, f (26). The differential method for measuring dichroism is much more accurate than the conventional method. A quantitative comparison of the two techniques has recently been published by Read (27).

The orientation data of this study is correlated in terms of the uniaxial orientation function, f. The maximum value of the function f is unity, corresponding to perfect axial orientation into the stretch direction. $f = 0$ in the absence of orientation and $f = -1/2$ for perfect orientation transverse to the stretch direction. Implicit in the use of this orientation function is the assumption that the uniaxial model is sufficient to represent the state of orientation.

A. Hard Segment Orientation Mechanisms

The orientation mechanism of the hard segment is controlled primarily by its length. Three distinct mechanisms have been identified on the basis of studies using the differential method. Chemical parameters for the materials used are shown in Figure 4 and Table I. Results are presented only for polyether-urethanes, with the understanding that similar conclusions apply to the polyester-urethanes.

The lowest diisocyanate content material studied was ET-24-1, in which the hard blocks average only slightly greater than one MDI unit in length (Table I). This material is mechanically quite soft and exhibited the orientation function-elongation curves for the NH, CH_2 and free and bonded carbonyl groups shown in Figure 5. Orientation in the hard block may be slightly higher than that in the soft, though it is extremely low in both. This results from the fact that most of the molecular alignment induced by the deformation process is able to relax during the time scale of the test. The lower level of mechanical properties and orientability of ET-24-1 may be ascribed to the inability of this material to form an interlocking domain structure. A more detailed description of this situation may be found elsewhere (28).

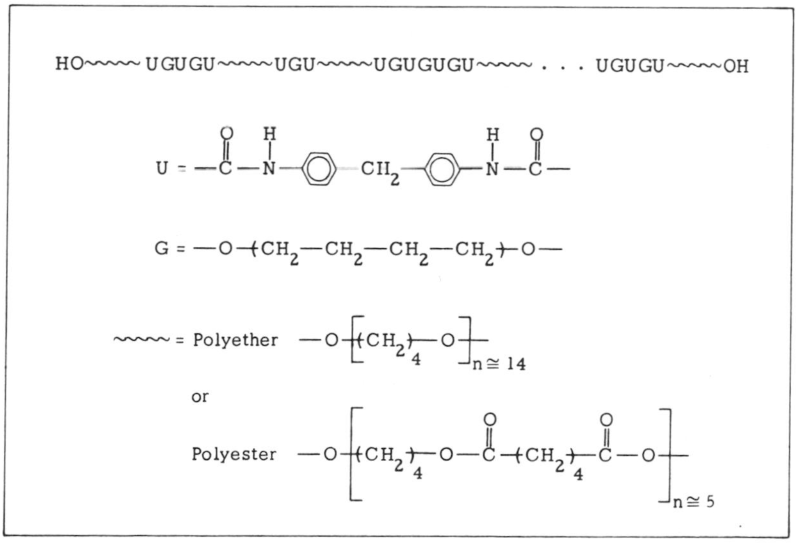

HO~~~~ UGUGU~~~~UGU~~~~UGUGUGU~~~~ . . . UGUGU~~~~OH

U = —C(=O)—N(H)—⟨C₆H₄⟩—CH_2—⟨C₆H₄⟩—N(H)—C(=O)—

G = $—O—(CH_2—CH_2—CH_2—CH_2)—O—$

~~~~ = Polyether $—O—[(CH_2)_4—O]_{n \cong 14}$

or

Polyester $—O—[(CH_2)_4—O—C(=O)—(CH_2)_4—C(=O)—O]_{n \cong 5}$

Figure 4 Chemical Structure of Polyurethane Block Polymers.
~~~~

TABLE I CHEMICAL PARAMETERS OF EXPERIMENTAL MATERIALS

Designation	Soft Seg. Mol. Wt.	Wt. % MDI	Wt. % Hard Seg.	x + 1	$\bar{M}_w x10^{-3}$	$\bar{M}_n x10^{-3}$	$\bar{M}_w/\bar{M}_n$
ES-24-1	901	24.0	26.3	1.39	89.4	20.3	3.41
ES-28-1	901	27.4	30.6	1.49	138	24.0	5.73
ES-31-1	901	31.3	37.0	1.87	98	22.1	4.44
ES-35-1	901	34.5	41.0	2.19	151	34.1	4.43
ES-38-1	901	38.6	47.6	3.11	236	32.4	7.28
ES-38-2	1962	37.9	54.0	7.	147	26.2	5.62
ES-38-5	4545	38.0	54.0	15.	NA	NA	NA
ET-24-1	980	23.7	25.2	1.3	206	79.6	2.59
ET-28-1	980	27.3	30.9	1.6	228	35.3	6.47
ET-31-1	980	31.2	36.6	2.0	274	107.0	2.58
ET-35-1	980	34.4	41.5	2.47	169	23.2	7.27
ET-38-1	980	37.7	46.3	2.86	252	87.3	2.88
ET-38-2	2000	38.0	50.0	6.	NA	NA	NA

x + 1: average number of diisocyanate residues per hard segment

NA: not available

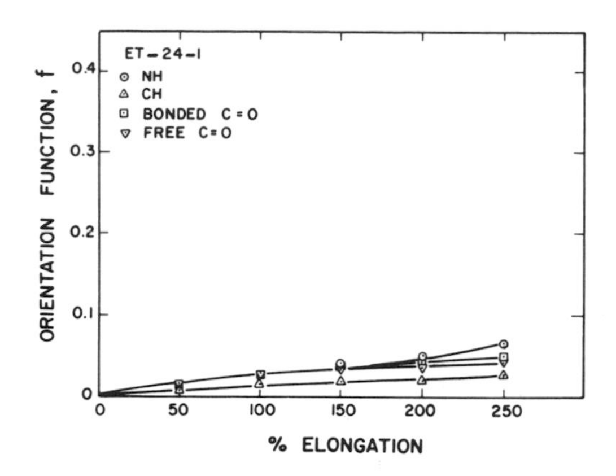

Figure 5 Orientation Function-Elongation Curves for ET-24-1.

A significant change in orientation behavior is seen on raising the MDI content to 28%. Orientation function-elongation curves for ET-28-1 are shown in Figure 6. Overall there is a much higher degree of orientation and a much greater distinction between the behavior of the NH and CH_2 groups. This suggests an important change in morphology occurs on going from 24% to 28% MDI.

While some orientation of the flexible soft segment relaxes, resulting in a comparatively low value of the C-H orientation function, much more orientation is held in the hard blocks (bonded C=O and NH curves).

An interesting difference is also seen in comparing the free and bonded carbonyl groups. The bonded C=O are associated with other hard segments and thus are representative of the behavior of the segments within the domains. Free carbonyl, on the other hand, could exist in either of two situations. It may be located at some sort of domain interface, where its degree of orientation would be expected to be at most comparable to that of the hard segments within the domain. Alternatively, a free C=O group can arise when the hard segment to which it is attached is dispersed in a soft segment region. Under these conditions, it should orient more like the CH_2 group. A combina-

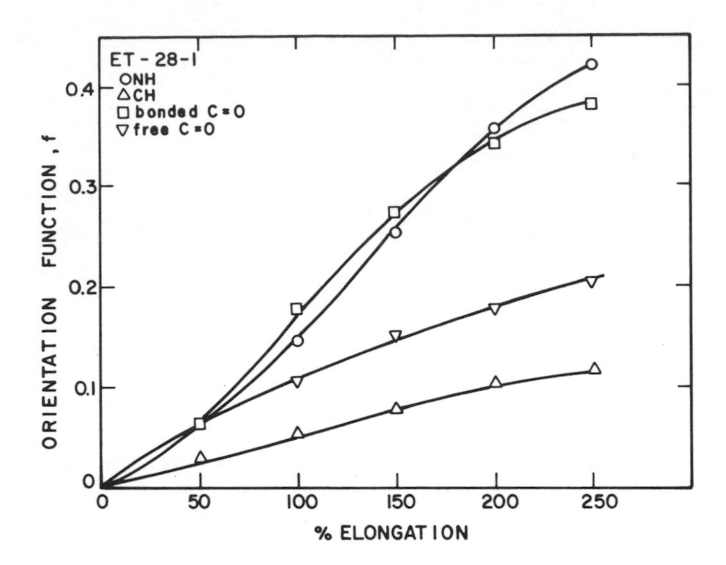

Figure 6 Orientation Function-Elongation Curves for ET-28-1.

tion of these effects could produce the observed orientation level intermediate between those for the bonded C=O and CH_2 groups.

The hard segment orientation behavior shown in Figure 6 is characteristic of materials with compositions between ET-28-1 and ET-38-1, there being only minor differences between them. This behavior is summarized in Figure 7, where the NH orientation function curves for ET-28-1, ET-31-1 and ET-38-1 all lie within the same band. The DSC curves for these materials show that only short range hard segment order is present (17).

Introducing microcrystallinity into the hard block distinctly changes the orientation behavior. DSC (17) and x-ray diffraction (6) show that the materials whose NH orientation functions fall into the lower band on Figure 7 contain micro-crystalline hard segments. ET-38-2 is normally crystalline and crystallinity may be introduced into ET-38-1 by prolonged annealing at 150°C (16,17).

No f values are shown at very low strains (below 100% in ET-38-2 and 50% in annealed ET-38-1) as the differential dichroic spectra show that two types of orientation are present at these strain levels. Some representative raw data for ET-38-2 are shown in Figure 8.

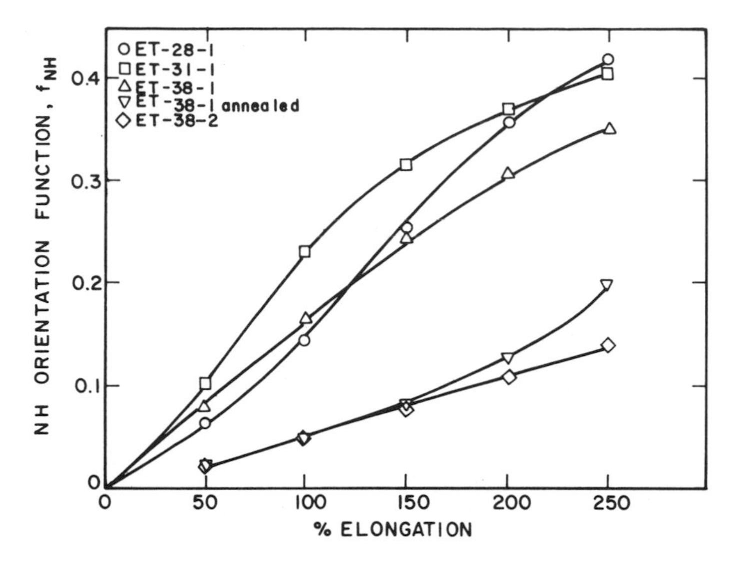

Figure 7 Summary of NH Orientation Functions for Polyether-Urethanes.

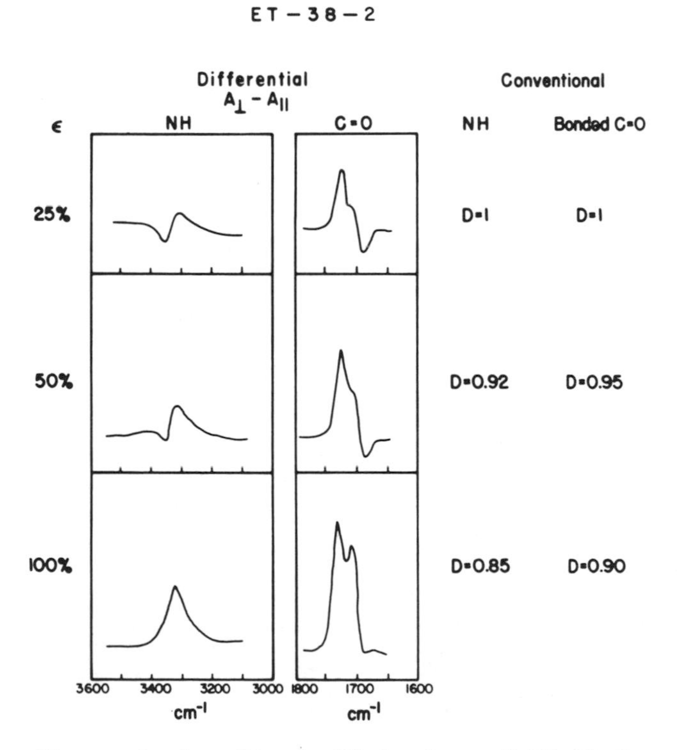

Figure 8 Low Strain Dichroism of ET-38-2.

The dichroic difference $A_{\perp} - A_{\parallel}$ at low strain levels is shown for the NH and C=O regions in this figure. Upward deviation from the baseline corresponds to positive values of the orientation function, i.e. orientation of the segments into the stretch direction. It can be seen that the NH and bonded C=O peaks contain both positively and negatively orienting components. The splitting decreases with increasing strain, until at 100% elongation only positive orientation is observed. The CH_2 region orients positively at all elongations, as did all groups in materials with non-crystalline hard segments. Also shown in Figure 8 are the values of the dichroic ratio D which were measured with the conventional single polarizer technique. The value of $D = 1$ at 25% elongation indicates that the conventional measurement would lead to the conclusion that there is no orientation. In actuality there are both positively and negatively orienting structures whose optical anisotropies tend to cancel each other in the conventional dichroism measurement.

The peak splitting observed in the NH and bonded C=O regions is observed only in samples with partially crystalline hard segments. By analogy with the data already discussed, the non-crystallized hard segments may be expected to orient positively, implying that it is the crystallized hard segments which are orienting negatively. Such a result would be predicted if the crystalline regions themselves orient as a body into the stretch direction. Since the hard segments are aligned approximately normal to the long axis of the crystalline regions (8,9,15,21), this would result in a negative chain backbone orientation.

Orientation functions for the NH and bonded carbonyl regions have been calculated only when purely positive orientation is observed. f values that would be calculated for the split peaks are of doubtful value, as it is not clear that the peak absorbance is a true measure of the maximum. It may be noted that the NH and bonded carbonyl orientation functions are significantly lower than those observed in the samples discussed previously.

As can be seen in Figure 8 by 100% elongation no negative peak can be detected. This does not mean that the crystalline domains have been totally disrupted. If the crystalline regions were totally disrupted and oriented, one would expect the hard segment f values to be equivalent to those of the non-crystalline polyurethanes. Since this does not occur, it appears that the disruption of the crystalline domains is a continuous process, occurring over the range of strain levels.

The IR dichroism results indicate that the crystalline hard segments do not orient in the same manner as the non-crystalline hard segments. Rather, the orientation of the crystalline regions occurs in two steps. Initially the long dimension of the crystalline domain orients into the stretch direction, resulting in a transverse segment orientation. The crystalline regions gradually break up with increasing strain and the segments then begin to orient into the stretch direction. The initial transverse orientation, which may be partially retained even at higher elongations, serves to lower the overall value of the orientation function.

B. Soft Segment Orientation Mechanism

Figure 9 summarizes the soft segment orientation for the materials ET-28-1 through ET-38-2. Soft segment orientation is always into the stretch direction and is largely unaffected by the composition. The exception to this may be noted in Figure 5, where extensive relaxation of all segments occurred. Thus, soft segment orientation behavior depends only on whether an interlocking phase morphology has developed.

Residual orientation functions for the two segments of ET-38-1 are plotted in Figure 10. Residual orientation functions are

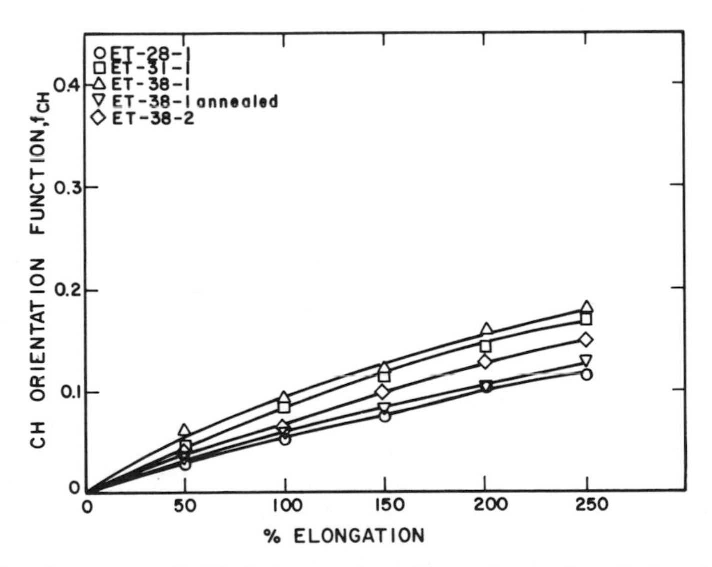

Figure 9 Summary of CH Orientation Functions for Polyether-Urethanes.

obtained by stretching the sample to a given elongation, allowing it to relax for five minutes at zero load, then measuring the orientation. The abscissa in Figure 10 is the strain level to which the samples were extended prior to relaxation; hence, it is labelled prestrain. It is significant that when relaxed, the hard segments (NH orientation) exhibit substantially larger orientation functions than the soft segments. Both ET-38-1 and ES-38-1 show this behavior. Since the elastomers creep considerably (65-70 percent following straining to 300 percent), it might be thought that the residual orientation is due to the residual strain. This could be true in part, but cannot explain the entire residual hard segment orientation after high extension, for at 70 percent initial strain the orientation function for NH alignment is only about 0.1. It appears that the differences in the residual orientation functions are due to different retractive stresses acting on the separate domains.

To further investigate the connection between strain history and residual orientation, orientation functions were measured on films prestrained to various elongations. The results of these experiments are shown in Figure 11 along with comparable data for a control sample. It is evident that the soft segment orientation is essentially independent of strain history. However, as expected

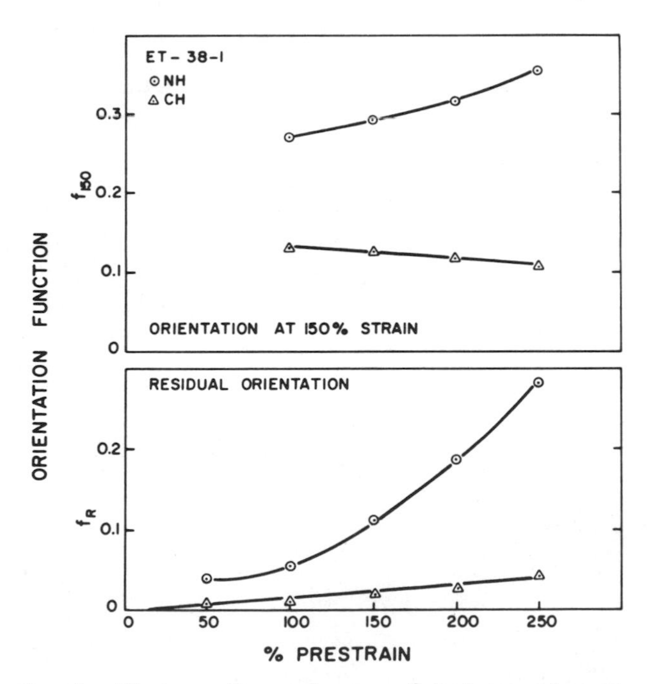

Figure 10 Strain History Dependence of Orientation Functions for ET-38-1.

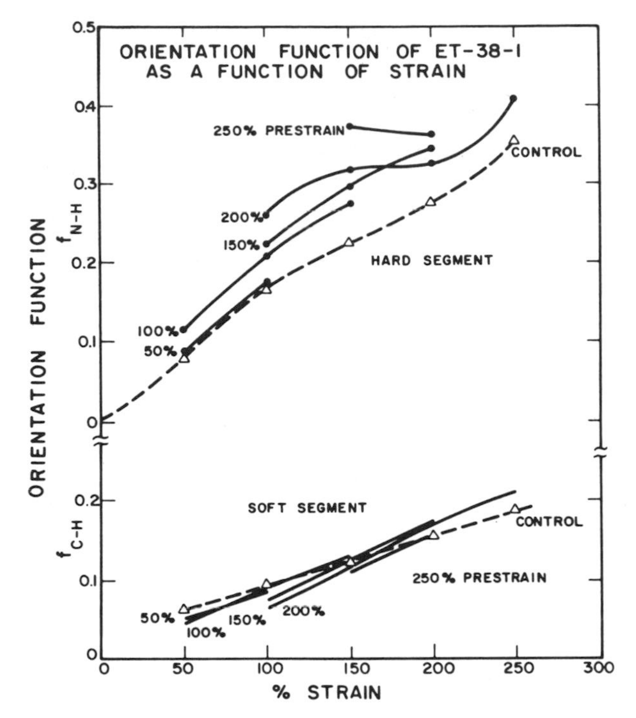

Figure 11 Orientation Hysteresis in ET-38-1.

from the residual orientation functions, the hard segment orientation shows a marked prestrain effect.

The phenomenon of stress-softening may then be described in terms of two separate mechanisms which are important during the second straining. One is operative at strain levels between that corresponding to the permanent set of the sample, ε_1, and that strain level, ε_2, which during initial extension would correspond to the residual hard segment orientation. A second is operative between ε_2 and the prestrain level ε_3. Between ε_1 and ε_2 lower stresses are required because only chains in the soft domains need to be oriented and aligned. No deformation of the hard domains is required. At ε_2 and above deformation of both soft and hard domains is required, but the hard domains have been subjected to this degree of deformation during the first straining. The previous deformation has softened the hard domains and thus lowered their capacity to act as the reinforcement for the soft domains. Hence, stress-softening is observed. Such stress-softening is observed up to the prestrain level, ε_3, at which point both soft and hard domains become "tense". Straining beyond ε_3 causes the hard domains to undergo additional plastic deformation.

CONCLUSIONS

The value of infrared dichroism in the study of orientation of block polymer systems has been clearly established. It does not suffer the ambiguity in interpretation of birefringence, nor does it require crystalline ordering, as does x-ray diffraction.

Application of the technique of differential infrared dichroism to a well characterized system of polyurethane block polymers has resulted in a more detailed understanding of their response to deformation. It has been pointed out that there is a drastic change in orientation behavior in going from ET-24-1 to ET-28-1. Very little orientation is held in either segment in ET-24-1. In ET-28-1, ET-31-1 and ET-38-1, relatively high degrees of orientation are maintained, particularly in the hard segments. It is suggested that this is connected with the development of interlocking hard segment domains in ET-28-1 which do not exist in ET-24-1. In ET-24-1, the independent hard domains allow extensive relaxation of the orientation in both segments. In ET-28-1, relaxation of orientation is severely restricted by the interconnecting morphology in the hard segments. The flexible soft segments can relax to a greater degree, but they too are restrained, by covalent bonding to the hard segments.

The first change in orientation behavior with composition thus occurs at the point where the morphology changes from independent hard domains to an interlocked structure. Once the interlocked non-crystalline hard regions are formed, only minor changes in orientation behavior due to the changing volume fraction of hard segment occur. The basic mechanism for these materials is an increasing orientation of both segments into the stretch direction (positive orientation) at all elongations. A similar conclusion based upon low angle x-ray studies on comparable non-crystalline segmented polyurethanes was made by Wilkes (29).

A fundamental change in orientation mechanism is observed when crystallinity is introduced into the hard blocks. While the non-crystalline segments still present orient as before, the crystalline regions show a transverse (negative) orientation of the segments at low elongations. This may be visualized as an orientation into the stretch direction of the long axis of crystalline lamellae causing chain orientation transverse to the stretch direction. Low strain level hysteresis experiments show that this is a nearly reversible process. As the elongation is increased, however, the original crystallites are broken up and the segments orient positively. Orientation of the crystalline hard segments thus occurs in two steps. First the positive orientation of the crystalline region itself, followed by disruption of the region and increasing positive orientation of the segments.

Finally, it has been found that the stress-softening which is very pronounced in the results of cyclic stress-strain experiments can be attributed to a nonrecoverable deformation and distortion of the hard domains during the initial application of strain. This residual hard segment deformation is evidenced by the high levels of segmental orientation which remain in the reinforcing phase after the application and removal of stress.

EXPERIMENTAL

The materials investigated were described in Figure 4 and Table I. Differential infrared dichroism was measured on a Perkin-Elmer Model 180 Infrared Spectrophotometer. The sample was placed at an intermediate focus in the source beam before splitting at an angle of 45° to the slits. Wire grid polarizers were placed in the sample and reference beams at a setting of 45°. This results in a crossed polarizer configuration because of an unequal number of beam reflections between the source and polarizer location in the two beams. The 45° setting was chosen to minimize machine polarization effects. In this configuration, the recorded output is the dichroic difference $A_{\perp} - A_{\parallel}$.

Data were recorded in linear absorbance using programmed slits and automatic gain control. The slit program was set for 1 mm at 3550 cm^{-1}. Most data were recorded using 5x ordinate expansion and a time constant setting of three. Sample stretching was accomplished in a jig designed to elongate the specimen from both ends simultaneously. Strain was measured using bench marks.

Orientation of the soft segment has been followed using the asymmetric CH_2 stretch vibration. This group resides primarily in the soft segment (21). Hard block orientation has been characterized by the NH group, located entirely in the hard segment. In addition, orientation of the carbonyl group of the urethane linkage has been studied in polyether-urethanes. This band is clearly split into free and hydrogen bonded components (7).

Orientation function-strain curves were generated by sequentially straining a single sample. Sufficient time was allowed between application of the strain and measurement of the spectrum for the orientation to reach a constant value. The time required varied with diisocyanate content and block length, the maximum being about 30 minutes for a 50% strain increment in the lowest diisocyanate content materials.

Baselines for unoriented samples were drawn according to the recommendations of Estes (22). A single baseline through the NH and CH regions was generally used for the differential spectra.

In calculating orientation functions, transition moment directions of 90° were assumed for the NH and CH vibrations and 79° for the carbonyl.

ACKNOWLEDGEMENT

The authors wish to acknowledge the assistance of Mr. Jeffrey T. Koberstein in carrying out some of the experimental work. The materials used in this study were kindly supplied by Dr. E. A. Collins of the B. F. Goodrich Chemical Co. We are also grateful to the National Science Foundation for support of this research through Grant GH-31747.

BIBLIOGRAPHY

1. G. M. Estes, S. L. Cooper and A. V. Tobolsky, J. Macromol. Sci., Rev. Macromol. Chem., C4(1), 167 (1970).

2. J. A. Koutsky, N. V. Hien and S. L. Cooper, Polymer Letters, 8, 353 (1970).

3. S. B. Clough, N. S. Schneider and A. O. King, J. Macromol. Sci., B2, 641 (1968).

4. S. L. Cooper and A. V. Tobolsky, Textile Res. J., 36, 800 (1966).

5. S. L. Cooper, D. S. Huh and W. J. Morris, Ind. and Engr. Chem. Prod. Res. and Dev., 7(4), 248 (1968).

6. D. S. Huh and S. L. Cooper, Polymer Engr. and Sci., 11, 369 (1971).

7. R. W. Seymour, G. M. Estes and S. L. Cooper, Macromolecules, 3, 579 (1970).

8. R. Bonart, J. Macromol. Sci.-Phys., B2(1), 115 (1968).

9. R. Bonart, L. Morbitzer and G. Hentze, J. Macromol. Sci.-Phys., B3(2), 337 (1969).

10. R. D. Andrews and T. J. Hammack, Polymer Letters, 3, 655 (1965).

11. E. P. Otocka and F. R. Eirich, J. Polymer Sci., A-2, 6, 895 (1968).

12. W. E. Fitzgerald and L. E. Nielsen, Proc. Royal Soc., A282, 137 (1964).

13. A. V. Tobolsky and M. C. Shen, J. Phys. Chem., 67, 1886 (1963).

14. K. Ogura and H. Sobue, Polymer J., 3, 153 (1972).

15. I. Kimura, H. Ishihara, H. Ono, N. Yoshihara and H. Kawai, XXIII IUPAC Preprints, 525 (1971).

16. R. W. Seymour and S. L. Cooper, Polymer Letters, 9, 689 (1971).

17. R. W. Seymour and S. L. Cooper, Macromolecules, 6, 48 (1973).

18. D. S. Huh, Ph.D. Thesis, University of Wisconsin (1971).

19. G. M. Estes, R. W. Seymour and S. L. Cooper, Polymer Engr. and Sci., 9, 343 (1969).

20. G. M. Estes, D. S. Huh and S. L. Cooper, in Block Polymers, (S. L. Aggarwal, ed.), Plenum Press, New York, (1970), p. 225.

21. G. M. Estes, R. W. Seymour and S. L. Cooper, Macromolecules, 4, 452 (1971).

22. G. M. Estes, Ph.D. Thesis, University of Wisconsin (1970).

23. M. J. Folkes and A. Keller, Polymer, 12, 222 (1971).

24. R. S. Stein, Polymer Letters, 9, 747 (1971).

25. R. Gotoh, T. Takenaka and N. Hayama, Kolloid Z. u.z. für Polymere, 205, 18 (1965).

26. R. W. Seymour, A. E. Allegrezza and S. L. Cooper, Macromolecules, in press.

27. B. E. Read, D. A. Hughes, D. C. Barnes and F. W. M. Drury, Polymer, 13, 485 (1972).

28. R. W. Seymour, Ph.D. Thesis, University of Wisconsin (1973).

29. C. E. Wilkes and C. S. Yusek, J. Macromol. Sci., B7(1), 157 (1973).

BUTADIENE-STYRENE AB TYPE BLOCK COPOLYMERS

G. Kraus and H. E. Railsback

Phillips Petroleum Company
Research and Development Department
Bartlesville, Oklahoma 74004

I. INTRODUCTION

Few recent developments in polymer science have created more interest than the discovery of the so-called "thermoplastic elastomers", block copolymers of the general structure $A_nB_mA_n$, wherein A is a monomer whose polymer is glassy at ordinary temperatures, while monomer B forms a rubbery center block. These materials owe their unique properties to immiscibility of the two kinds of block sequences. The result is a two-phase domain structure in which the glassy domains assume the role of both crosslinks and reinforcing filler particles for the rubbery matrix composed of the center blocks. Thus, at ambient temperature the polymers resemble vulcanized, filler reinforced rubbers in reversible extensibility and strength. Above the glass transition of the A-blocks flow becomes possible - - the polymers are true thermoplastics. The first, and most extensively investigated, polymers of this kind are those in which A is styrene and B is butadiene or isoprene.

While much has been published in the past five years on the morphology and the mechanical behavior of styrene-butadiene-styrene (SBS) and styrene-isoprene-styrene (SIS) block polymers, relatively little has been said about the corresponding two-sequence block polymers (SB or SI). Clearly, such polymers cannot form networks without further (chemical) crosslinking. Still, they are highly interesting and useful materials, which have already attained a significant degree of technological and commercial importance as specialty rubbers.

There are many similarities between SB and SBS block polymers. The thermodynamic incompatibility of long styrene and butadiene block sequences is fundamentally independent of the details of the arrangement of the blocks along the chain and both types of polymers exhibit domain structures of similar morphology. Also, the Tg of the domains are similar so that low temperature flexibility of a crosslinked SB polymer is quite comparable to that of an SBS thermoplastic elastomer. The polystyrene domains also stiffen and reinforce a crosslinked SB rubber, although high tensile strength is not attained with pure gum vulcanizates of SB block polymers. This, however, is not as great a handicap as it may seem, for fillers are used in compounding of both SBS and SB block polymers. These fillers have an equalizing effect; they further reinforce SB block polymers, but tend to lower the tensile strength of SBS polymers.

SBS block polymers form highly viscous melts and are prepared to relatively low molecular weights to ensure processing. This is also true of SB polymers, but to a lesser degree - - at equal molecular weight and styrene content the viscosity of SB polymer is significantly lower. SB block polymers are thus exceptionally well suited for injection molding. (Of course, crosslinking agents and high mold temperatures are required and scrap cannot be reprocessed.)

A major field of application of SB block copolymers is in blends with other rubbers. Because the second rubber requires vulcanization, there is no incentive for using the structurally more complicated and costly SBS block polymers in such applications. Advantages derived from SB block polymers in blends are, among others, higher hardness, improved low temperature flexibility, better mold flow and smoother extrusions.

The present report is a brief discussion of the preparation, structure, properties and selected applications of SB block polymers of the type currently in commercial use.

II. SYNTHESIS OF SB BLOCK POLYMERS

The preparation of styrene-butadiene block polymers has been described in the literature and reviewed in some detail by Zelinski and Childers(1). Accordingly, we only outline the principles underlying the two general methods available. The first of these employs sequential monomer addition in anionic polymerization. For example, polymerization of butadiene (in hydrocarbon solution) is initiated with n-butyllithium. After all butadiene is converted to polymer, styrene is added to the living polybutadienyl lithium anions and polymerized to quantitative conversion. The result is a BS block polymer with compositionally pure sequences. The second method is even simpler, although the product has a somewhat more complicated structure. If both monomers are charged simultaneously, polymer formed at low conversion will be rich in polybutadiene, simply as the result of the relative magnitudes of the monomer reactivity ratios. As polymerization proceeds the chain becomes gradually enriched in styrene. When all butadiene has polymerized, the remaining styrene adds on as a pure block. The polymer thus consists of a random copolymer block richer in butadiene than the monomer charge ratio and a pure polystyrene block. The random copolymer block is "tapered" in composition, becoming richer in styrene toward the junction point between the two blocks.

It has been found that rubbery block polymers prepared by the copolymerization method differ little in their fundamental mechanical behavior from those made by the sequential addition technique. The reason for this becomes clear upon examination of the composition distribution. (This distribution is readily calculated from the composition vs conversion curve.) An actual example is shown in Figure 1. The polymer contains a total of 30% styrene, 22% as a pure styrene block (shown in the figure as weight per cent between 95 and 100% styrene content). The remaining 8% styrene is concentrated almost entirely in portions of the molecule containing less than 20% styrene and there is very little material of intermediate styrene content situated between these two main blocks. The presence of 8% styrene in the "rubbery" block raises its Tg from -98 C to about -85 C and does not render it miscible with polystyrene. The polymer thus resembles an SB polymer of 22% styrene content with pure sequences, except that Tg of the "rubbery" block is raised 13°. This has very little effect on mechanical properties at ordinary service temperatures ($T \gg Tg$). For this reason we shall refer to block polymers made by <u>either</u> method of synthesis as "SB", keeping in mind the distinction with regard to compositional purity of the "B"-block. Because of its simplicity and the near-equivalence of the resulting products, the copolymerization technique is the preferred method for commercial manufacture of SB block polymers.

III. MOLECULAR STRUCTURE

The anionic polymerization techniques described above produce linear polymers of narrow molecular weight distribution. The microstructure of the combined butadiene units is roughly 50% trans, 40% cis and 10% vinyl. The polystyrene domains are amorphous. A fairly complete description of molecular structure thus consists of the weight average molecular weight (M_w), total styrene content and block styrene content. Butadiene microstructure and number average molecular weight (M_n) serve to further define small differences.

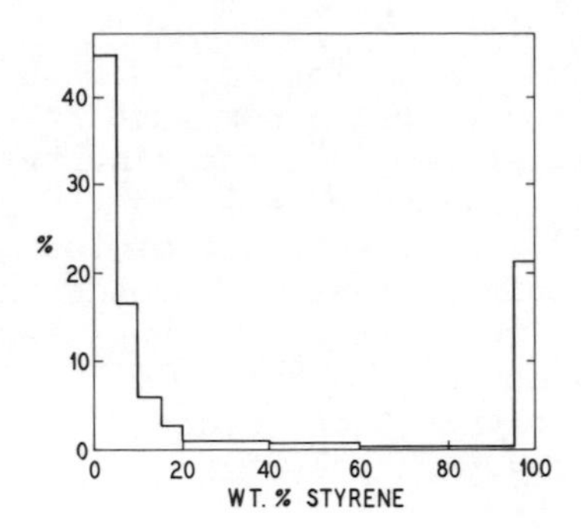

FIGURE 1
COMPOSITION DISTRIBUTION OF 70/30
BUTADIENE-STYRENE BLOCK COPOLYMER
PREPARED BY COPOLYMERIZATION METHOD

In polymers made by the copolymerization technique the distribution of styrene in the rubbery block is of interest. Composition distributions such as Figure 1 can be obtained from copolymerization data. Since initiation is fast compared to propagation and there is no chain transfer or termination, the compositional variation is almost entirely _intra_-molecular and plots like Figure 1 represent the variation of styrene concentration along the chain.

Lack of _inter_-molecular compositional heterogeneity can be established by density gradient sedimentation equilibrium measurements. A typical result is illustrated in Figure 2 for a polymer of 25% total styrene content, 18% block styrene. The Schlieren pattern shows a single, symmetrical boundary indicative of one composition only. Inter-molecular compositional homogeneity is an advantage of block polymers made by the copolymerization method over those made by sequential monomer addition. In the latter case it is difficult to avoid terminating some of the polymer chains by adventitious impurities in the second monomer increment. The final block polymer is thus frequently "contaminated" by a small amount of homopolymer of the monomer introduced as the first increment.

The NMR spectra of SB block polymers made by the copolymerization technique are too complex to permit detailed sequence length distributions to be established. However, the two ortho-protons of blocked styrene units of sequence length $n \geqq 4$ give rise to an absorption at 6.1-6.8 ppm, while all the aromatic protons in isolated styrene units ($n = 1$) and the remaining three protons of blocked units absorb at 6.8-7.4 ppm.* If only styrene sequences of $n = 1$ and $n \geqq 4$ are present, it is easily shown that the fraction of styrene units in blocks is

$$S_B = (5/2)A_B \qquad (1)$$

where A_B is the fraction of the total aromatic area of the NMR spectrum between 6.1 and 6.8 ppm. If all styrene blocks in a polymer are long, equation (1) produces excellent agreement with estimates of block styrene from the chemical degradation method(2) or, in sequential addition block polymers, from the monomer charge ratio.

The NMR spectrum of the polymer of Figure 1 is shown in Figure 3. In this example $A_B = 0.315$, $S_B = 0.79$. Since the total styrene content is 30%, block styrene = 0.79 (30) = 23.6%. This compares with 21.8% found by the chemical method and ca. 22% from analysis of the copolymerization data (Figure 1). The generally good agreement shows that styrene in the rubbery block must be present almost exclusively in very short sequences (n = 1 to 3).

*Spectra in CCl_4, 8% concentration with TCE internal standard.

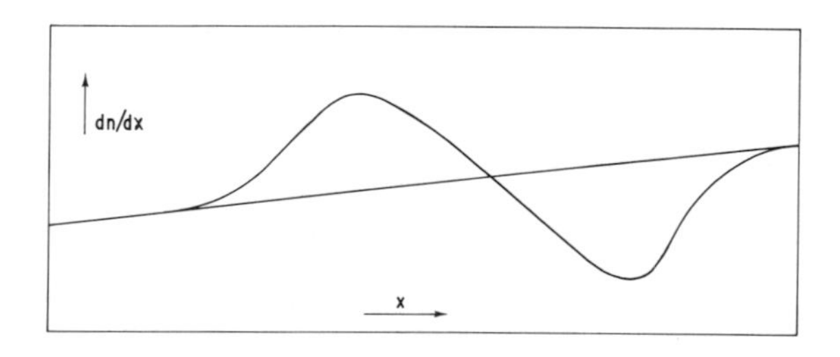

FIGURE 2
SEDIMENTATION EQUILIBRIUM OF SB-BLOCK POLYMER (25% STYRENE, 18% BLOCK STYRENE) IN CYCLOHEXANE (75 VOL.%) - CARBON TETRACHLORIDE (25 VOL.%)

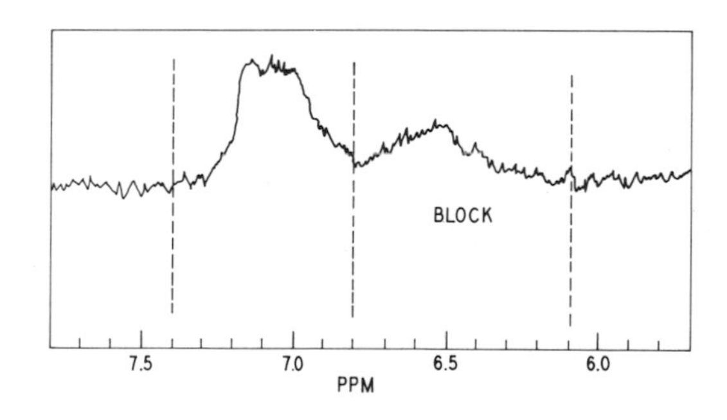

FIGURE 3
AROMATIC REGION OF NMR SPECTRUM FOR SB BLOCK POLYMER (30% STYRENE, 22% BLOCK STYRENE)

IV. MORPHOLOGY

The domain morphology of block copolymers has been studied by many authors. In particular, Matsuo et al(3,4) investigated SB and SBS block polymers, while Kawai and associates(5-7) studied the corresponding isoprene-styrene polymers. These investigations show that, in general, the domain structure is governed by the overall composition, not the sequence arrangement. Spherical domains are formed by the phase present in minor amount, provided the second phase is present in large excess. Near 50:50 composition the morphology is lamellar. When one phase is present in moderate excess, the second phase forms rod-like domains. One distinction between SBS and SB block polymers is that the domain structure of the latter is coarser, provided that the comparison is made at equal molecular weight and composition.

V. MELT RHEOLOGY

Below the glass transition of the polystyrene domains an SB block polymer (excess B), unlike an SBS polymer of comparable molecular weight and composition, can still undergo significant viscous flow, for the association of styrene blocks into glassy domains does not lead to formation of an infinite network. Nevertheless, the association does produce structures of very large molecular weight, consisting of a central polystyrene domain from which emanate many polybutadiene chains. The high molecular weight causes the viscosity at small shear rates to become very large. However, the polymers have strongly non-Newtonian characteristics and some flow can occur at larger shear rates. Kraus and Gruver(8) compared SB and SBS block polymers at 80 C, finding highly non-Newtonian behavior for the SB copolymer, but no significant flow for the SBS polymer.

More important from a practical view is the behavior above Tg (polystyrene), i.e., above 100 C. The transition of the polystyrene domains from the glassy to a rubbery state is not accompanied by realization of molecular compatibility and the resulting polymer melt maintains its two-phase structure. The glass transitions of polybutadiene and polystyrene are roughly 200 C apart. Hence T-Tg and segmental mobility are much greater for the polybutadiene phase than for the polystyrene domains. If the butadiene phase is present in excess and forms the continuum, flow can occur in it with little disruption of the more viscous polystyrene inclusions and the situation resembles that below Tg (polystyrene): The limiting zero shear viscosity is very great and the flow is strongly shear rate dependent. It is interesting to compare the flow behavior of SB and SBS block polymers with that of a random copolymer of similar composition and molecular weight (Table I). Although styrene contents are not exactly matched, it is obvious that the SBS polymer is by far the most viscous, as its flow requires disruption (and re-formation) of the domain structure, but even the SB block polymers greatly exceed the random copolymer in viscosity.

There is evidence that under the influence of high shear stresses disruption of polystyrene domains occurs also in SB block polymers, accentuating the shear rate dependence of the apparent viscosity. As temperature is increased the viscosity of the polystyrene domains falls more rapidly than that of the polybutadiene matrix and disruption of the domain structure becomes easier. As a result the viscosity of the block polymer melts also undergoes rapid change.

Temperature dependence of non-Newtonian viscosity is best evaluated at constant shear stress, for only this manner of comparison can lead to a meaningful activation energy for flow. For block polymers even this activation energy is only an apparent one, since the system is thermorheologically complex and it is not clear what elementary processes are reflected in the overall value of E_a and how. Nevertheless, as shown in Table II, some insight is provided by a comparison of the apparent activation energies for flow of block polymers with flow activation energies of polybutadiene, polystyrene and a random copolymer. In all cases E_a was calculated assuming Arrhenius' law to apply between 130 and 160 C. This is a good approximation for polybutadiene and for the random copolymer (Tg are -98 C and -64 C, respectively), but not for polystyrene (Tg = 100 C) or for the block polymers. For the latter polymers E_a becomes simply a convenient way to express the temperature dependence of viscosity. It is at once obvious that all the block polymers have substantially higher E_a than polybutadiene or random butadiene-styrene copolymer of the composition corresponding to the styrene content of either the rubbery block alone or of the entire block polymer. It is also evident that E_a for the block polymers is shear stress dependent. This means that simple shear rate vs temperature superposition of SB block polymer viscosity data is impossible, a characteristic also of SBS block polymers(8,11). The temperature and shear dependencies of the apparent viscosity are thus entirely consistent with a flow mechanism involving shear and temperature-induced changes in the polystyrene domain structure.

As seen from the data of Tables I and II, which include block polymers made by both the sequential addition and copolymerization routes, there is no fundamental difference in kind in the flow behavior of the two types of rubbers.

TABLE I

Non-Newtonian (Apparent) Viscosity of Block and Random Copolymers of Butadiene and Styrene

	Styrene, %				$\eta \times 10^{-5}$, poise			
					130 C		160 C	
Structure	Total	Block	Mw/ 1000	Mn/ 1000	$\dot{\gamma}$ = 1 sec^{-1}	$\dot{\gamma}$ = 10 sec^{-1}	$\dot{\gamma}$ = 1 sec^{-1}	$\dot{\gamma}$ = 10 sec^{-1}
SB	25	25	78	63	1.80	0.90	0.48	0.30
SB	25	17	88	69	2.10	0.98	0.77	0.41
SBS	30	30	86	75	9.5	2.6	2.4	1.2
Random	30	0	86	74	0.14	0.14	.059	.059

TABLE II

Temperature Dependence of Non-Newtonian Flow

Styrene, %		Mw/ 1000	Mn/ 1000	$\tau = 10^5$ dynes/cm^2			$\tau = 10^6$ dynes/cm^2		
Total	Block			η(130 C) poise	η(160 C) poise	E_a kcal/mole	η(130 C) poise	η(160 C) poise	E_a kcal/mole
25	25	68	58	1.04	0.175	20.8	0.51	0.111	17.7
25	25	78	63	2.05	0.40	19.0	0.85	0.20	16.8
25	17	88	69	2.50	0.71	14.6	0.94	0.26	14.9
30	22	70	59	1.73	0.33	19.3	0.74	0.185	16.1
40	22	67	57	0.78	0.139	20.1	0.39	0.088	17.3
40	24	80	66	1.43	0.28	19.0	0.68	0.165	16.5
50	22	57	50	0.35	0.074	18.1	0.23	0.052	17.3
50	24	92	79	1.50	0.232	21.7	0.75	0.160	18.0
30[a]	0	$\gg M_c$[d]	$\gg M_c$[d]	-	-	10.1	-	-	10.1
0[b]	0	↓	↓	-	-	5.2	-	-	5.2
100[c]	0	↓	↓	-	-	59.8	-	-	59.8

a - Random copolymer, authors' data.

b - Polybutadiene, E_a calc. from data of Gruver and Kraus(9).

c - Polystyrene, E_a calc. from data of Plazek(10).

d - Critical molecular weight for entanglement coupling.

VI. TRANSITION BEHAVIOR

To illustrate similarities and differences between SB block polymers made by the sequential addition and copolymerization methods, temperature scans of tan δ (loss tangent) are compared in Figure 4 for polymers of equal total styrene content (25%) and nearly equal molecular weight ($\bar{M}_w \approx 85000$). The dynamic measurements were performed in the tensile mode at 35 Hz. Both polymers were lightly crosslinked with dicumyl peroxide to permit reliable determination of the upper transition temperature. The results show very similar behavior. The block polymer made by the copolymerization method contained 17% block styrene, the remainder of the styrene being distributed over the rubbery block. This causes the lower maximum in tan δ, corresponding to Tg of the rubbery block, to be shifted toward higher temperature. The height of this maximum is raised, as the transition involves a greater amount of material (83% vs 75% of the total). The slight inward shift of the upper maximum in tan δ seems to suggest partial compatibility. This interpretation, however, is questionable in view of other, more detailed, observations. Thus, Kraus et al(12) found copolymer blocks in polymers of the present type to have the same Tg (measured dilatometrically) as random copolymers of identical styrene content and butadiene microstructure. In this study the styrene content of the rubbery blocks did not exceed 20%. Results of a similar set of experiments, with styrene contents of the copolymer block ranging from zero to 40%, are shown in Figure 5. Here the temperature of the maximum in tan δ at 35 Hz is plotted against Tg calculated from composition and microstructure, using the Gordon-Taylor type equation(13) developed by Kraus, Childers and Gruver(12). The equation gives the dilatometric value of Tg of the uncrosslinked random copolymer. As expected, the temperature of the 35 Hz maximum in tan δ of lightly crosslinked samples is substantially higher. However, the difference remains a constant 15° up to about 20% styrene content, thereafter increasing very gradually to 25° at 40% styrene. This is good evidence that copolymer blocks not exceeding 20% styrene are strictly incompatible with polystyrene blocks of sufficient length ($>$12000 molecular weight units for all polymers of Figure 5). Even up to 40% styrene, partial miscibility cannot be extensive.

The position of the polystyrene loss maximum has been shown to be sensitive to polystyrene block length(12) and to the morphology of the domain structure and its orientation with respect to the direction of the stress(14). For the block polymers of Figure 5 the upper loss maximum (35 Hz) varies rather unsystematically between 108 to 123 C. For block length $>$12000 the molecular weight effect should be small and at least a portion of the variability must be ascribed to orientation effects. This was confirmed by measuring tan δ on one of the

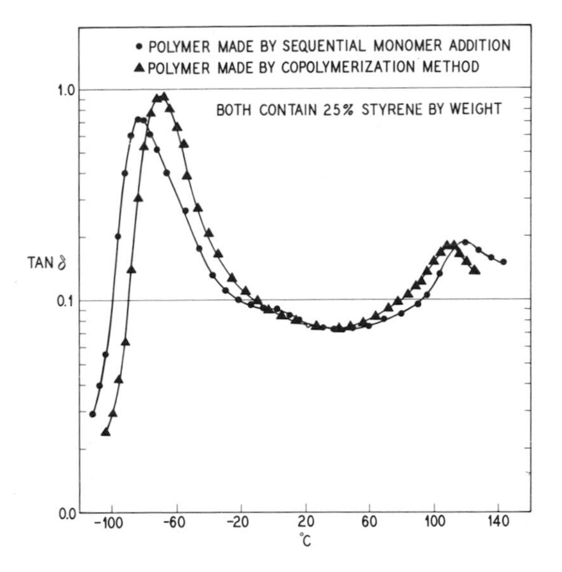

FIGURE 4
LOSS TANGENTS (35Hz) FOR BLOCK POLYMERS OF BUTADIENE AND STYRENE

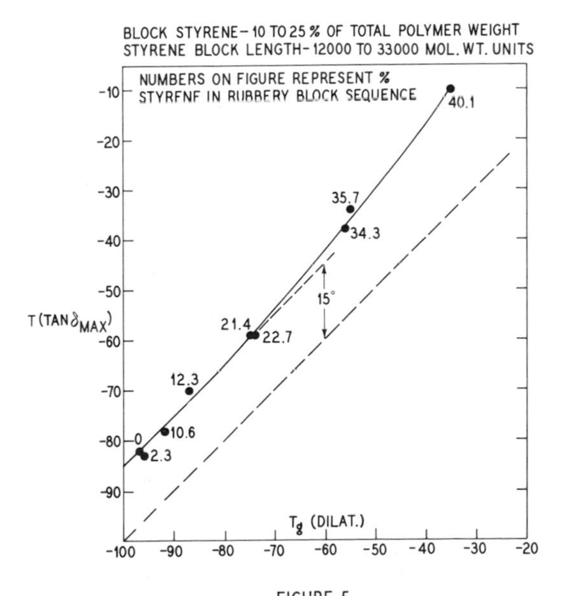

FIGURE 5
POSITION OF LOW TEMPERATURE LOSS MAXIMUM OF SB BLOCK POLYMERS MADE BY COPOLYMERIZATION METHOD

polymers (50% total styrene, 22% block styrene) both parallel and perpendicular to the direction of mold flow, observing a difference of 8° in the position of the loss maximum. It is thus highly probable that the polystyrene domains are substantially pure homopolymer and the minor shifts in $T(\tan \delta_{max})$ should not be interpreted as being due to limited solubility of the copolymer portion of the molecule in polystyrene.

Several investigators have proposed the existence of a mixed interlayer between butadiene and styrene domains of SBS block polymers(15,16). It has also been suggested that occasionally observed secondary loss maxima between the two principal transitions do, in fact, represent the glass transition of this interlayer(15). It would appear that formation of an interlayer would be favored in block polymers made by the copolymerization method. This idea is not supported by Figure 4; it is the data for the sequential addition polymer which show evidence of weak secondary dispersion (near 0°C).

In many applications (see below) SB polymers are blended with other rubbers. The resulting polyblends are almost without exception three-phase systems, as is apparent from Figure 6. In each of these blends the polybutadiene and polystyrene glass transitions are detectable. Of course, resolution of the two low temperature transitions suffers when the two Tg are close

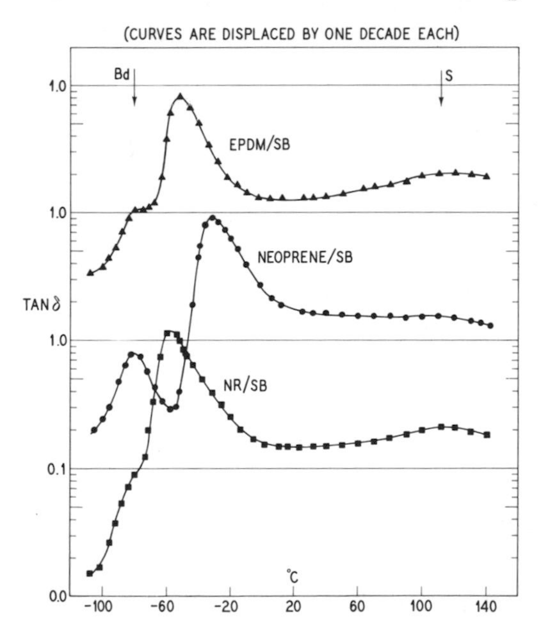

FIGURE 6 LOSS TANGENTS (35 Hz) OF CROSS-LINKED POLYBLENDS OF 30% SB-BLOCK POLYMER (25% STYRENE) WITH 70% OF ANOTHER RUBBER

together. In practical formulations containing fillers and plasticizers the transition behavior will in general be modified, but the basic multiphase morphology of the blends is usually preserved.

VII. STRENGTH PROPERTIES

Uncrosslinked butadiene-rich SB block polymers exhibit negligible tensile strength at room temperature. Fully crosslinked, however, they behave as filler-reinforced networks which, in fact, they are. When the tensile strength is followed as a function of the degree of crosslinking or of the quantity of crosslinking agent used, as in the data of Figure 7, a curve exhibiting two maxima is observed. The second of these clearly marks the point of optimum crosslink density for a fully formed network. This maximum in tensile strength occurs in all unfilled and filled rubbers. At the first maximum of Figure 7, however, the polymer still contains a rather large sol fraction (ca. 25%). The relatively high strength of these very poorly formed chemical networks must be explained by the completion of the network by the polystyrene domains. Coupled with the high breaking elongation, it is clearly reminiscent of the behavior of SBS block polymers in which the entire network is formed by polystyrene domains. Furthermore, the high strength of SBS polymers (ca. 300 kg/cm^2) is very sensitive to dilution with SB block polymer; an example given by Morton(17) shows a 24% reduction in tensile strength with an addition of only 5% SB. A tensile strength of 84 kg/cm^2 seems, therefore, quite reasonable for a sample containing 25% sol, most of which is undoubtedly composed of the original SB block polymer.

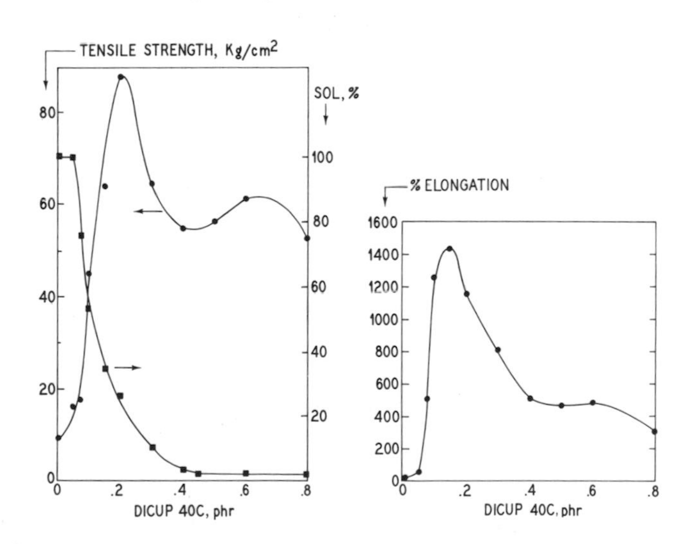

FIGURE 7
EFFECT OF CHEMICAL CROSS-LINKING ON AN SB BLOCK COPOLYMER (25% STYRENE)

The results of Figure 7 suggest that the mechanism of reinforcement by polystyrene domains is different in networks formed entirely (SBS) or predominantly (lightly crosslinked SB) by these domains on one hand, and in fully developed chemical networks on the other. In the latter the distribution between SB and SBS block polymers is, of course, largely lost.

The effect of additional filler reinforcement of the SB block polymer of Figure 7 is shown in Table III. At the low peroxide level the polymer is not reinforced by silica, which is again reminiscent of the behavior of SBS block polymers. At the higher peroxide level excellent response to silica reinforcement is observed.

TABLE III

ADDITION OF REINFORCING SILICA TO AN SB-BLOCK POLYMER[a]

Crosslinking Agent,[b] phr	Silica,[c] phr	Tensile Strength kg/cm^2
0.2	0	87
	10	115
	20	53
	40	15
0.6	0	61
	10	104
	20	177
	40	220

a - 25% styrene, 18% block styrene, M_w = 85000.
b - Supported dicumyl peroxide, 40% active material (DiCup - 40 C).
c - Precipitated silica of specific surface area approximately 100 m^2/g.

VIII. APPLICATIONS

A block polymer of the SB type, although not specifically identified as such, was first disclosed by Crouch and Short in 1961 under the name Philprene® X-40(18); further identification and a discussion of the structure and mechanical properties were reported later(19). This rubber was first produced commercially in December 1962(20). Characteristic mechanical properties were recognized early in the development and the polymer was quickly accepted by rubber technologists. The current U.S. annual consumption of this rubber alone (Solprene® 1205, a 75/25 butadiene-styrene block copolymer having 18 per cent block styrene) is in excess of 5000 metric tons. Solprene 303, a 52/48 butadiene/styrene copolymer with 11 per cent block styrene and Solprene 410, a 52/48 polymer with 32 per cent block styrene are also marketed.

In most applications these polymers are blended with other rubbers(21) to take advantage of specific properties conferred by their structure and morphology, as well as other characteristics inherently provided by the solution polymerization process. Some of these are:

Resistance to low temperature stiffening
Thermoplasticity
Moldability
Low mill shrinkage and smooth calendering
Low extrusion die swell
Glossy, smooth surface finish
Very light or no color
High purity
Low ash and water absorption

A few examples of applications are given in succeeding sections of this report showing how SB block polymers can be used to improve one or more important properties of a given compound without sacrificing others essential to the intended use.

A. Modification of Processing Characteristics

A common rubber processing problem is excessive shrinkage of the compound during processing, resulting in rough milled sheets, uneven extruded profiles and inconsistencies in the size and shape of raw blanks for vulcanization. In the example of Table IVA replacement of only small amounts of premasticated natural rubber (No. 1 RSS) with an SB block polymer in a mechanical goods formulation reduced mill shrinkage significantly without detriment to other important properties. Similarly, the extrusion rate, volume and appearance of conventional SBR, made either by emulsion or solution processes, can be improved appreciably by replacing part of the random copolymer with block polymer (Table IVB).

B. Injection Molded Neoprene Gasket Stock

Although many elastomers are developed for special uses such as resistance to heat, light, solvents and other environmental factors, these rubbers seldom provide all of the desired characteristics; e.g., ease of processing, mold flow and resistance to low temperature stiffening. The blending of polychloroprene rubber (Neoprene) with SB block copolymers provides polyblends of both improved processing and performance characteristics. Table V shows how, in an injection molded compound, replacement of 30 per cent of the Neoprene W with SB block polymer reduces injection time significantly and lowers the brittle point of the vulcanizate far below that ordinarily obtained. Ozone resistance was not impaired at this level of blending under the conditions tested. The remarkable influence of the SB block copolymer on transfer mold flow in this compound is apparent from the diagrams in Figure 8.

TABLE IV

MODIFICATION OF PROCESSING CHARACTERISTICS OF GENERAL PURPOSE RUBBERS

A. Control of Mill Shrinkage[a]

Premasticated #1 RSS[b]	100	95	90	80
SB Block Copolymer[c]	-	5	10	20
Mill Shrinkage, %	35	16	11	9.5
Cured 16 Minutes at 160 C				
300% Modulus, kg/cm^2	43	38	41	40
Tensile Strength, kg/cm^2	102	115	107	112
Elongation, %	490	550	540	550
Hardness, Shore A	57	59	62	63

B. Control of Extrusion[a]

Solution SBR Random Copolymer[d]	100	50	25	-
SB Block Copolymer[c]	-	50	75	100
Garvey Die Extrusion at 121 C				
Rate, cm/minute	167	193	218	258
grams/minute	138	148	159	160
Appearance Rating (12 best)	6	7	11	11
Cured 30 Minutes at 153 C				
300% Modulus, kg/cm^2	66	70	76	73
Tensile Strength, kg/cm^2	119	119	122	105
Elongation, %	530	620	640	670

a - All compounds contain 150 parts mineral filler and the usual rubber chemicals.
b - Premasticated to 80 ML-4 at 100 C.
c - Solprene 1205
d - Solprene 1204

TABLE V

INJECTION MOLDED NEOPRENE BLEND[a]

Neoprene W	100	70
SB Block Copolymer	-	30
Injection Time, seconds		
Pressure, 122 kg/cm^2	4.0	2.5
Cured 60 Seconds at 232 C		
300% Modulus, kg/cm^2	144	118
Tensile Strength, kg/cm^2	162	150
Elongation, %	340	350
Shore A Hardness	60	61
Brittle Point, ASTM-D746, °C	-38	-49
Ozone Resistance (0 - no cracks)	0	0

a - Recipe: Rubber - 100, N330 carbon black - 40, hard clay - 25, naphthenic oil - 25, stearic acid - 0.5 and 2 respectively, zinc oxide - 5, magnesium oxide - 4, Arkoflex CD - 2, NA-22 - 0.5, Thionex - 0.5, DOTG - 0.5 and 0.75 respectively, sulfur - 1 and 1.5 phr, respectively.

MOLDING TEMPERATURE 153°C
ORIFICE 3.14 cm
PRESSURE 210 Kg/cm^2

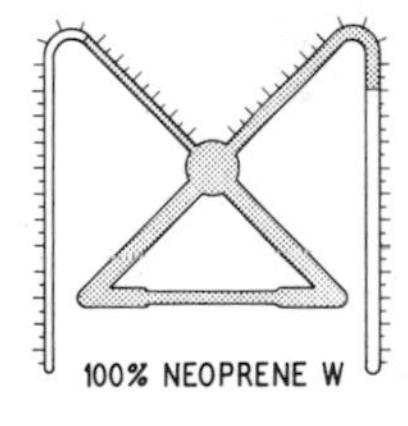

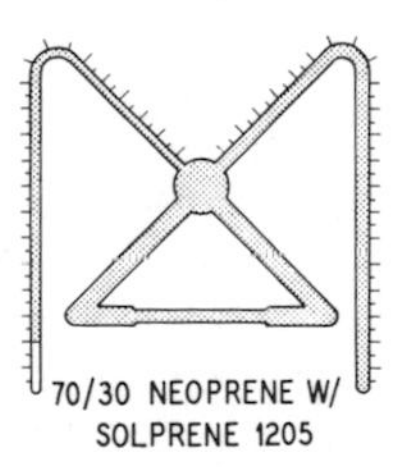

FIGURE 8
MOLD FLOW TEST

C. Electrical Insulation Compound

Emulsion SBR rubbers used for electrical insulation are usually specially finished to obtain light color, low ash and good electrical properties. Smooth extrudate appearance and good low temperature flexibility are desired characteristics in the compounded stock and in the vulcanizates. The data in Table VI show how low viscosity, fast extruding compounds with low shrinkage and excellent appearance can be obtained by using SB block copolymer as a partial replacement for SBR 1503 (specially finished emulsion copolymer) or as a replacement for SBR 1018 (high Mooney crosslinked emulsion SBR) in a blend with SBR 1503. This improvement is realized while maintaining modulus, tensile strength and electrical resistivity at the levels of the controls. Additionally tear strength, water absorption and resistance to low temperature stiffening or embrittlement are improved.

D. Microcellular Soling

Block polymers of butadiene and styrene have found ready acceptance in shoe soling formulations since they normally provide a thermoplastic base for incorporation of the commonly used high levels of filler, facilitate expansion in sponge stocks, and impart high hardness to the final vulcanizates. An example of the reduction in cure time, higher hardness and significantly lower shrinkage given by Solprene 303 (see above) in microcellular soling, compared to low Mooney emulsion SBR, is shown in Table VII. The lower level of blowing agent required by the more thermoplastic high styrene block copolymer, equal expansion and low shrinkage are especially noteworthy since the stock was reduced even further in cost by increasing the filler level and by addition of oil. The shrinkage after vulcanization permits a significant reduction in post-cure time and gives less tendency to warpage or curling of the shoe after finishing. The photographs in Figure 9 show the more uniform cell structure given by the block polymer in sponge.

E. Other Applications

A recent development is the improvement in the quality of pressure sensitive adhesives realized through the use of linear SBS and multichain $(S\text{-}B)_x$ block ("radial teleblock") polymers(22). By replacement of a part of the base rubber with an SB block polymer it is possible to tailor pressure sensitive adhesives of low cost, excellent shear (creep) resistance and high tack. Comparative data (Table VIII) show that use of an SB block polymer leads to a marked reduction in formulation viscosity with little effect on adhesive properties. Either adhesive is superior to conventional formulations based on SBR or natural rubber in creep resistance (50 vs approx. 0.1 hr) and failure temperature (above 100 C vs 25 to 35 C).

TABLE VI

ELECTRICAL INSULATION[a]

SBR 1503[b]	100	50	25	25
SBR 1018[c]	-	-	75	-
SB Block Copolymer[d]	-	50	-	75
Mooney Viscosity @ 100 C	47	42	63	41
Extrusion @ 82 C, Garvey Die				
g/min	102	114	79	119
Rating(12 best)	9	11	9	12
Extrusion Shrinkage, %	53	43	36	34
Cured 30 Minutes at 153 C				
300% Modulus, kg/cm^2	46	55	52	57
Tensile Strength, kg/cm^2	126	113	113	102
Tear Strength, kg/cm	29	32	26	34
Water Absorption, mg/cm^2	0.54	0.38	0.32	0.31
Electrical Resistivity 180 V, ohm-cm	3.2×10^{14}	3.4×10^{14}	2.9×10^{14}	2.9×10^{14}
Low Temperature				
Gehman Freeze Point, °C	-50	-54	-51	-63
Brittle Point, °C	-44	-58	-43	-62
Shore Hardness	77	71	77	72

a - Recipe: High styrene resin - 20, mineral rubber - 30, CCO Whiting - 25, zinc oxide - 15, stearic acid - 2, antioxidant - 2.5, sulfur - 1.25, litharge - 1, Zenite - 1.25, Monex - 1.

b - Specially finished, 41 F emulsion SBR.

c - Specially finished, 122 F crosslinked SBR.

d - Solprene 1205

TABLE VII

POLYMER REPLACEMENT IN MICROCELLULAR SOLING

SBR[a]	75	-
S/B Block Copolymer[b]	-	75
Rubber Dust	25	25
High Styrene Resin	5.8	10
Filler	75	110
Oil	-	3
Blowing Agent	5.5	3
Pound-Volume Cost, $	.1790	.1620
Cure Time, min at 153 C	12	8
Density, grams/cc	0.55	0.56
Split Tear, kg/cm	4.8	5.9
Shrinkage, % - 24 hrs at 100 C	3.5	0.8
Shore A Hardness (center)	29	38

a - SBR 1509, low Mooney 41 F emulsion SBR.
b - Solprene 303, 48% total styrene, 11% block styrene.

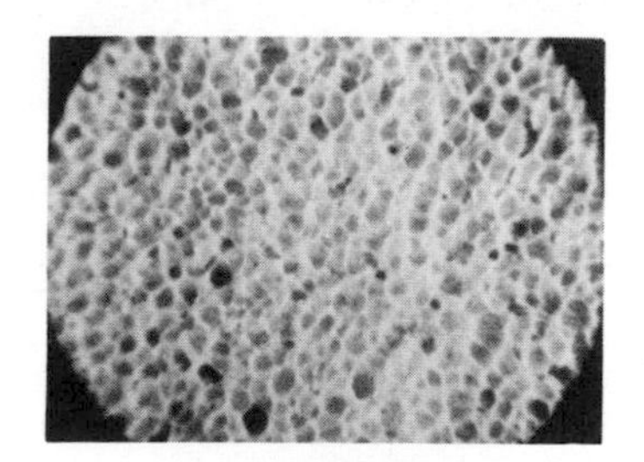

Solprene 303

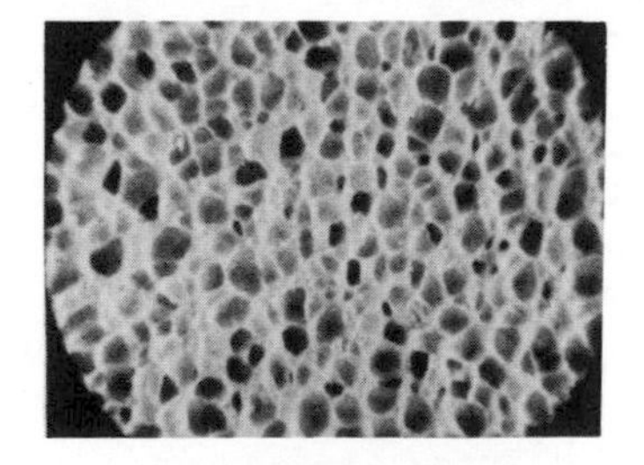

Low Mooney SBR

FIGURE 9 - COMPARISON OF CELL STRUCTURE

TABLE VIII

PRESSURE SENSITIVE ADHESIVE

Radial Teleblock Polymer[a]	100	75
SB Block Polymer[b]	-	25
Tackifier[c]	100	100
Formulation Viscosity, poise[d]	120	54
Tack, grams	1150	1100
Peel Strength, kg/cm	0.8	0.8
Failure Temperature, .14 kg/cm^2, C	$>$100	$>$100
Shear Creep Resistance, hrs to 0.16 cm slip	72	50

a - Solprene 411, 70/30 butadiene-styrene, 30% block styrene.
b - Solprene 1205, 75/25 butadiene-styrene, 18% block styrene.
c - Piccolyte α-115, alpha-pinene resin.
d - Solvent: 466 parts naphtha (80%) - toluene (20%) mixture.

TABLE IX

AUTOMOBILE SEAM SEALER[a]

SB Block Copolymer (Solprene 1205)	100
Hard Clay	200
Silica	15
Naphthenic Oil	300

Cured 15 Minutes at 135 C

Slump	$<$1.59 cm on vertical panel at 149 C
Flammability	Self Extinguishing
Weldability	Good (no resulting fire)

a - Recipe also contains resins, zinc oxide, chlorinated paraffins, antimony trioxide and curatives.

TABLE X

RUBBERIZED ASPHALT SEALER

SB Block Copolymer[a]	100
Airblown Asphalt	470
Highly Aromatic Oil	100
Tensile Strength, kg/cm^2	14
Elongation, %	1000
Recovery after 100% Elongation, %	60

a - Solprene 1205

A formulation for an automobile weld-through seam-sealer is shown in Table IX with pertinent physical properties. This recipe contains SB block copolymer as the base rubber with very high levels of mineral fillers and oil. The compound gives minimal slump, is self-extinguishing and permits welding through without ignition.

The asphaltic sealer formulation in Table X is typical of a variety of applications where SB block copolymers can be used effectively. Not only can such rubbers be used to make adhesives and sealants but also caulking compounds and for the modification of asphalt to control the viscosity.

LITERATURE CITED

(1) R. P. Zelinski and C. W. Childers, Rubber Chem. Tech., 41, 161 (1968).
(2) I. M. Kolthoff, T. S. Lee and C. W. Carr, J. Polymer Sci., 1, 429 (1946).
(3) M. Matsuo, T. Ueno, H. Horino, S. Chujo and H. Asai, Polymer, 9, 425 (1968).
(4) M. Matsuo, S. Sagae and H. Asai, Polymer, 10, 79 (1969).
(5) T. Inoue, T. Soen, T. Hashimoto and H. Kawai, J. Polymer Sci. A-2, 8, 1283 (1969).
(6) T. Uchida, T. Soen, T. Inoue and H. Kawai, J. Polymer Sci. A-2, 10, 101 (1972).
(7) T. Soen, T. Inoue, M. Miyoshi and H. Kawai, J. Polymer Sci. A-2, 10, 1757 (1972).
(8) G. Kraus and J. T. Gruver, J. Appl. Polymer Sci., 11, 2121 (1967).

(9) J. T. Gruver and G. Kraus, J. Polymer Sci. A, 2, 797 (1964).
(10) D. J. Plazek, J. Phys. Chem., 69, 3480 (1965).
(11) G. Holden, E. T. Bishop and N. R. Legge, J. Polymer Sci. C, 26, 37 (1969).
(12) G. Kraus, C. W. Childers and J. T. Gruver, J. Appl. Polymer Sci., 11, 1581 (1967).
(13) M. Gordon and J. S. Taylor, J. Appl. Chem., 2, 493 (1952).
(14) G. Kraus, K. W. Rollman and J. O. Gardner, J. Polymer Sci. A-2, 10, 2061 (1972).
(15) J. F. Beecher, L. Marker, R. D. Bradford and S. L. Aggarwal, J. Polymer Sci., C, 26, 117 (1969).
(16) D. H. Kaelble, Trans. Soc. Rheol., 15, 235 (1971).
(17) M. Morton, ACS Advances in Chemistry Series, 99, 490 (1971).
(18) W. W. Crouch and J. N. Short, Rubber and Plastics Age, 42, 276 (1961).
(19) H. E. Railsback, C. C. Biard, J. R. Haws and R. C. Wheat, Rubber Age, 94, 583 (1964).
(20) Rubber Age, 93, No. 2, 303 (1963).
(21) H. E. Railsback and C. Porta, Materie Plastiche et Elastomeri, 35, 63 (1969).
(22) O. L. Marrs, F. E. Naylor and L. O. Edmonds, Journal of Adhesion, 4, 211 (1972).

NEW DEVELOPMENTS IN BLOCK COPOLYMER APPLICATIONS

G. HOLDEN

SHELL DEVELOPMENT COMPANY

TORRANCE, CALIFORNIA 90509

During the last ten years, block copolymers have developed into important commercial materials. Most commercial applications have been based on polymers which have alternating blocks of hard and rubberlike materials, such as poly(styrene-b-butadiene-b-styrene) (S-B-S) and poly(styrene-b-isoprene-b-styrene) (S-I-S). These polymers are thermoplastic elastomers, that is, they can be molded like thermoplastics to give products similar in many properties to normal vulcanizates. Also, they can be dissolved in common solvents and applied as solutions. When the solvent evaporates, the polymer regains its physical properties.

It is now generally accepted that those unique properties are the result of a phase separation,[1,2] in which the polystyrene end segments aggregate to form hard regions or "domains". When the polystyrene segments are less than about one-third of the total polymer, the domains are probably spherical[3,4] and often form regular arrays.[5,6] Since the polystyrene is a hard glasslike solid at room temperature, these spherical domains "tie down" the ends of the flexible midsegments and act both as crosslinks and as reinforcing filler particles, thus giving the product properties similar to those of a vulcanized rubber. As the polystyrene content is increased, the form of the domains changes. At first they form rods and then lamellae. This changes the polymer from a rubberlike to a harder, more rigid material which draws on extension.[1,7] When the polymer is heated the polystyrene domains soften, and when a stress is applied the material is able to flow.

Because of the two phase structure, S-B-S and S-I-S block copolymers show two distinct glass transition temperatures, one corresponding to that of polystyrene and one to that of the elastomer phase.[1] The second transition takes place at a low temperature, and these polymers remain flexible even under very cold conditions.

Surprisingly, the viscosities of S-B-S and S-I-S polymers of the same molecular weight are quite different,[7] with that of the S-B-S being about ten times greater at low shear rates (Figure 1). Other differences between S-B-S and S-I-S block copolymer are related to the differences between polybutadiene and polyisoprene. For example, in the presence of air, polyisoprene degrades by chain scission but polybutadiene crosslinks. Thus, during aging S-I-S polymers become soft and sticky, whereas S-B-S polymers become hard and resinous. These effects can, of course, be retarded by the use of suitable antioxidants.

ESTABLISHED USES FOR S-B-S AND S-I-S POLYMERS

These can be divided into three groups and it should be noted that all involve compounded products rather than pure polymer.

Molded and Extruded Products

Since these products should have a nontacky surface, they are usually based on S-B-S polymers. They can be used for extrusion, blow molding and injection molding.[7,8] Special products intended for pharmaceutical and food contact applications and for footwear applications are available. The commercial advantages include short cycle times, absence of vulcanization residues and easy recycling of scrap. One important example of the use of these polymers as extruded and molded product is their applications in shoes. The soles of tennis shoes can be directly molded onto the canvas upper, and

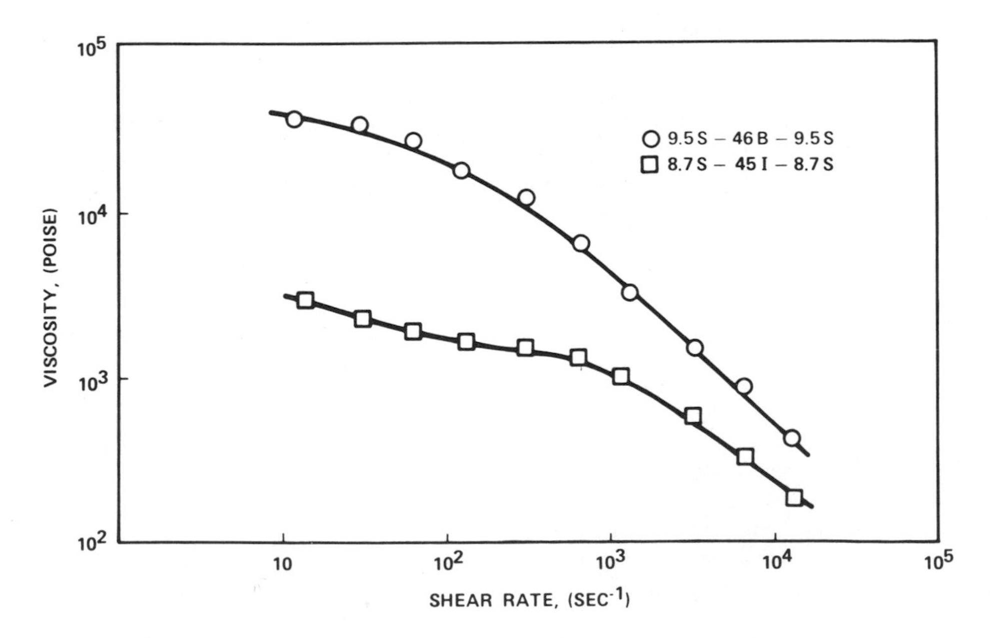

Figure 1. Viscosities of an S-B-S and an S-I-S Block Copolymer

both soles and heels of leather shoes can be cut from extruded sheet and then stitched or adhesively bonded to the body of the shoe. At equal hardness, the frictional properties of the thermoplastic elastomers are better than those of flexible thermoplastics such as PVC or EVA.[7] In addition, the coefficient of friction generally increases with increasing slip rate, whereas that of similar PVC blends falls, as shown in Figure 2.

Adhesives

Block copolymers are used to formulate both hot melt and solution adhesives. Tackifying and reinforcing resins are added to achieve a desirable balance of properties, as are oils and fillers. Principles have been described governing the choice of these additives in terms of their compatibility with the two phases in the block copolymer.[9,10,11] Oils and resins which associate with the center polybutadiene or polyisoprene segments give softer, stickier products, while resins which associate with the end polystyrene segments increase hardness and strength. Generally, oil and soft resins which associate with the polystyrene segments are to be avoided, since they plasticize the domains and allow the polymer to flow under stress. An exception to this would be in sealants where a controlled amount of stress relaxation or plastic flow may be desirable.

Protective Coatings

Solutions of block copolymer formulations can be coated onto

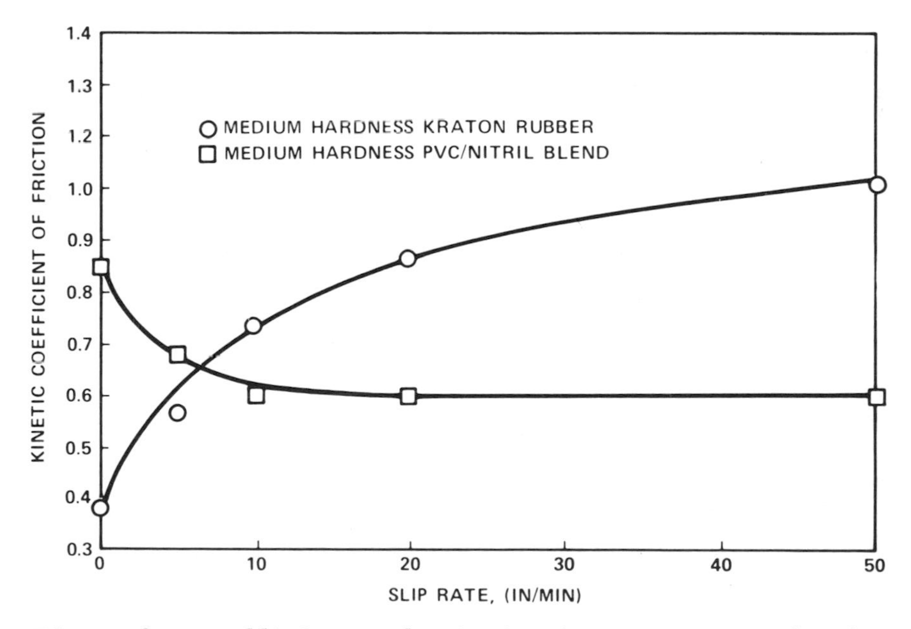

Figure 2. Coefficients of Friction Against Wet Vinyl Tile

metal or other parts. In the process of chemical milling, the polymer coating is removed from certain areas of the part, which is then dipped in an etchant bath. The unprotected areas of the metal are dissolved away, but the remainder is left unchanged.

In these last two applications the ability to dissolve in common solvents and to form a tough elastomeric film when the solvent is evaporated give S-I-S and S-B-S block copolymer unique advantages. The effect of the solubility parameter of solvents on their ability to dissolve these block copolymers and on the properties of the solutions have been discussed.[7] Generally, solvents for these block copolymers are also solvents for both polystyrene and polybutadiene or polyisoprene. A list of some solvents and their solubility parameters is given in Table I.

Table I. Solubility Parameters of Various Liquids and Polymers

n-Pentane[a,b]	7.0
n-Hexane[a,b]	7.4
Diethyl ether	7.4
POLY(ETHYLENE-BUTYLENE)	7.8
POLYISOPRENE	8.2
Cyclohexane	8.2
Butyl chloride	8.4
POLYBUTADIENE	8.4
Methyl isobutyl ketone	8.4
Amyl acetate	8.5
Carbon tetrachloride	8.6
Cellosolve® acetate	8.7
Ethyl chloride	8.8
m- and p-Xylenes	8.8
Toluene	8.9
O-Xylene	9.0
Butyraldehyde	9.0
Ethyl acetate	9.1
Methyl ethyl ketone[b]	9.1
POLYSTYRENE	9.1
Benzene	9.2
Chloroform	9.3
1,1,2,2,-Tetrachloroethane	9.7
Nitrobenzene[b]	10.0
Acetone[a,b]	10.0
Bromobenzene[b]	10.3
Isopropyl-alcohol[a,b]	11.9

a) Non-solvents for poly(styrene-b-diene-b-styrene) materials.

b) Non-solvents for poly(styrene-b-ethylene-co-butylene-b-styrene) materials.

Table II. Compounding Block Copolymers

COMPONENT	EFFECT ON				
	HARDNESS	PROCESSABILITY	OZONE RESISTANCE	COST	OTHER
Oils	Decreases	Increases	-	Decreases	Decreases U.V. Resistance
Polystyrene	Increases	Increases	Some increase	Decreases	-
Polyethylene	Increases	-	Increases	Decreases	Often gives satin finish
Polypropylene	Increases	-	Increases	Decreases	Improves high temperature properties. Best incorporated as powder
EVA	-	-	Increases	-	-
Fillers	Some increase	-	-	Decreases	Often improves surface

NEW DEVELOPMENTS IN S-B-S AND S-I-S

Extruded and Molded Products

As previously noted, almost all applications of block copolymers involve blends with substantial amounts of other materials such as oil, other polymers and fillers. Some guidelines for these blends are given in Table II. It has been found that polystyrene helps the processability of the product and makes it harder. Oils also help processability but lower the hardness. Oils with low aromatic content must be used, since other oils plasticize the polystyrene domains and weaken the product. Large amounts of oil and polystyrene can be added to these block copolymers, up to the weight of the polymer in some cases.

In addition to or instead of polystyrene and oils, polymers such as polypropylene, polyethylene or ethylene-vinyl acetate copolymer can be blended with S-B-S or S-I-S block copolymers. The blends usually show greatly improved ozone resistance and some solvent resistance, both produced by an interpenetrating continuous network of the added polymer, which is developed when the polymer mixture is sheared and then quickly cooled. For this reason, compression molded samples do not show the same improvements as injection molded or extruded ones. Surprisingly, even oils which themselves are stable to UV radiation reduce the stability of the blends. Oil-free blends of S-B-S or S-I-S block polymers with these other polymers have improved resistance to sunlight, particularly when absorptive or reflective pigments are also present. Oil-free blends of S-B-S with other polymers are sometimes rather hard and difficult to process. Thus, blends based on S-I-S may be particularly useful, since this block copolymer tends to be softer and easier processing than S-B-S. One example of the properties which can be achieved is shown in Table III

Large amount of inert fillers such as whiting, talc and clays can be used in these compounds. Frequently, up to 200 parts are used in compounds intended for footwear applications. Reinforcing fillers such as carbon black are not required and, in fact, large quantities of such fillers make the final product stiff and difficult to process.

Hot Melt Extrudable Adhesives

Solvent free adhesive formulations for coating onto tape or product assembly are another new application for S-I-S and S-B-S block copolymers, and three such formulations are given in Table IV.

These polymers can easily be tackified by the addition of oils, synthetic polyterpenes and similar resins. Since S-I-S and S-B-S block polymers have unsaturated backbones, they are susceptible to both oxidative and shear degradation. Thus, it is important that compositions containing them be adequately stabilized. S-I-S block copolymers are often preferred for tape coating because of their

Table III. Composition and Properties of an S-I-S/EVA Blend

Composition	
KRATON® 1107 Thermoplastic Rubber[a]	100
Ultrathene UE-631[b]	25
Carbon Black	1
Ethyl 330[c]	0.5
Plastanox LTDP[d]	0.5
Physical Properties (measured on injection molded slabs)	
300% Modulus (psi)	250
Tensile Strength (psi)	950
Elongation at Break (%)	> 1000
Shore A Hardness	43
Melt Flow Condition E (gm/10 min)	3.1
Ozone Resistance	
Bent Loop using 1/4" thick slab	No change after 1028 hrs in 50 pphm ozone at 90°F

a) S-I-S block copolymer, ex Shell Chemical Co.
b) Ethylene-vinyl acetate copolymer, ex U.S.I. Industries.
c) Antioxidant, ex Ethyl Corp.
d) Antioxidant, ex American Cyanamid Corp.

Table IV. Hot Melt Adhesives Based on S-I-S and S-B-S Block Copolymers

Pressure Sensitive	
KRATON 1107 Thermoplastic Rubber[a]	100
CUMAR LX-509[c]	30
PERMALYN' XA 100[d]	100
TUFFLO 6056[g]	60
BUTAZATE[h]	2
Pressure Sensitive	
KRATON 1102 Thermoplastic Rubber[b]	100
CUMAR LX-509[c]	20
ESCOREZ 5820[e]	100
TUFFLO 6056[g]	60
BUTAZATE[h]	2
Assembly	
KRATON 1102 Thermoplastic Rubber[b]	100
CUMAR LX-509[c]	150
SUPERSTATAC 80[f]	50
TUFFLO 6056[g]	50
BUTAZATE[h]	5

a) S-I-S block copolymer, ex Shell Chemical Co.
b) S-B-S block copolymer, ex Shell Chemical Co.
c) Coumarone Indene resin, ex Neville Chemical Co.
d) Cyclo aliphatic hydrocarbon resin, ex Hercules.
e) Polyolefin resin, ex EXXON Chemical Co.
f) Polyolefin resin, ex Reichhold Chemical Co.
g) Oil, ex ARCO.
h) Stabilizer, ex Uniroyal.

low viscosity. In addition, if degradation of S-I-S does take place, the polyisoprene segment is cleaved and lower molecular weight molecules which are relatively harmless are produced.

In many cases it is desirable to nitrogen blanket hot melt adhesives both during mixing and in the applicator.

Toughening Polystyrene

S-B-S block copolymers blended with high impact polystyrenes give a super high impact product. The more viscous S-B-S polymers give significant increases in Izod and falling weight impact resistance, with minimum loss of hardness. With crystal grade polystyrenes, S-B-S block copolymers act as flexibilizers rather than impact improvers.

BLOCK POLYMERS WITH OLEFIN RUBBER MIDSEGMENTS

Some new block copolymers have recently been introduced under the trade name KRATON® G Thermoplastic Rubbers. The difference in these new polymers is that the polydiene midsegments have been replaced by a polyolefin rubber which is an ethylene-butylene copolymer. Thus, they are designated S-EB-S. Similar polyolefin rubbers (such as EPR) have better stability than polydiene rubbers (such as IR or BR), and so the new products have much more resistance to oxidation, ozone attack and UV radiation. An example of their excellent stability is shown in Figure 3. The two GPC curves define the molecular weight-distribution of an S-EB-S block copolymer before and after 15 min. exposure to a temperature of 300°C. Little if any change can be seen. Similar results were obtained when the same polymer was exposed to pure oxygen at 300 psi and 70°C for 2000 hours.

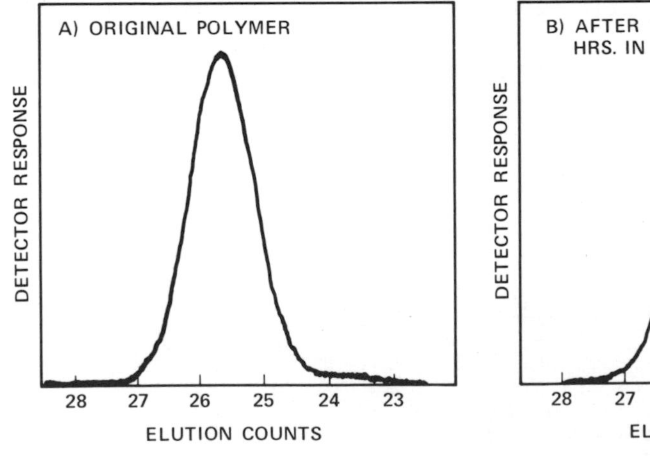

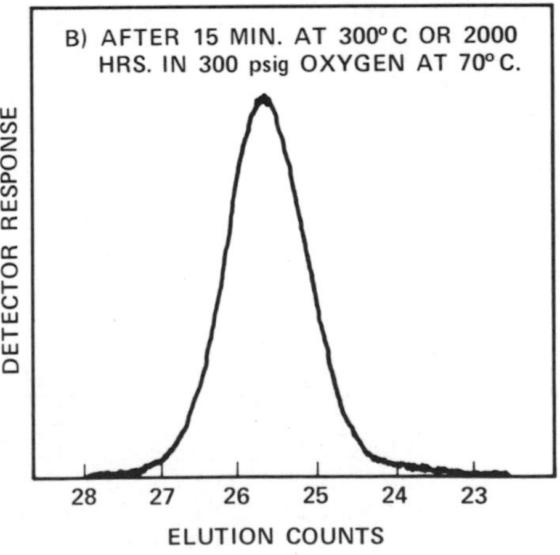

Figure 3. GPC Curves of an S-EB-S Block Copolymer

Table V. Adhesive Formulations Based on an S-EB-S Block Copolymer

Type	Sealant	Assembly Hot Melt Adhesive	Case Sealing Hot Melt Adhesive
KRATON® G Rubber GX-6500[a]	100	100	100
Wingtack 95[b]	200	100	100
Hercolyn D[c]	100	-	-
Piccotex 120[d]	-	150	50
Tufflo 6056[e]	25	50	50
SHELLWAX® 234[f]	-	-	150
Asbestine 3X[g]	400	-	-
Irganox 1010[h]	0.5	2	2
Cyclohexane	100	-	-

a) S-EB-S block copolymer, ex Shell Chemical Co.
b) Synthetic polyterpene resin, ex Goodyear Tire & Rubber Co.
c) Hydrogenated rosin ester, ex Hercules, Inc.
d) Vinyl toluene copolymer, ex Pennsylvania Industrial Chem. Co.
e) Oil, ex ARCO.
f) Petroleum wax, ex Shell Chemical Co.
g) Talc, ex International Talc Co.
h) Antioxidant, ex Ciba-Geigy.

Table VI. Resins Used with S-EB-S Block Copolymers

Resin	Type	Softening Point (°C)	Phase Association[a]
Piccotex 120[b]	Vinyl toluene copolymers	120	E
Piccovar 130[b]	Alkyl aromatic	130	E
Cumar R-16[c]	Coumarone-Indene	104	E
Wingtack 95[d]	Polyterpene	100	M
Super Statac 80[e]	Polyolefin	80	M
Foral 85[f]	Hydrogenated rosin ester	85	M
Hercolyn D[f]	Hydrogenated rosin ester	Liquid	B
Picco LTP-100[b,g]	Terpene phenolic	100	M

a) E designates a resin which primarily associates with the endblock phase.
M designates a resin which primarily associates with the midblock phase.
B designates a resin which associates with both phases.
b) Ex Pennsylvania Industrial Chemical Corp.
c) Ex Neville Chemical.
d) Ex Goodyear Tire and Rubber.
e) Ex Reichhold Chemical.
f) Ex Hercules, Inc.
g) Improves bond to metal surfaces.

Applications of Block Polymers With Olefin Rubber Midsegments

Molded or Extruded Goods

S-EB-S block copolymers can be formulated to give moldable or extrudable products which can then be formed by normal plastics processing equipment. Formulation principles are similar to those previously described for S-B-S and S-I-S polymers (see Table II). In compounds based on S-EB-S polymers, polypropylene is a particularly valuable additive, serving both to improve the processability and to extend the upper service temperature. Presumably, at least some of the improved service temperature can be attributed to the high crystal melting temperature (~ 165°C) of the polypropylene network. Blends of S-EB-S block copolymer, paraffinic or naphthenic oils and polypropylene are transparent. This is probably due to the fact that the refractive index of an S-EB-S polymer/oil blend almost exactly matches that of crystalline polypropylene.

Another advantage of S-EB-S polymers is that because of their lower midsegment solubility parameter, they are very compatible with paraffinic or naphthenic oil. As much as 200 parts of some oils can be added without bleedout. Again, properties of these blends depend on the formation of interpenetrating networks during injection molding or extrusion. The articles have excellent heat stability. In one case an injection molded sheet was exposed for one week at 300°F with no loss of properties. In addition, the resistance of pigmented stocks to outdoor exposure is very good.

Adhesives and Sealants

When compounded with suitable resins and oils, S-EB-S block copolymers can be used to make adhesive and sealants which have improved stability during long term or extreme temperature exposure. In addition, hot melt adhesives formulated from these new polymers are more stable under melt compounding conditions and can be applied at very high temperatures. Nitrogen blanketing is not required. Low cost adhesives and sealants with large amounts of waxes, oils and resins can be formulated, as shown in Table V. Table VI lists some resins which have been used and also gives details of their compatibility with the end or center segments in the S-EB-S block copolymer.

REFERENCES

1) G. Holden, E. T. Bishop, N. R. Legge, J. Poly Sci C.26,37 (1969).
2) G. Kraus, C. W. Childers and J. T. Gruver, J. Appl. Poly Sci, 11, 1581 (1967).
3) D. J. Meier, J. Poly Sci. C.26, 81 (1969).
4) H. Hendus, K. H. Hillers & E. Ropte, Kolloid ZS and ZS Polymers, 216-217, 110 (1967).
5) E. Fischer, J. Macromol Sci-Chem A2 (6), 1285 (1968).
6) P. R. Lewis and B. Price, Nature 223, 494 (1969).

7) Chapter 6, Block and Graft Copolymerization, R. J. Ceresa Ed. John Wiley and Sons.
8) W. R. Hendricks, Modern Plastics Encyclopedia, McGraw-Hill Inc., 1971-72 edition, p. 132.
9) J. T. Bailey, J. Elastoplastics 1, 1 (1969).
10) L. A. Petershagen, Adhesives Age 14, (2), 20 (1971).
11) J. T. Harlan, W. B. Luther, L. A. Petershagen and W. J.Robertson, Adhesives Age 15 (12), 30 (1972).

POLYSTYRENE-POLY (α-METHYL STYRENE) AB BLOCK COPOLYMERS AND ALLOYS

L. M. Robeson*, M. Matzner*, L. J. Fetters**, J. E. McGrath*

*Research and Development
Union Carbide Corporation
Bound Brook, New Jersey 08805

**Institute of Polymer Science
The University of Akron
Akron, Ohio 44325

INTRODUCTION

An in-depth investigation of a variety of block copolymers has been in progress in our laboratories for several years.[1-20] One of the most intriguing aspects of block copolymers lies in the fact that mechanical compatibility (i.e. polysulfone-Nylon 6)[20] or even true polymer-polymer solubility can be achieved.

In polystyrene-polybutadiene-polystyrene ABA block copolymers, Meier[21] reported that block weights of >5000 - 10,000 for polystyrene and >50,000 for polybutadiene are necessary for microphase separation. In polysulfone-poly (dimethyl siloxane) $(AB)_n$ block copolymers[6], microphase separation occurs at block molecular weights as low as ∿2000.

Important differences in the phenomenon of phase separation are found with physical blends versus their corresponding block or graft copolymers.[22] In this paper AB block copolymers of poly (α-methyl styrene) and polystyrene were prepared and studied with regard to microphase separation per se and in blends with their respective homopolymer constituents.

Blends of poly (α-methyl styrene) and polystyrene are known to exhibit two-phase behavior at high molecular weight, $>10^5$.[23] Baer, however, found that tri-block (ABA) copolymers of polystyrene and poly (α-methyl styrene) exhibited one-phase behavior even at number average molecular weights above 10^5.[23] Blends of the ABA block copolymers with the homopolymer constituents were not reported. Therefore, further study of this very interesting system was indicated.

Phase separation of a blend containing an AB polystyrene-polyisoprene block copolymer with the constituent homopolymers has been studied by Inoue et al.[24] The results indicated that when the molecular weight of the added homopolymer is less than, or equal to, that of the corresponding block segment, the added homopolymer can be solubilized into the domain. At higher molecular weights of the added homopolymer, a phase of the block copolymer and a phase of the homopolymer are observed.

Theoretical considerations of microphase separation in block copolymers have been reported by Krause.[22, 25, 26] Parameters under consideration include block molecular weight, number of blocks per molecule, and the interaction parameters between the homopolymers. The results show that covalent bonding in a block copolymer has a significant effect in solubilizing its segments. This is in agreement with the experimental results reported by Baer.

Block copolymer blends with their homopolymer constituents could exhibit various morphologies and viscoelastic responses. Since we will be considering a one-phase block copolymer, the discussion will be limited to the possibilities that could occur in that case.

1) Addition of homopolymer to a one-phase block copolymer results in a one-phase system with a single Tg (glass transition temperature) intermediate between block copolymer and the added homopolymer.

2) Addition of homopolymer to the block copolymer results in phase separation and two transitions corresponding to each homopolymer.

3) Two-phase system resulting with addition of homopolymer to the single phase block copolymer. This case is different from case 2 in that the Tg's of the homopolymer and block copolymer are observed (i.e., the homopolymer does not enter the block phase of equivalent composition).

4) Heterogeneous systems (infinite number of different phases) in which a broad transition exists between the Tg of the single phase block copolymer and the added homopolymer. A single Tg cannot be ascribed to this particular system.

In the case of both homopolymers added to the two-constituent block polymers even more possibilities will exist. One of these cases involves preferential solubility of the block copolymer in one of the added homopolymers.

The mechanical loss spectra of the four cases listed above are shown in Figure 1, and will form the basis for the interpretation of the experimental results that are presented.

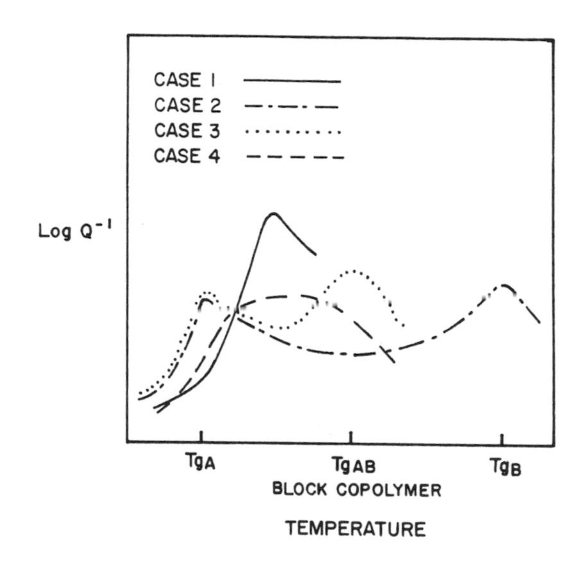

Figure 1. Hypothetical Mechanical Loss (Q^{-1}, defined on p. 300) Data for Homopolymer A Mixtures with Block Copolymer AB.

EXPERIMENTAL

The polymers utilized in this study are listed in Table I. All were prepared by the anionic sequential polymerization as described(18, 27) elsewhere. Briefly, this involved very rigorous purification of the monomers, tetrahydrofuran (THF), glassware and the utilization of high vacuum preparative conditions. The reaction is illustrated in equation (1). The "random" copolymer (CP-I) was prepared by distilling styrene into the stirred tetrahydrofuran-α-methyl styrene solution, which was held at -78°C. This technique was utilized since several preliminary runs demonstrated that with a mixture of these monomers, the styrene polymerized first with the near exclusion of α-methyl styrene. The distillation method, however, permitted the incorporation of

TABLE I

Polymer Characterization[1]

Polymer Sample	Description	$M_n^{(3)} \times 10^{-4}$ gm mole^{-1}	$M_w^{(4)} \times 10^{-4}$ gm mole^{-1}	Tg[(5)] °C
Polystyrene	SMD-3500[(2)]	11.0 (±5%)	27.0	100
Polystyrene		-	3.6	
Poly-α-methyl-styrene		-	31.0	171
Poly-α-methyl-styrene[(6)]		16.0 (±5%)	16.5	
CP-I	Styrene/α-methyl-styrene random copolymer	27.0 (±6%)	30.0	138
BP-I	Styrene/α-methyl-styrene AB block copolymer	7.4 (±4%)	8.0	134
BP-IIA	Styrene/α-methyl-styrene AB block copolymer	14.8 (±5%)	15.0	132
BP-III	Styrene/α-methyl-styrene AB block copolymer	39.0 (±7%)	42.0	130
PPO[(7)]	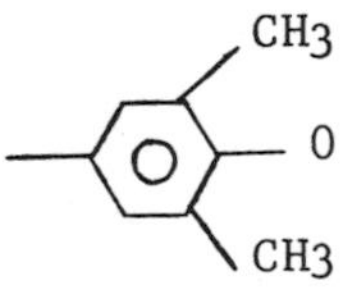 CH_3 O CH_3	-	-	206

(1) All copolymers contained 50 mole percent of each constituent
(2) Union Carbide Corporation
(3) Membrane Osmometry
(4) Via gel permeation chromatography
(5) Dynamic Mechanical Loss Peak
(6) Utilized only in ternary blends with BP-IIA
(7) Obtained from General Electric Company

α-methyl styrene into the chain along with styrene. When all of the styrene was distilled the reaction was immediately terminated by the addition of ethanol. The yield was >99% of copolymer.

$$nCH_2{=}CH(C_6H_5) \xrightarrow[\text{sec-butyl lithium}]{\text{THF } -78^\circ C} S\text{-}C_4H_9 - (CH_2 - CH(C_6H_5))_n Li \xrightarrow{CH_2{=}C(CH_3)(C_6H_5)}$$

polystyrene poly (α-methyl styrene)

(1)

The number-average molecular weights were determined with a Hewlett-Packard 503 osmometer. The solvent used was toluene at 37° while S and S-08 membranes were used. The number-average molecular weights were determined from the usual $(\pi/c)^{0.5}$ vs. c plots.

A Waters Ana-Prep gel permeation chromatograph was used at a flow rate of 0.25 ml min^{-1} with 0.25 wt. % solutions in tetrahydrofuran at a temperature of 45°. An ultraviolet detector was used with the following seven, 4 ft., columns of polystyrene gel: 7-50 (10^5) Å, 1.5-7 (10^5) Å, 5-15 (10^4) Å, 1.5-5 (10^4) Å, two columns with 5-15 (10^3) Å, and 2-5 (10^3) Å. The theoretical plate number for these seven columns was 730.

The GPC columns were calibrated with 19 polystyrene samples. Of these samples, four possessed M_w/M_n ratios >2 while the remaining samples possessed M_w/M_n ratios <1.2. In addition, the GPC columns were calibrated with a series of 11 samples of poly-α-methylstyrene. These materials were synthesized and characterized in these laboratories and all possessed M_w/M_n ratios of <1.1. The calibration generated by the plot of log M_w vs. elution volume was linear between the molecular weights of 2.4×10^3 and 2.5×10^6.

The GPC chromatograms of the three copolymer materials are shown in Figure 2. It should be noted that the interval between each count corresponds to a volume of 5 mls.

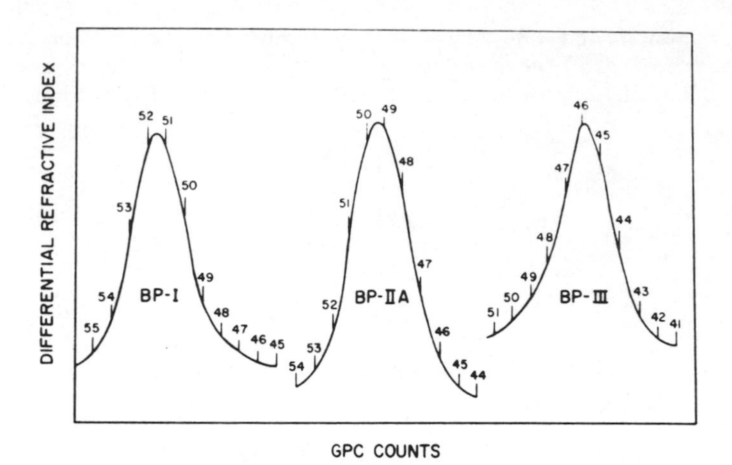

Figure 2. Gel Permeation Chromotography Curves for BP-I, BP-IIA, and BP-III

Blends of the block copolymers with the constituent homopolymer were made by dissolving in methylene chloride and devolatilizing in a vacuum oven at 140°C. The resultant material was compression molded into specimens for dynamic mechanical testing. The blends of the polystyrene-poly (α -methyl styrene) AB block copolymer with PPO were similarly prepared except that chloroform was used. Removal of the last traces of chloroform was very difficult and required temperatures of > 180°C for > 1 hour under vacuum. All samples were molded at 210°C.

The dynamic mechanical testing was done using a torsion pendulum similar to the design of Nielsen.[28] The frequency of the free vibration torsion pendulum analysis varies with modulus. All sample dimensions were chosen such that in the glassy state (i.e. room temperature) the frequency was between 1.5 and 2.5 cycles/sec.

RESULTS

The expected two-phase behavior with alloys of 25/75 polystyrene 270,000 M_w and poly (α -methyl styrene) 310,000 M_w (Figure 3) was observed. The characteristic glass transition temperatures of the homopolymers are clearly seen, thus demonstrating complete phase separation. Similar results were also obtained for a 75/25 blend of a poly (α -methyl styrene) of 165,000 M_w and a polystyrene of 36,000 M_w. However, broader transitions and shifts in Tg were observed indicating slight mutual solubility.

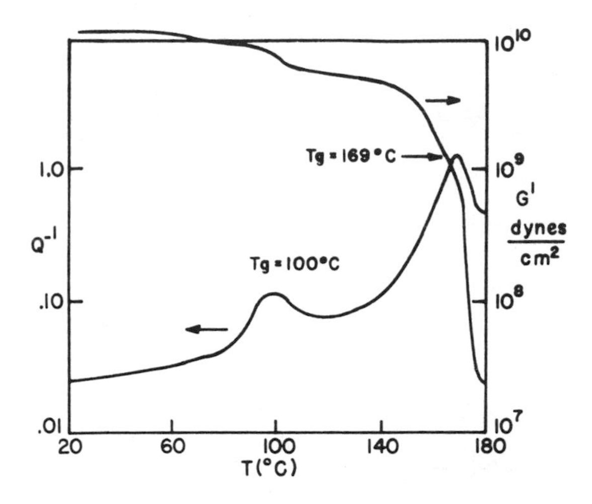

Figure 3. Mechanical Loss and Shear Modulus Temperature Data on 75/25 Poly (α-methyl styrene) (M_W=310,000)/ Polystyrene (M_W=270,000) Blend

The AB block copolymer of polystyrene and poly (α-methyl styrene), however, exhibits true solubility even up to molecular weights of 200,000 for each block. This is illustrated in Figure 4 for BP-I and BP-IIA and in Figure 5 for BP-III. For each block copolymer a single Tg is observed without evidence of any transitions characteristic of polystyrene or poly (α-methyl styrene). Note that the transition region broadens as molecular weight increases and a monotonic reduction in Tg occurs from the random copolymer CP-I to the high molecular weight block copolymer BP-III. These results show the dramatic effect that one covalent bond has on the mutual solubility characteristics of the subject polymers.

Blends of the block copolymers were made with polystyrene. Mechanical loss and shear modulus curves for 50/50 (wt %) polystyrene/BP-I and polystyrene/BP-IIA (Figure 6) show single phase behavior with intermediate Tg's. With BP-III, however, phase separation occurs. A blend of 70/30 (by wt) of BP-III and polystyrene exhibits a pronounced shoulder in the mechanical loss curve at 100°C and the characteristic BP-III transition at 130°C (Figure 7).

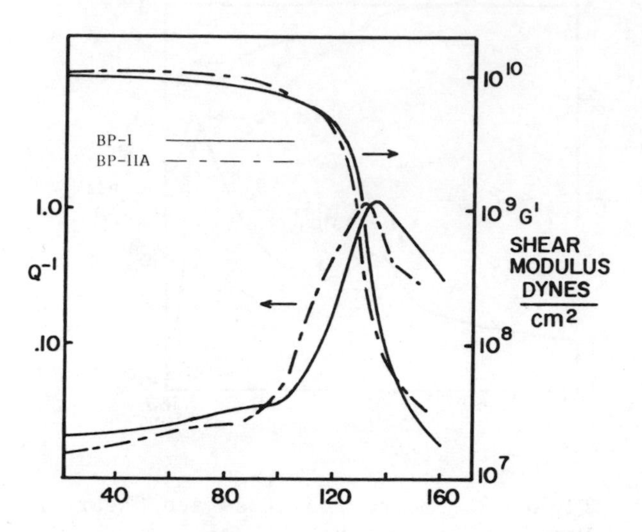

Figure 4. Mechanical Loss and Shear Modulus Temperature Data on BP-I and BP-IIA (Polystyrene/Poly (α-methyl styrene) AB Block Copolymers

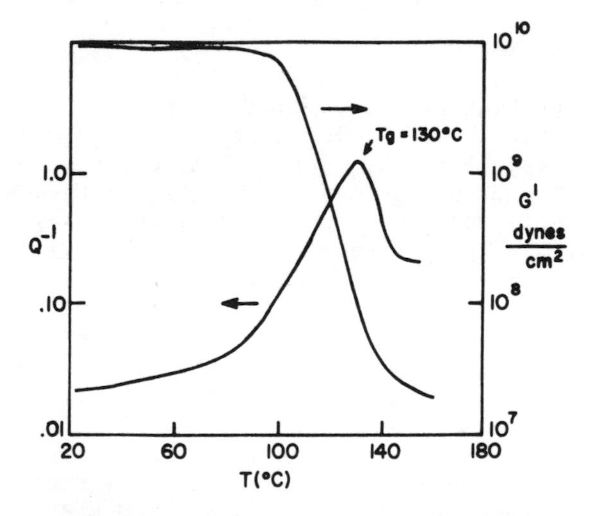

Figure 5. Mechanical Loss and Shear Modulus Temperature Data on BP-III (Polystyrene/Poly (α-methyl styrene) AB Block Copolymer

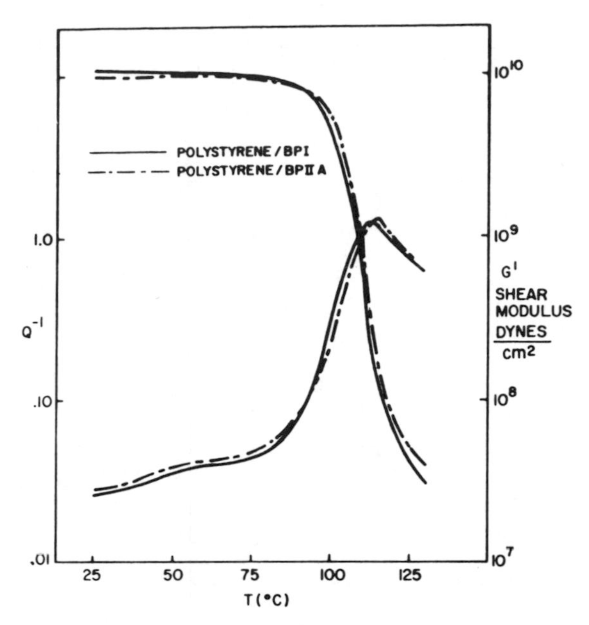

Figure 6. Mechanical Loss and Shear Modulus Temperature Data for Polystyrene (270,000 M_W)/BP-I 50/50 (by wt.) (Tg=114°C) and Polystyrene (270,000 M_W)/BP-IIA 50/50 (by wt.) (Tg=115°C)

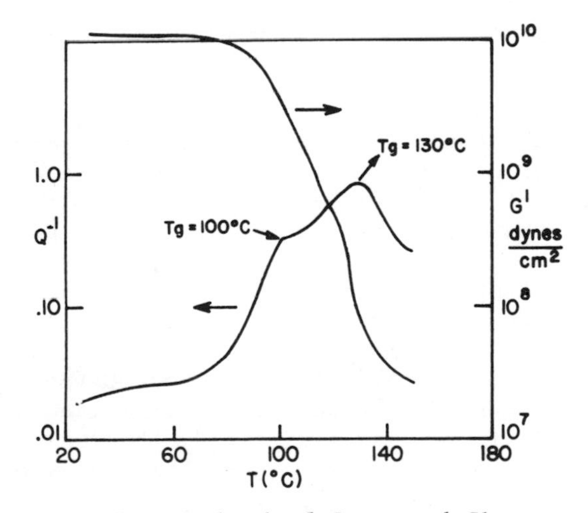

Figure 7. Mechanical Loss and Shear Modulus Temperature Data on 70/30 (by wt.) BP-III/Polystyrene (270,000 M_W) Blend

Similar results were observed on compression molded specimens of a 50/50 (by wt.) blend of BP-III and polystyrene (Figure 8).

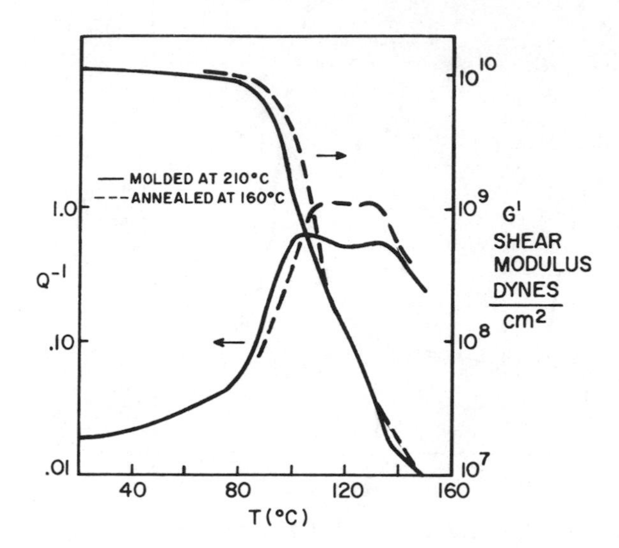

Figure 8. Mechanical Loss and Shear Modulus Temperature Data on Polystyrene (270,000 M_w)/BP-III 50/50 (by wt.) Blend

In this latter case, the transitions are obscured by the proximity of the constituent Tg's and equivalent phase continuity. An interesting observation was made during these investigations. It was found that subjecting the blends of homopolymers, as well as the polystyrene blends with the block copolymer BP-III, to higher temperatures (>160°C) resulted in decrease of transparency. This may be due to further microphase separation and/or the refractive index temperature coefficient difference between polystyrene and poly (α-methyl styrene). Note that our studies on polymer-polymer solubility have shown generally increased solubility with decreasing temperature. Where borderline solubility occurs, lower critical solution temperatures are observed. This behavior was also theoretically predicted by McMaster, of these laboratories.[29] Because of these previous observations, the 50/50 (by wt.) blend of BP-III and polystyrene (which was initially molded at 210°C) was annealed at 160°C for 30 minutes. The mechanical loss data after annealing shows a trend towards one-phase behavior as shown in Figure 8.

A blend of the styrene-α-methyl styrene random copolymer (CP-I) and polystyrene at a 70/30 (by wt.) ratio shows two-phase behavior (Figure 9). Thus both the random and block copolymers of styrene-α-methyl styrene exhibit similar characteristics with polystyrene homopolymer.

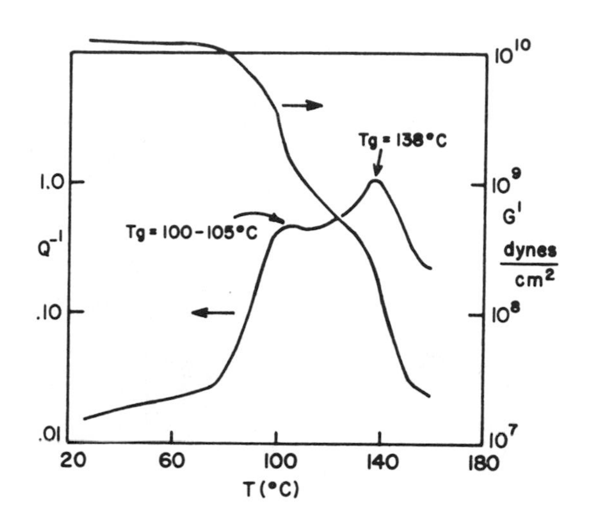

Figure 9. Mechanical Loss and Shear Modulus Temperature Data on 70/30 (by wt.) CP-I/Polystyrene (270,000 M_w) Blend

A 50/50 blend (by wt.) of poly (α-methyl styrene) and BP-III, however, yielded a single Tg at 142°C as shown in Figure 10. Similar results (also Figure 10) were obtained with a 75/25 (by wt.) blend of poly (α-methyl styrene) and BP-III with a single Tg at 156°C.

Since poly (2,6 dimethyl 1,4 phenylene ether) PPO is soluble in polystyrene,[30] the blends of PPO and BP-III were also investigated. The 75/25 (by wt.) blend of PPO and BP-III (Figure 11) as well as a 50/50 (by wt.) blend of PPO and BP-III are one-phase single systems with Tg's at 177.5°C and 158°C, respectively.

These results raised the question of PPO/poly (α-methyl styrene) solubility. The mechanical loss data for a 50/50 wt. blend of the two polymers yielded a single Tg (182°C; Figure 12) although the transition was broad indicating some heterogeneous character. In contrast, similar data obtained on polystyrene/PPO blends shows nearly ideal solubility.

Ternary blends of BP-IIA, polystyrene, and poly (α-methyl styrene) were prepared. Note that the molecular weight of the poly (α-methyl styrene) used in this part of our study was lower

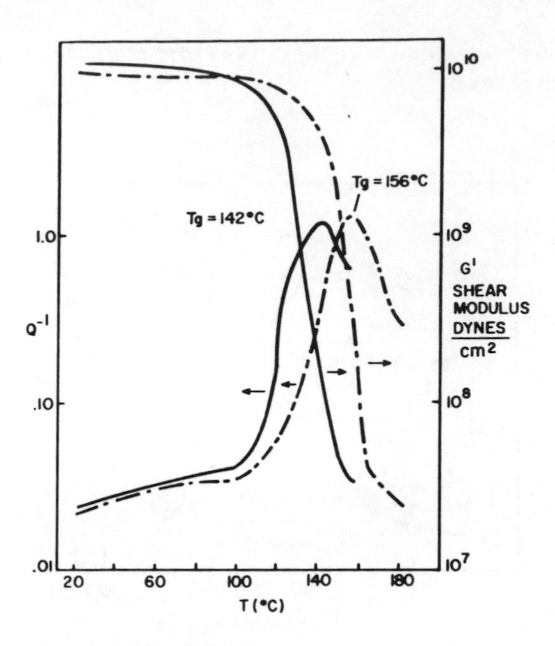

Figure 10. Mechanical Loss and Shear Modulus Temperature Data on BP-III/Poly (α-methyl styrene) (310,000 M_w) Blends (50/50 by wt. ———) (75/25 by wt. — · — · — ·)

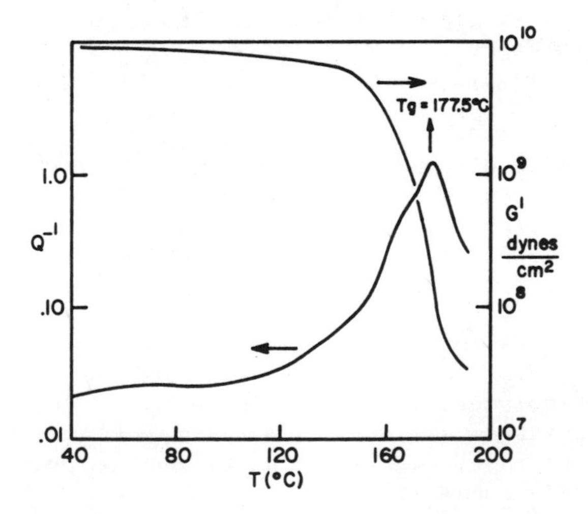

Figure 11. Mechanical Loss and Shear Modulus Temperature Data on 75/25 (by wt.) PPO/BP-III Blend

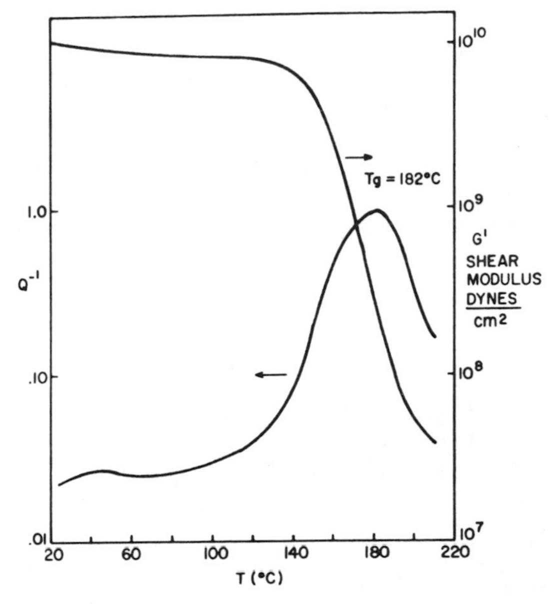

Figure 12. Mechanical Loss and Shear Modulus Temperature Data on 50/50 PPO/Poly (α-methyl styrene) (310,000 M_W) Blend

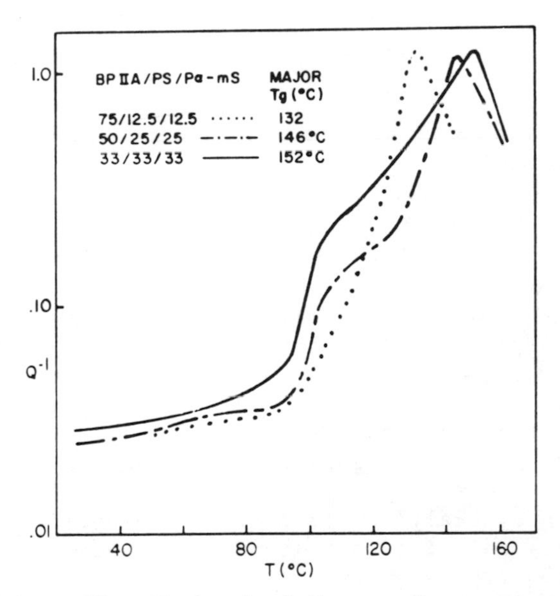

Figure 13. Mechanical Loss - Temperature Data for Ternary Blends of BP-IIA/Polystyrene (270,000 M_W)/Poly (α-methyl styrene) (165,000 M_W)

than that used above. The mechanical loss curves of 75/12.5/12.5, 50/25/25, and 33/33/33 (by wt.) blends of BP-IIA, polystyrene, and poly (α-methyl styrene) are given in Figure 13. A single Tg at 132°C is observed for the 75/12.5/12.5 blend; however, phase separation is observed with the other two blends. The compatibilization efficiency of BP-IIA is, therefore, rather low. The shear modulus curves for these alloys are given in Figure 14.

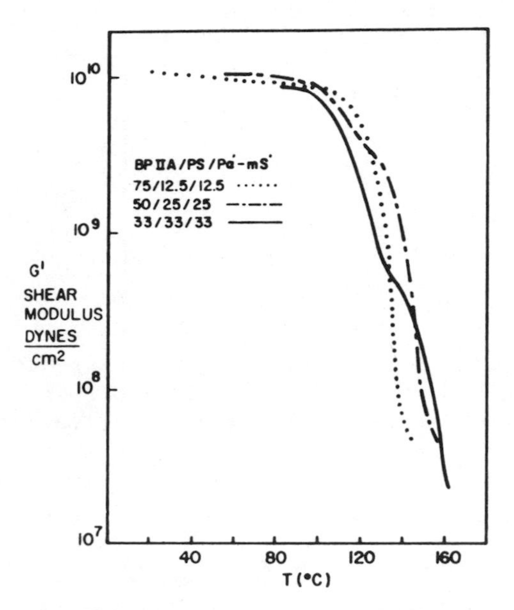

Figure 14. Shear Modulus - Temperature Data for Ternary Blends of BP-IIA/Polystyrene (270,000 M_w)/Poly (α-methyl styrene) (165,000 M_w)

A ternary blend of BP-III, poly (α-methyl styrene) and polystyrene (50/25/25) gave results that were similar to those obtained with BP-IIA. Here, however, the molecular weights of the block copolymer and the poly (α-methyl styrene) were again higher than in the previous ternary blends. The mechanical loss and shear modulus curves are given in Figure 15. The results show a polystyrene transition and a poly (α-methyl styrene)-BP-III transition. Thus, the previously observed higher solubility of poly (α-methyl styrene) relative to that of polystyrene in BP-III is again clearly demonstrated by the ternary blend data.

DISCUSSION

The single-phase behavior of block copolymers of poly (α-

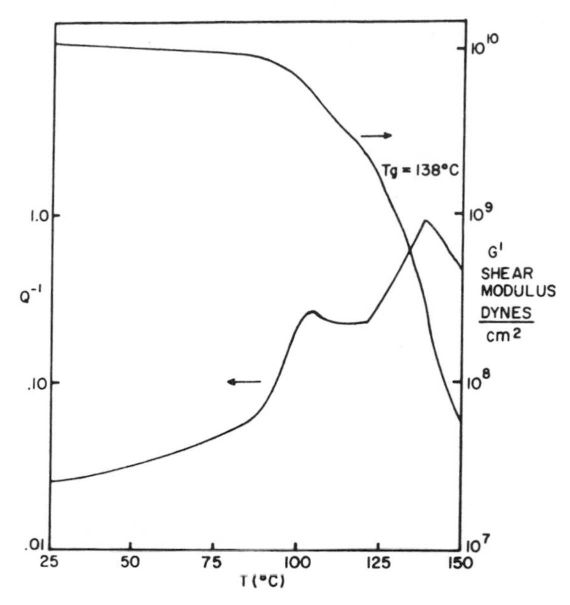

Figure 15. Mechanical Loss and Shear Modulus Temperature Data for a Ternary Blend of BP-III/ Polystyrene (270,000 M_w)/Poly (α -methyl styrene) (310,000 M_w)

methyl styrene) and polystyrene was reported previously by Baer(23) for the ABA tri-block copolymers (A = either polystyrene or poly (α -methyl styrene)). Our results show that one-phase behavior is also observed in an AB block copolymer at molecular weights as high as 200,000 for each segment. It is of interest to note that the Tg decreases and the transition broadens as one goes from the random copolymer CP-I to increasing block length in the AB block copolymer. Note that the higher Tg for the random copolymer is in contradiction with the results obtained by Baer.(23)

The theoretical treatment of phase separation in polymer blends and block copolymers was recently given by Krause(25) and Meier(31). It was predicted that even one chemical bond per chain may lead to solubility provided that the solubility parameters (or "interaction" parameters) be very similar for the two blocks. In fact, Krause predicted that phase separation in a tri-block polymer of styrene and α-methyl styrene would not occur up to molecular weights in the order of 10^6. Our results yield further credence to this theoretical treatment.

Using the same approach and assumptions outlined by Krause,

the critical interaction parameter $(\chi_{AB})_{cr}$ given as

$$(\chi_{AB})_{cr} = \frac{z\, V_r}{(z-2)\, V_B V_A{}^M\, n_B{}^C} \left[-\ell n\, (V_A{}^M)^{V_A{}^C} (V_B{}^M)^{V_B{}^C} - \frac{N_{HA}}{N_C}\, \ell n\, V_A{}^M + 2\,(m-1)\,\frac{S_{dis}}{R} - \ell n\,(m-1) \right]$$

where: V_B = volume of B repeat units

$n_B{}^C$ = number of B units in each block copolymer molecule

$\Delta S_{dis}/R$ = entropy lost when one segment of molecule is immobilized (assumed = 1.0)

$V_A{}^C$ and $V_B{}^C$ are volume fraction of monomer A and B in block copolymer molecule

$V_A{}^M$ and $V_B{}^M$ are volume fractions of monomers A and B in total mixture

z = lattice coordination number (assumed = 8)

V_r = volume of a lattice site

m = number of blocks per molecule (in this study $m = 2$)

$(\chi_{AB})_{cr}$ = .002 (based on analysis by Krause for polystyrene and poly (α - methyl styrene)

N_{HA}, N_C = number of homopolymer molecules; and total number of molecules

Since the block copolymers are on an equal molar basis, the volume of poly (α-methyl styrene) is slightly higher than that of polystyrene. For simplicity purposes calculations on an equal volume basis are reported. The results show that microphase separation will occur for the AB block copolymer at a total degree of polymerization of 7,200 which corresponds to a molecular weight of approximately 800,000. Results using the same assumption for the tri-block copolymer yield a total degree of polymerization of 10,700 with a total molecular weight of approximately 1,200,000.

When the calculations are done for homopolymer (50% by volume) added to the block copolymer, the term χ_{AB} increases. However, the calculations for BP-III and 50% V_f of either component of molecular weight of 200,000 give a value of $(\chi_{AB})_{critical}$ of 0.0030, thus predicting a one-phase system. In fact, even at infinite molecular weight of added homopolymer (50% V_f) $(\chi_{AB})_{critical}$ is only 0.0025, thus still predicting a one-phase system. While these results are consistent with the blends of (α-methyl styrene) and BP-III, they do not agree with the data on polystyrene and BP-III. The theory, of course, does not differentiate between assymetric phase solubility such as component A more soluble in component B than the reverse case. It would, therefore, be interesting to look at the phase diagram for polystyrene and poly (α-methyl styrene) at low and equivalent molecular weights. The data presented here would predict that the poly (α-methyl styrene) rich phase would contain more polystyrene than the polystyrene rich phase would contain poly (α-methyl styrene). This behavior can occur due to free volume effects resulting from non-zero volume change of mixing. Perhaps the Flory equation of state approach to polymer thermodynamics could rationalize this difference in solubility between the constituents in the block copolymer which is not predicted by Krause's[25] extension of the Flory-Huggins theory of polymer solution thermodynamics.

Note that in this study all block copolymer and poly (α-methyl styrene) samples were relatively monodisperse in that GPC data yielded $M_w/M_n < 1.1$. The polystyrene sample was prepared under free radical conditions and its measured M_w/M_n ratio was 2.5 (via GPC). However, since the M_w of polystyrene used in the BP-III blends was lower than the M_w of poly (α-methyl styrene), the higher solubility of poly (α-methyl styrene) than that of polystyrene in BP-III is therefore verified.

The apparent increase in solubility of the polystyrene/BP-III blend at a lower annealing temperature can be explained by recent work by McMaster[29] using Flory's equation of state thermodynamic approach. The theory shows that solubility will generally increase as temperature is lowered in polymer-polymer systems. This is due to free volume effects arising from differences in the pure component thermal expansion coefficients. This is, of course, in contrast to the conventional Flory-Huggins theory which predicts increasing solubility with increasing temperatures.

The experimental data with polystyrene and BP-III corresponds to the observation by Inoue(24) in studies relating to AB polystyrene/polyisoprene block copolymers.

The blends of the block copolymer with poly (2, 6 dimethyl 1, 4 phenylene oxide) yielded single Tg's as would be expected based on previous results with polystyrene and data presented here for the blends with poly (α-methyl styrene).

Ternary blends of the homopolymers and the AB block copolymers illustrate clearly the higher solubility of poly (α-methyl styrene) in the block copolymer. This is not only illustrated with BP-III but also a lower molecular weight block copolymer BP-IIA which exhibited solubility with polystyrene. With BP-IIA, the solubilizing effect is quite limited as only the 75/12.5/12.5 (by wt.) BP-IIA/polystyrene/poly (α-methyl styrene) system exhibited a single phase while the corresponding 50/25/25 (by wt.) blend exhibited phase separation.

In conclusion, contrary to the results with polystyrene, poly (α-methyl styrene) exhibits unique behavior in blends with its block polymer. It would appear reasonable to suspect that poly (α-methyl styrene) has a lower solubility in polystyrene than polystyrene in poly (α-methyl styrene).

REFERENCES

1. A. Noshay, M. Matzner and C. N. Merriam, U.S. Patent 3,539,657 (to Union Carbide Corp.), Nov. 10, 1970.

2. A. Noshay, M. Matzner and C. N. Merriam, J. Polymer Sci., Part A-1, 9, 3147 (1971).

3. M. Matzner and A. Noshay, U.S. Patent 3,579,607 (to Union Carbide Corp.), May 18, 1971.

4. A Noshay, M. Matzner, G. Karoly and G. B. Stampa, Polymer Preprints, 13, 292 (1972).

5. M. Matzner, A. Noshay, L. M. Robeson, C. N. Merriam, R. Barclay and J. E. McGrath, NASA Conf. on Polymers for Unusual Service Conditions, Nov. 30, 1972, Appl. Polymer Symposia, in press 1973.

6. L. M. Robeson, A. Noshay, M. Matzner and C. N. Merriam, Die Angewante Makromoleculare Chemie, 29/30, 47 (1973).

7. M. Matzner, A. Noshay and J. E. McGrath, Polymer Preprints, 14, No. 1, 68 (1973).

8. M. Matzner and J. E. McGrath, U.S. Patent 3,659,385 (to Union Carbide Corp.), April 18, 1972.

9. L. J. Fetters and M. Morton, Macromolecules, 2, 453 (1969).

10. M. Morton, R. F. Kammereck and L. J. Fetters, Br. Polymer J., 3, 120 (1971).

11. M. Morton, R. F. Kammereck and L. J. Fetters, Macromolecules, 4, 11 (1971).

12. L. J. Fetters, B. Meyer and D. McIntyre, J. Appl. Polymer Sci., 16, 2079 (1972).

13. E. Compos-Lopez, D. McIntyre and L. J. Fetters, Macromolecules, 6, 415 (1973).

14. M. Morton, L. J. Fetters, F. C. Schwab, C. R. Strauss and R. F. Kammereck, "4th Synthetic Rubber Symposium", Rubber and Technical Press, London, 1969, p. 70.

15. L. J. Fetters, Chap. 5 in Block and Graft Copolymerization, edited by R. J. Ceresa, John Wiley and Sons, London, (1973).

16. L. J. Fetters, J. Elastoplastics, 4, 32 (1972).

17. M. Morton, J. E. McGrath and P. C. Juliano, J. Polymer Sci., Part C, No. 2699 (1969).

18. L. J. Fetters, J. Polymer Sci., Part C, No. 26, 1 (1969).

19. M. Morton, Encyclo. Polymer Tech., 15, 506 (1971).

20. J. E. McGrath, M. Matzner, L. M. Robeson, Polymer Preprints, 14, No. 2, 1032 (1973).

21. D. J. Meier, J. Polymer Sci., Part C, No. 26, 81 (1967).

22. S. Krause, "Colloidal and Morphological Behavior of Block and Graft Copolymers", edited by G. E. Molau, Plenum Press, New York (1971).

23. M. Baer, J. Polymer Sci., Part A, 2, 417 (1964).

24. T. Inoue, T. Soen, T. Hashimoto and H. Kawai, Macromolecules, 3, 87 (1970).

25. S. Krause, 19th Sagamore Conference on Block and Graft Copolymers, Sept. 5-8, 1972; to be published by Syracuse University Press.

26. S. Krause, Macromolecules, 3, 84 (1970).

27. L. J. Fetters, J. Research NBS, 70A, 421 (1966).

28. L. E. Nielsen, Rev. Sci. Instr., 22, 690 (1951).

29. L. P. McMaster, paper to be published in Macromolecules.

30. E. P. Cizek, U.S. Patent 3,383,435 (to General Electric Company), May 14, 1968.

31. D. J. Meier, 19th Sagamore Conference on Block and Graft Copolymers, Sept, 5-8, 1972; to be published by Syracuse University Press.

The mechanical loss, Q^{-1}, is defined as the ratio of the energy dissipated to the energy stored during a single cycle of oscillation. In the torsion pendulum free vibration output this translates to:

$$Q^{-1} = \frac{\ln \frac{A}{B}}{N\pi}$$

where A represents the magnitude of cycle i and B represents the magnitude of cycle i + N. See reference 28 for a more detailed description.

SECTION III. Introduction

The several papers in this section present a discussion of the synthesis and characterization of novel graft copolymers and polyblends. First, Kennedy and Smith discuss the synthesis, characterization and some properties of graft copolymers of polystyrene grafted onto ethylene-propylene rubber substrate. A novel cationic initiator/coinitiator system composed of triethylaluminum/ethyl chloride results in 85% graft efficiency which results in well defined materials with useful properties for many potential applications. The graft copolymers are well characterized for graft molecular weight, molecular weight distribution of the grafts, and the number of graft polystyrene chains per ethylene-propylene molecule. The graft copolymers possess appreciable tensile strength in the absence of chemical vulcanization. The stress-strain properties of the graft copolymers range from elastomeric to tough plastic depending on the graft polystyrene content. The mechanical properties of the ethylene-propylene-graft-polystyrene copolymers resemble the styrene-butadiene-styrene thermoplastic elastomer block copolymers. The significance and potential value of the ethylene-propylene-graft-polystyrene copolymers are in superior weather resistance and stability in sunlight.

Reed, Bair and Vadimsky discuss the origin of surface roughness resulting from processing commercially important ABS. Two kinds of surface roughness are observed in injection molded ABS parts. The first, haze, appears as a whitening of the surface and is attributed to a differential thermal contraction between the SAN glass phase and the butadiene rubber particles in ABS. Haze in molded parts can be reduced to an acceptable level by selecting molding conditions which provide adequate pressure within the mold cavity. The other surface blemish, pitting, appears as pinhole-size defects and is caused by the escape of volatiles from the plastic surface during molding. Pitting can be eliminated also by providing adequate cavity pressure during molding.

The next two papers discuss the very interesting and timely topic of interpenetrating polymer networks (IPN's). The term IPN refers to a unique type of polyblend in which the component crosslinked polymers interpenetrate one another, however, not completely. Synthesizing IPN's is one technique of combining incompatible polymers. The IPN's are held together by chemical crosslinks which result in the polymer chains of each component becoming entangled in the other. The morphology is governed by the crosslink density and the sequence of preparation. Dynamic mechanical and electron microscopy studies reveal the existence of two phases; the phase domain dimensions being on the order of hundreds of Angstroms. IPN systems could produce new materials from incompatible as well as compatible polymers. The paper by Donatelli, Thomas, and Sperling discusses the relationship of IPN's to graft copolymers, osmium tetraoxide staining behavior, and morphology of IPN's. The morphology of IPN's and graft copolymers are similar. K. Frisch, Klempner, H. Frisch, and Ghiradella report on the tensile properties and glass transition temperature of IPN's. The IPN's produced from polyurethanes with polyepoxides, polyesters and polyacrylates exhibit one Tg intermediate in temperature to the Tg's of the components. This result suggests at least partial phase mixing. These IPN's exhibit a maximum in tensile strength significantly higher than the tensile strength of either component.

The final paper in this section by Kaplan and Tschoegl presents a treatment of time-temperature superposition principle in two-phase polyblends. The time and temperature dependences of the mechanical properties of the poly(methylmethacrylate)/poly(vinylacetate) blend changes from dominance by the softer phase at low temperatures to dominance by the harder phase at higher temperatures. The complex dynamic mechanical response to time and temperature resulting from the two-phase nature of the polyblend results in lack of superposition.

James F. Kenney

THE SYNTHESIS, CHARACTERIZATION AND PHYSICAL PROPERTIES OF EPM-g-POLYSTYRENE THERMOPLASTIC ELASTOMERS

J. P. Kennedy and R. R. Smith

Institute of Polymer Science
The University of Akron
Akron, Ohio 44325

SUMMARY

The synthesis, characterization and some property evaluation of a new series of graft copolymers, the EPM-g-polystyrene system, was investigated. A novel cationic initiator/coinitiator system based on certain alkylaluminum compounds was employed for graft synthesis. A systematic examination of the effect of reaction parameters such as temperature, solvent polarity, and nature of the alkylaluminum coinitiator on grafting efficiency was carried out. Reaction conditions have been optimized to achieve grafting efficiencies greater than 85%. Evaluation of the tensile properties of solution cast films of the grafts indicates that these materials possess appreciable strength in the absence of chemical vulcanization. The stress-strain properties of the grafts ranged from elastomeric to tough plastic depending on the polystyrene content. Compared to poly(styrene-*b*-butadiene-*b*-styrene) copolymers, a higher weight percent polystyrene must be incorporated into the graft copolymers to achieve comparable strengths.

INTRODUCTION

The application of cationic polymerization techniques to graft copolymer synthesis has received little attention despite the potentially large number of monomers polymerizable by this process. Initial investigation[1-7] of this grafting technique using conventional Friedel-Crafts halides (e.g. $AlCl_3$, BF_3, $SnCl_4$, etc.) met with little success and discouraged further attempts. The main problems encountered were: (1) the inability to control initiation which led to large quantities of homopolymer along with graft copolymer and (2) the occurrence of side reactions such as gelation and backbone degradation which led to ill-defined product mixtures. For example, attempted grafting of styrene to poly(vinyl chloride) using either $AlCl_3$ or $TiCl_4$ resulted in appreciable degradation and discoloration of the PVC and very low yields of graft copolymer[4].

Recent fundamental research by Kennedy[8,9] on cationic polymerization initiation has provided a better understanding of the details of the initiation mechanism. As a result of these studies, a series of cationic coinitiators based on certain alkylaluminum compounds were found to be useful for the efficient synthesis of graft copolymers.

The present study concerns the grafting of styrene onto chlorinated ethylene-propylene rubber (Cl-EPM) using the alkylaluminum initiator systems. The main objectives were: (1) to demonstrate successful graft synthesis, (2) to study the effect of synthesis conditions on grafting efficiency, graft composition, graft properties, etc., (3) to characterize pure graft copolymer and (4) to examine the tensile properties of solution-cast films of the graft copolymers.

PREVIOUS RESEARCH ON CATIONIC GRAFTING

Haas and coworkers[1] used BF_3 or $SnCl_4$ for the alkylation of poly(p-methoxystyrene) with growing polystyrene chains. While graft copolymer was obtained in this particular system, no graft formed with vinyl butyl ether, vinyl ethyl ether, isobutylene, etc., perhaps due to the incompatibility of the polymeric species involved. Similarly, Overberger and Burns[2] prepared graft copolymers by polymerizing styrene with $SnCl_4$ in the presence of poly(2,6-dimethoxystyrene). In this study, the amount of grafting and grafting efficiency were found to increase with increasing temperatures. Jaacks and Kern[3] produced poly(vinyl acetate-g-trioxane) by utilizing the pendant ester linkage as transfer agents in the cationic polymerization of trioxane.

All these cationic "grafting onto" methods are inefficient for the preparation of graft copolymers because of the large amounts of homopolymer formed along with graft copolymer.

A number of workers have utilized the "grafting from" approach. Kockelberg and Smets[4] obtained low yields of graft copolymers by reacting chloromethylated polystyrene with $AlBr_3$ and subsequently with isobutylene.

Minoura and coworkers[5] investigated the polymerization of styrene by the chlorobutyl/$SnCl_4$ system and thus obtained a mixture of graft copolymer and homopolystyrene. These workers also investigated the effect of several reaction variables on the rate of polymerization, per cent grafting and grafting efficiency. The highest grafting efficiency achieved was only about 50% indicating that conventional Friedel-Crafts halides produce large amounts of homopolymer due to initiation in the presence of protogenic impurities. Of interest to the present study, these investigators noted that either decreasing the polymerization temperature or increasing the polar solvant level (i.e. 0 to 25% nitrobenzene) gave improved grafting efficiencies.

Plesch[6] treated poly(vinyl chloride) and poly(vinyl chloride-co-vinylidene chloride) with $AlCl_3$ or $TiCl_4$ to initiate the polymerization of styrene, indene, indole and trans-stilbene. Although few details were given, the presence of graft copolymer was indicated by solubility data. In addition, severe degradation and discoloration of the PVC was observed using $AlCl_3$. The extent of disturbing side reactions was somewhat reduced by using $TiCl_4$ but appreciable homopolymer formation and degradation was still evident. Solomon and coworkers[7] grafted N-vinylcarbazole from PVC using $AlCl_3$ at 20°C in nitrobenzene. The highest grafting efficiency reported was 44%. Although no reference was made to PVC degradation, this side reaction would be expected on the basis of Plesch's work.

Gaylord and Takahashi[10] claimed the modification of PVC by grafting to it 3-10% cis-1,4-polybutadiene branches with Et_2AlCl/cobalt stearate catalyst. The modified PVC exhibited improved thermal properties which was attributed to the presence of cis-1,4-polybutadiene acting as an antioxidant. Recent work by Kennedy et al.[11], however, indicates that the improved thermal properties of the grafted PVC were more likely due to the removal of labile chlorines from the PVC by Et_2AlCl and that cis-1,4-polybutadiene was unnecessary for improved thermal properties. Moreover, research in our laboratories has shown that a variety of alkylaluminum/polymer halide systems are uniquely suited for the efficient synthesis of various novel graft copolymers[12].

PREVIOUS ATTEMPTS AT EPM MODIFICATION AND GRAFTING BY ALKYLALUMINUM COMPOUNDS

Graft modification of EPM or EPDM has been attempted by a variety of techniques, i.e. free radical polymerization[13-17], anionic polymerization[18], peroxidation[19], irradiation[20] and comastication[21]. Unfortunately, none of these methods proved to be satisfactory for a variety of reasons (gelation, backbone degradation, copious homopolymer formation, etc.). As an extension of our fundamental mechanism studies with alkylaluminum/alkyl halide initiation systems, we have examined the possibility of grafting polystyrene from chlorinated EPM under mild conditions to obtain "clean" EPM-g-polystyrene by the use of alkylaluminum compounds.

Our approach is based on the recent finding that polymerization-initiation can be controlled by using certain dialkylaluminum halides or trialkylaluminum compounds in conjunction with suitable alkyl halides[8]. For example, the Et_2AlCl/t-BuCl system was found to be particularly effective in giving rise to high molecular weight polymer under mild conditions with either isobutylene or styrene:

$$Et_2AlCl + t\text{-}BuCl \rightleftharpoons t\text{-}Bu^{\oplus}/Et_2AlCl_2^{\ominus} \xrightarrow{M} \text{homopolymer}$$

Significantly, Et_2AlCl alone is unable to initiate the polymerization and initiation occurs only upon the addition of the t-BuCl. In this sense, the alkylaluminum is the coinitiator and its role is to assist in generating the true initiating entity the t-butyl carbocation.

It was anticipated and found[22] that replacement of t-BuCl with a suitable polymeric halide (PCl) in this system would lead to graft copolymer:

$$Et_2AlCl + PCl \rightleftharpoons P^{\oplus}/Et_2AlCl_2^{\ominus} \xrightarrow{M} \text{graft copolymer}$$

In this manner, graft initiation is controllable and the synthesis of essentially pure graft copolymers by cationic techniques should be possible.

In the present study, we examined the feasibility of grafting polystyrene from EPM by the scheme shown in Figure 1.

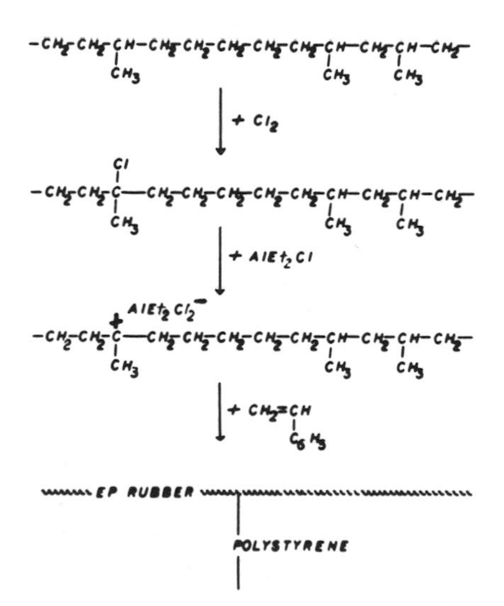

Figure 1. Reaction scheme used in synthesizing EPM-g-Polystyrene

EXPERIMENTAL

Materials and Procedures

Highest grade commercial reagents were used and were purified by conventional methods. The alkylaluminums (Texas Alkyl Co.) were freshly distilled prior to use. Et_2AlCl was stirred with NaCl for 2 hours at 80^0C prior to distillation. The ethylene-propylene copolymers were supplied by the Enjay Polymer Laboratories (experimental copolymer A, courtesy Dr. E. N. Kresge) and the B. F. Goodrich Company (Epcar 505). Both copolymers were purified by several reprecipitations.

All material handlings and polymerizations were performed in a stainless steel enclosure under nitrogen[23].

Chlorination of EPM

Photochlorination of EPM was carried out in benzene solution in an all glass reactor at $20-30^0$ using a 275 watt General Electric sun lamp. Reaction conditions for the chlorination experiments and characterization results from the EPM's and Cl-EPM's are summarized in Table I. The EPM copolymers were selected on the basis of their high solubility, relatively low molecular weight and low degree of unsaturation (i.e. both copolymers were reported to contain very little if any unsaturation). Only the chlorinated EPM's obtained from Enjay EPM were used for grafting.

Previous workers have shown that the relative amount of primary, secondary and tertiary chlorines obtained in the chlorination of EPM was dependent on temperature and the nature of the solvent[24,25]. Thus, a high ratio of labile, tertiary chlorines was obtained by chlorinating in aromatic solvents at low temperatures. Accordingly, we have chlorinated EPM samples in benzene at approximately ambient temperature (Table I). The chlorinated copolymers were purified (reprecipitation) and characterized with respect to total chlorine content,

TABLE I

PREPARATION AND CHARACTERIZATION OF Cl-EPM COPOLYMERS

Chlorination Conditions			Cl-EPM-1	Cl-EPM-2	Cl-EPM-3	Cl-EPM-4	Cl-EPM-5	Cl-EPM-6
EPM, Type/g.			Enjay/35	Enjay/225	Enjay/70	Enjay/225	Epcar/215	Epcar/185
Benzene, l.			0.9	9.0	2.5	9.0	9.0	9.0
Chlorine, g.			0.7	4.5	4.9	13.5	4.3	18.5
Cl_2 addition time, min.			20	50	60	60	50	110
Time UV light, min.			30	65	75	75	60	130
Polymer Characterization	Enjay	Epcar						
Propylene, mole %	51	53	-	-	-	-	-	-
Cl cont., wt. %	-	-	1.1	1.6	3.0	3.0	1.1	5.1
$M_n \times 10^{-3}$	60	57	66	60	64	74.5	-	-
$[\eta]$ Tol. $70^{\circ}C$	1.9	2.3	-	-	-	-	-	-
Toluene Sol. wt. % (R.T.)	98-100	98-100		98-100	98-100	-	-	88-90

number average molecular weight and solubility.

The chlorine content of the copolymer was found to be directly related to the amount of chlorine in the charge (Table I). In contrast, the number average molecular weights and solubilities of the Cl-EPM's did not vary substantially with the amount of chlorine charged except at the highest chlorine level (Cl-EPM-6, containing 5.1 wt. % chlorine). The decreased solubility of Cl-EPM-6 is due to crosslinking which is probably a result of the increased free radical concentration during chlorination with increasing amounts of chlorine.

Polymerization Techniques

Prior to grafting experiments, control experiments were performed to monitor the purity of the reagents. Styrene, solvent and alkylaluminum were combined in the same proportions used for grafting and agitated at the same temperature and time required for grafting reactions. After termination and precipitation, the absence of any polymer in the controls indicated the suitability of the reagents for grafting.

Styrene homopolymerizations were carried out by charging styrene, hexane, ethyl chloride and t-butyl chloride to a test tube, cooling to the desired temperature and adding the appropriate amount of alkylaluminum coinitiators. Polymerizations were usually rapid, the final conversion being attained within a few minutes. After 10 minutes, a few drops of methanol were added and the polymers isolated by addition to excess methanol.

Degradation studies were carried out on the Cl-EPM backbones by adding ethyl chloride to a hexane solution of the chlorinated copolymer to simulate reagent concentrations used in grafting. After temperature equilibration, the required amount of alkylaluminum coinitiator was added. The reactions were terminated with methanol and the polymers isolated by precipitation into excess methanol. After drying the number average molecular weights were determined.

Graft copolymerizations were carried out in a 500 ml., 3-neck flask equipped with low temperature thermometer, stirrer and stopper. The flask was charged with a hexane solution of the Cl-EPM, styrene and cosolvent (ethyl chloride or methyl chloride) at the desired temperature. Alkylaluminum coinitiator was added dropwise over 2 minutes. The styrene conversion could be approximated with experience by the turbidity of the medium. The reactions were terminated by introducing chilled methanol and the product immediately precipitated into a 10-fold excess of acetone. After refluxing about 2 hours in the acetone-solvent mixture, the precipitated polymer was filtered, dried and redissolved in benzene. Two precipitations into acetone were found to be sufficient to give graft copolymers of constant composition by infrared. The homopolystyrene was recovered by concentrating and drying the acetone-soluble fractions. The weight percent styrene in the graft was determined by infrared spectroscopy and the grafting efficiency calculated as the ratio of the grafted polystyrene over the total polystyrene formed.

Polymer Characterization

The selective precipitation technique used to isolate the graft copolymers is described in the Results and Discussion section.

The overall composition of the graft was determined by quantitative infrared spectroscopy using a Perkin-Elmer 521 Grating Spectrophotometer and films cast on NaCl plates from carbon tetrachloride solution. The relative concentrations of polystyrene and EPM were measured by characteristic bands at 1491 cm^{-1} and 1378 cm^{-1} respectively. The infrared spectra of polystyrene and EPM films are shown in Figure 2. Since polystyrene shows a slight absorption at 1378 cm^{-1}, overlap correction was effected by plotting the absorbance ratio 1378 cm^{-1}/1491 cm^{-1} versus the 1491 cm^{-1} absorbance for polystyrene films of various thickness.

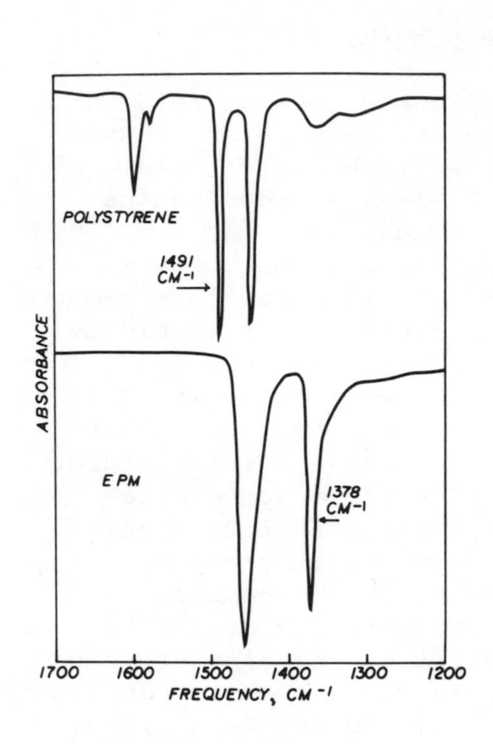

Figure 2. Infrared spectra of polystyrene and EPM.

Figure 3. Infrared calibration curve for EPM/polystyrene blends.

The correct absorbance for EPM was shown to be:

$$\underset{(\text{corr.})}{\text{Abs.}1378\ \text{cm}^{-1}} = \underset{(\text{obs.})}{\text{Abs.}1378\ \text{cm}^{-1}} - \frac{(\text{Abs.}1491\ \text{cm}^{-1})^2}{4.0}$$

The composition of graft copolymers was obtained directly from a calibration curve (Figure 3) constructed by using a series of polystyrene/EPM blends of known composition and the absorbance at 1378 cm^{-1} (corr.) and 1491 cm^{-1}. An infrared spectrum of a typical EPM-g-polystyrene copolymer is shown in Figure 4 with the appropriate baselines indicated for the determination of the 1378 cm^{-1} and 1491 cm^{-1} absorbances.

Polystyrene contents of several graft copolymers were also determined by NMR spectroscopy (Varian T-60) of polymer solutions (5-10%) in carbon tetrachloride. The relationship used was:

$$\text{wt. \% polystyrene} = \frac{104A}{104A + 70\left(\frac{B - 0.6A}{2}\right)}$$

where: A = Integrated area from aromatic protons

B = Integrated area from aliphatic protons

Number average molecular weights ($\overline{M}_n$) were determined using toluene solutions and a Mechrolab 503 high speed membrane osmometer at 37°C.

Gel permeation chromatograms were determined on a Waters Associates Ana-Prep Instrument using dilute polymer solutions (0.125%) in tetrahydrofuran at 37°C and a flow rate of 1 ml/minute. The chromatograph was equipped with five (pore size: $5x10^4$ - $1.5x10^4$ to $2x10^3$ - $5x10^3$Å) or seven (pore size: $7x10^5$ - $5x10^6$ to $2x10^3$ - 5×10^3Å) columns in series. Calibration of the GPC instrument was carried out using well-characterized polystyrene samples of known M_n and M_w. The

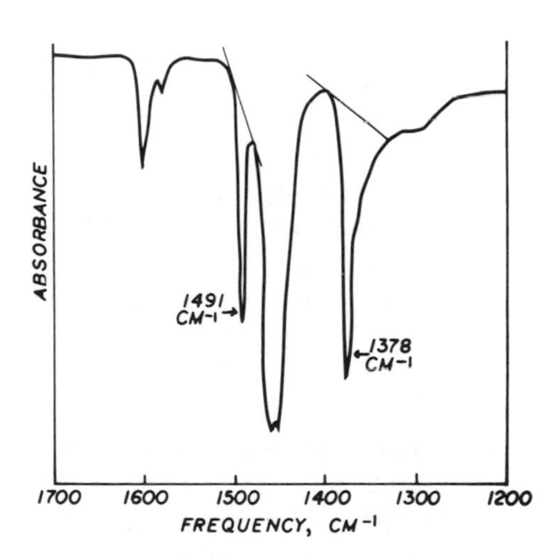

Figure 4. Infrared Spectrum of Typical EPM-g-polystyrene

EPM and EPM-g-polystyrene samples were run using five columns whereas the homopolystyrenes were run with seven columns. Column plugging was encountered during analysis of the graft copolymers so that only limited GPC studies were attempted with these materials.

Physical Property Measurements

Physical property measurements were carried out with solution-cast films. The polymers were dissolved in benzene (5% w/v) and centrifuged for 1-2 hours to remove any extraneous solids prior to casting the solutions on taut cellophane surfaces prepared by stretching wet cellophane sheets over an embroidery hoop. A glass cylinder (9.5 cm. dia.) was placed on the taut cellophane film and the polymer solutions cast inside the cylinder. Solvent evaporation rate was controlled by covering the cylinder. After 2-3 days, the polymer films were removed from the cellophane and dried in vacuum at ambient temperature for 2-3 days. A polymer sample of 2.5 g was found to give a film with a thickness of about 0.025 cm. No attempt was made to examine the effect of sample preparation (i.e. molded films vs. cast films) on physical properties.

Blends of homopolystyrene with graft copolymers were prepared by combining 5% benzene solutions of polymers and casting films.

Stress-strain data were obtained on micro-dumbbells stamped from the solution-cast films. Measurements were performed at ambient temperature using an Instron Tester at a crosshead speed of 5 cm./minute. Five specimens were tested per sample and the average value is reported.

The glass transition temperature, Tg, was determined by differential thermal analysis (DTA) using a DuPont 900 DTA instrument. Thermograms were obtained over a temperature range from -100^0 to $+150^0$C. The transition temperatures were obtained from the point of intersection of the tangent lines.

RESULTS AND DISCUSSION

Preliminary Experiments and Demonstration of Grafting

Preliminary experiments have been carried out prior to detailed grafting studies to demonstrate by chemical and physical evidence that grafting has been achieved.

Chemical evidence for grafting: styrene polymerization with macromolecular initiator. Styrene polymerizations were examined using various aluminum-based coinitiators under conditions representative of grafting (Table II). These experiments were first conducted

TABLE II

STYRENE POLYMERIZATION WITH VARIOUS ALUMINUM COMPOUNDS

Aluminum Coinitiator	Results
$AlCl_3$, $EtAlCl_2$, Me_2AlCl	Immediate polymerization upon coinitiator addition (no t-BuCl)
Et_2AlCl, Me_3Al, Et_3Al	No polymerization after 60 minutes upon coinitiator addition (no t-BuCl); immediate polymerization upon introduction of 1 drop of initiator (t-BuCl).

Conditions: styrene = 1.0 m/l, aluminum compound = $2x10^{-2}$ M , ethyl chloride = 50% vol., hexane = 38.5% vol., total volume = 35 ml., temperature - -30^0C.

by mixing styrene, solvent and aluminum compound in the absence of initiator (cationogen or protogen). Homopolystyrene could not be detected in the experiments using Et_2AlCl, Et_3Al or Me_3Al. However, upon addition of one drop of cationogen (e.g., t-BuCl), rapid polymerization ensued. In contrast, initiation did not require the purposeful addition of an initiator with $AlCl_3$, $EtAlCl_2$

and Me_2AlCl. It is possible that traces of impurities were sufficient to induce polymerization with these compounds. These experiments indicate that only the former group of coinitiators is suitable for efficient cationic graft synthesis and that with these materials, polymerization must commence at the polymeric backbone leading to graft copolymers.

Blend separation by selective precipitation. Separation of EPM/polystyrene and EPM-g-polystyrene/polystyrene blends has been examined to demonstrate the effectiveness of selective acetone precipitation for removal of homopolystyrene from the graft copolymers. These experiments involved: a) blend preparation, b) separation by acetone extraction and c) analysis of the acetone soluble and insoluble fractions by gravimetric and spectroscopic methods.

In the first series of experiments, solution blends of polystyrene and EPM were prepared using the reagents and concentrations given in Table III. The polystyrenes used were prepared by anionic polymerization and were characterized by GPC prior to blending. The blends were slowly precipitated at ambient temperature into a 10-fold excess of acetone followed by refluxing for 2 hours. The precipitated polymer was filtered and dried in vacuum at 50^0C. The acetone soluble fraction was recovered by evaporation and drying. Each fraction was examined by infrared and gravimetric analyses and the results given in Table III. The blends were prepared using either benzene or a mixture of hexane and 1,1-dichloroethane solvents. The latter solvent-mixture was used to simulate the solvent system used in grafting. Gravimetric analyses of both the soluble and insoluble fractions show essentially quantitative polymer recovery. Infrared analysis confirms the essentially quantitative separation of the two polymers. The distribution of the polymers between the soluble and the insoluble fractions varies slightly depending on the solvent used for blend preparation. A somewhat better separation was achieved using benzene (Expt's. A-1 and A-2) compared to the hexane/CH_3CHCl_2 mixture (Expt's. A-3 and A-4). This

TABLE III

SEPARATION OF EPM/POLYSTYRENE BLENDS

	Exp.A-1	Exp.A-2	Exp.A-3	Exp.A-4
Blend Preparation				
Polystyrene,g.	1.00	1.00	1.00	1.00
$M_n \times 10^{-3}$	33	100	33	100
EPM, g.	1.00	1.00	1.00	1.00
Solvent, l.	0.10[1]	0.10[1]	0.10[2]	0.10[2]
Separation Results				
Acetone, l.	1.0	1.0	0.5	0.5
(a) Acetone soluble fraction, g.	1.00	1.00	0.86	0.78[3]
(b) Acetone insoluble fraction, g.	1.03	1.05	1.13	1.10
EPM content of (b), % by I.R.	~100	~100	~95	~95

1. Benzene
2. Hexane/CH_3CHCl_2 mixture (3/1 v/v).
3. Slight polymer loss during work-up.

is probably due to the higher solvating ability for polystyrene of a benzene/acetone mixture (1/10 vol.) compared to that of a hexane/CH_2CHCl_2/acetone (3/1/20 vol.) system.

In the 33,000 to 100,000 $\bar{M}_n$ range, the number average molecular weight of the polystyrene does not appear to affect the separation. Tate and coworkers[26] have studied the problem of polystyrene extraction from poly(butadiene-g-styrene) and poly(isoprene-g-styrene) systems using acetone. They found that polystyrene of 50,000 M_n is entirely soluble in acetone. In the present study, polystyrene of 100,000 M_n gave a turbid solution at room temperature but a clear solution at 40°C in acetone/benzene (10/1) mixtures.

In the second series of experiments, graft copolymers, previously purified to constant composition by acetone precipitation were blended in benzene solution with anionically prepared polystyrenes.

Separations were effected by precipitation into acetone as described above. All fractions were analyzed gravimetrically and by infrared spectroscopy and the acetone soluble fractions by GPC. The results, given in Table IV and Figure 5, indicate the applicability

TABLE IV

SEPARATION OF POLYSTYRENE/EPM-g-POLYSTYRENE BLENDS

	Exp.A-5	Exp.A-6	Exp.A-7	Exp.A-8
Blend Preparation				
Polystyrene,g.	0.50	1.00	0.50	1.00
$M_n \times 10^{-3}$	33	33	100	100
EPM-g-polystyrene,g.	0.50	1.00	0.50	1.00
% wt. styrene	27	48	27	48
Benzene, l.	0.05	0.10	0.05	0.10
Separation Results				
Acetone, l.	0.5	1.0	0.5	1.0
Acetone soluble fraction, g.	0.54	1.17	0.52	1.15
% wt. styrene	>90	>90	~100	>95
Acetone insoluble fraction, g.	0.48	0.83	0.51	1.00
% wt. styrene	27	44	27	48

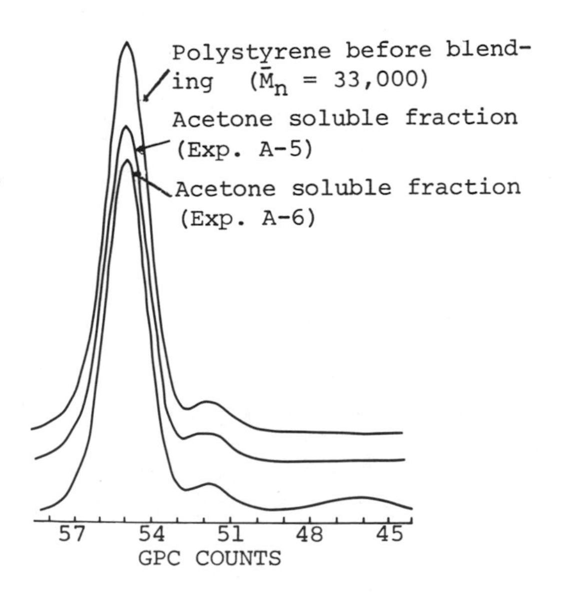

Figure 5. GPC curves of polystyrene before blending and acetone soluble fractions of experiments A-5 and A-6.

of acetone precipitation for graft isolation in our system. In all four experiments, the soluble fractions contained a quantity of polymer as great as or slightly greater than the amount of homopolystyrene charged to the blends. Infrared analysis of these fractions indicated essentially pure polystyrene. Likewise, the quantity and composition of the insoluble fractions were in good agreement with the amount and composition of graft copolymer charged to the blend. Only in experiment A-6 did the data appear to have a slightly larger deviation than could be accounted for by experimental error. This deviation will be discussed later in this section.

To provide further evidence for the separations attained in these experiments, gel permeation chromatograms of the soluble homopolystyrene fractions were compared with chromatograms of the original homopolystyrene charged to the blend (Figure 5). Both polystyrenes used in blend preparation had a bimodal molecular weight distribution indicating the presence of a small percentage (<10%) of a species having approximately twice the molecular weight of the predominant species.

Chromatograms of the acetone soluble fractions and those of the homopolystyrenes used to prepare the blends are essentially identical except in the case of experiment A-6 which shows a small, diffuse, new peak corresponding to a relatively high molecular weight species. Since the graft copolymers used in this study had been purified by 4 reprecipitations from benzene solution into acetone, it is unlikely that they contained any precipitable homopolystyrene impurities. Further studies to be described later indicate that homopolystyrene generated during grafting is of a substantially lower molecular weight than that of the species corresponding to the new peak in the chromatogram of experiment A-6. Thus, the new peak cannot be due to a high molecular weight polystyrene contaminant in the graft used for blend preparation. The new peak probably represents an EPM-*g*-polystyrene species of very high styrene content. The solubilization of this styrene-rich graft may have been enhanced by the blended homopolystyrene (M_n = 33,000) of approximately the same molecular weight as the grafted polystyrene branches. The deviation from ideal of the gravimetric and spectroscopic data of experiment A-6 (Table IV) is also explainable in terms of a small fraction of styrene-rich graft being soluble in acetone.

In summary, these blend separation studies demonstrate the feasibility of quantitatively separating EPM/polystyrene and EPM-g-polystyrene/polystyrene blends. This separation technique appears to be valid for polystyrenes of up to 100,000 M_n and blends containing as much as 50 wt. % polystyrene.

Degradation studies with chlorinated EPM. An important side reaction often encountered in cationic grafting is backbone degradation which renders graft characterization and correlation of graft properties with graft structure difficult. More importantly, degradation is unacceptable from a property standpoint.

Very little detailed work has been reported on the degradation of polymeric halides (backbones) under the influence of Friedel-Crafts halides. Recently, Kennedy and Phillips[27] showed that conventional coinitiators such as $AlCl_3$ and $EtAlCl_2$ severely and rapidly degrade chlorobutyl rubber, whereas breakdown is avoided with the less acidic Et_2AlCl and Et_3Al under mild conditions.

In this study, we examined the stability of Cl-EPM toward various alkylaluminum compounds under a variety of conditions chosen to simulate the grafting experiment. Results are given in Tables V and VI.

Initially, two Cl-EPM backbones with 1.1 and 3.0% chlorine and an EPM control were examined under the conditions shown in Table V. GPC and number average molecular weight data indicate that our EPM and Cl-EPM samples did not degrade in the presence of Et_2AlCl at -30^0C. The variation in number average molecular weights is considered to be within the 5-10% accuracy range of membrane osmometry.

TABLE V

DEGRADATION STUDIES WITH EPM AND Cl-EPM's

Prior to Attempted Degradation	Exp.B-1	Exp.B-2	Exp.B-3
Polymer Backbone, Cl (% wt.)	0	1.1	3.0
$\overline{M}_n \times 10^{-3}$	60	66	64
After Treatment with Et_2AlCl[a]			
$\overline{M}_n \times 10^{-3}$	67	64	71
Comparison of GPC Traces Before and After Et_2AlCl Treatment	no change	no change	no change

[a]Reaction Conditions: Et_2AlCl = 2 x 10^{-2} M, EPM or Cl-EPM = 1 wt. %, Hexane/MeCl = 16 vol. % MeCl, 1 hour at -30°C.

Additional experiments using a wider range of conditions have been carried out with a Cl-EPM containing 3.0% chlorine (Table VI). Molecular weight analyses indicate the absence of degradation under the conditions indicated in Table VI. Experiments with both Me_3Al and Et_3Al resulted in somewhat higher molecular weights as compared to that of the initial Cl-EPM. It appears as though the molecular weights increase with increasing alkylaluminum concentrations. Conceivably, this phenomenon might be due to the higher concentration of macromolecular carbenium ions formed from Cl-EPM in the presence of higher concentrations of R_3Al. The Cl-EPM very likely contains a small fraction of chains with terminal unsaturation resulting from the Ziegler-Natta polymerization utilized for the EPM preparation. Hence, the molecular weight increase observed in the degradation experiments with R_3Al is probably due to alkylation between the polymeric carbenium ions and the unsaturated chains. Alkylation in this case would lead to long chain branching.

Two facts should be emphasized. First, the highest concentration of R_3Al used in grafting experiments was 4 x 10^2 m/l. At this concentration, degradation experiments with Et_3Al showed only about a 10% increase

TABLE VI

DEGRADATION STUDIES WITH Cl-EPM-4[a]

Experiment No.	Alkylaluminum Type	Alkylaluminum moles/liter	EtCl/Hexane (V/V)	Time (min.)	Temp. (^oC)	$\bar{M}_n \times 10^{-3}$ After Al Cpd. Treatment
B-4	Me_3Al	$4x10^{-2}$	50/50	60	-30	104
B-5	Me_3Al	$4x10^{-2}$	50/50	60	-30	104
B-6	Me_3Al	$2x10^{-2}$	50/50	60	-30	98
B-7	Me_3Al	$2x10^{-2}$	50/50	15	-30	92
B-8	Me_3Al	$1x10^{-2}$	50/50	60	-30	75
B-9	Me_3Al	$2x10^{-2}$	50/50	60	-50	90
B-10	Me_3Al	$2x10^{-2}$	20/80	60	-30	112
B-11	Et_3Al	$2x10^{-2}$	50/50	60	-30	76
B-12	Et_3Al	$4x10^{-2}$	50/50	60	-30	81
B-13	Et_3Al	$8x10^{-2}$	50/50	60	-30	93
B-14	Et_3Al	$2x10^{-2}$	50/50	60	-50	77
B-15	Et_2AlCl	$2x10^{-2}$	20/80	5	-30	68
B-16	Et_2AlCl	$2x10^{-2}$	20/80	15	-30	68
B-17	Et_2AlCl	$2x10^{-2}$	20/80	60	-30	63
B-18	Et_2AlCl	$2x10^{-2}$	50/50	15	-30	73

[a]Initial $\bar{M}_n$ of Cl-EPM = 74,500 (Cl-EPM-4, Table I).
Cl-EPM concentration = 1 wt. %

in $\bar{M}_n$ (exp. B-12), whereas with Me_3Al under similar conditions (exp. B-4), a 40% increase was noted. Thus, under grafting conditions, alkylation or long chain branching might be more important with Me_3Al than with Et_3Al. Secondly, grafting is carried out in the presence of monomer (i.e. styrene), whereas the degradation experiments are performed in the absence of monomer. Therefore,in graft synthesis, this kind of alkylation would be kinetically and thermodynamically less favorable than polymerization (i.e. grafting) and indeed long chain alkylation might be entirely suppressed under grafting conditions.

In summary, experiments designed to test the stability of Cl-EPM in the presence of various alkylaluminum compounds used for grafting gave no indication for backbone degradation. A small increase in molecular weight observed in the presence of R_3Al has been attributed to long chain branch formation due to the presence of macromolecular carbenium ions and terminally unsaturated polymer.

Molecular weight versus composition of the EPM-g-polystyrene system. The number average molecular weight of EPM-g-polystyrene should be directly related to the polystyrene content of the graft. To examine this, a series of grafts was prepared with Et_2AlCl, Et_3Al and Me_3Al and their M_n's determined. Experimentally, grafting was initiated by introducing the above coinitiators to quiescent styrene-Cl-EPM systems and removing samples at various intervals to obtain copolymers with increasing polystyrene contents. The grafts were analyzed for composition (IR) and M_n(Figure 6).

The theoretical curves in Figure 6 represent the calculated M_n - composition relationships for the two Cl-EPM backbones used. The experimental points are in reasonably good agreement with the predicted values. This agreement between the predicted and observed M_n values of the graft copolymers is regarded as conclusive evidence for grafting. In a few instances, the observed M_n's are lower than the calculated M_n's by an amount greater than can be attributed to experimental error.

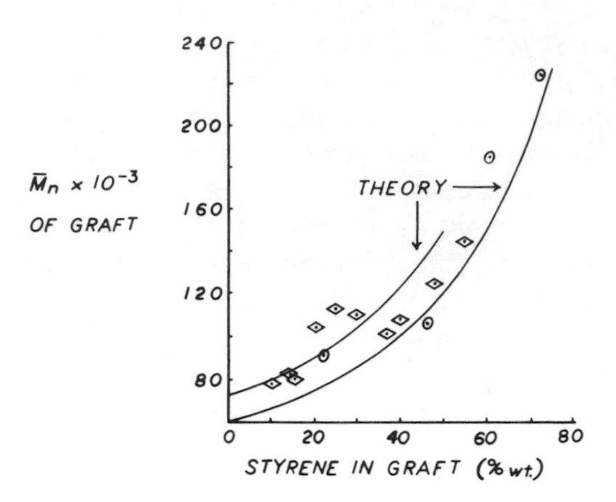

Figure 6. $\bar{M}_n$ of EPM-g-polystyrene vs. polystyrene content of the graft.

These discrepancies are usually noted for those grafts with relatively high polystyrene content (37-55%) and may be due to the removal of high molecular weight grafted species of high polystyrene content by the acetone extraction procedure.

The various experiments described in this section provide conclusive quantitative chemical and physical evidence that grafting is indeed the prevalent reaction in the Cl-EPM/styrene/alkylaluminum system under investigation.

The Effect of Synthesis Variables on the Grafting Reaction

Influence of Reaction Time. The effect of reaction time on the grafting reaction was studied by two experiments. Data are shown in Figures 7 and 8. In the first experiment, reactions were conducted in separate flasks and terminated at various times. Styrene conversion, grafting efficiency, M_n and styrene content increase with time. Unfortunately, the reaction times were too short to obtain high conversions and thus to ascertain the effect of both time and high styrene conversion on grafting efficiency.

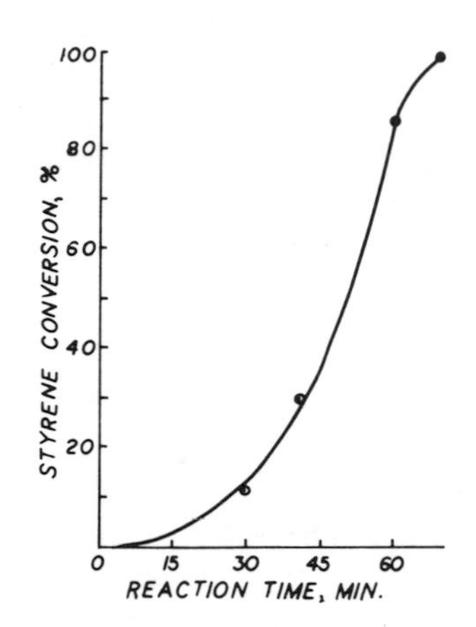

Figure 7. Conversion-time plot for a graft reaction.

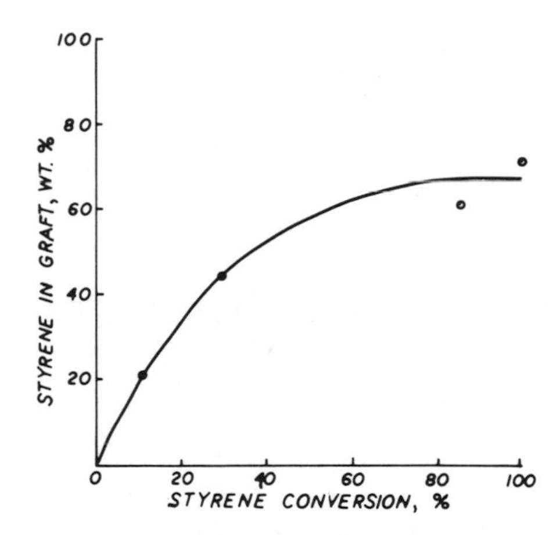

Figure 8. Styrene incorporation-conversion plot for graft reaction shown in Figure 7.

In a second experiment, samples were withdrawn from a single reaction flask at selected times followed by immediate quenching with methanol. In this experiment, nearly 100% styrene conversion was attained in about 70 min. (Figure 7). Grafting efficiency was 45 to 46% in the 11 to 29% conversion range. However, at high conversions (85-100%), the grafting efficiency appears to be reduced (Figure 8). Hence, to obtain optimum grafting efficiency, it is desirable to terminate grafting prior to about 30% conversion.

Evidently, the main mode of chain breaking is by transfer to monomer and new initiating sites are being created along the polymer backbone with time until all the potential grafting sites are utilized. The relative rates of graft initiation, chain propagation and chain transfer are relatively constant in the 11-29% styrene conversion range. Beyond about 30% conversion, the rate of graft initiation increasingly declines because the grafting sites are gradually used up. Thus, transfer becomes relatively more important and efficiency decreases. The apparent increase in grafting efficiency from 29 to 39% with increasing styrene conversion (85 to 95%) may be due to alkylation and could be a result of branched branch formation.

On the basis of these preliminary investigations, further synthesis studies were preferentially carried out in the 10-30% conversion range.

<u>Influence of styrene concentration on grafting.</u> The effect of styrene concentration on grafting was examined. Results are summarized in Table VII.

A critical styrene concentration appears to be necessary for grafting. Thus, in Series A, 0.5 and 1.0 molar styrene gave grafts containing 44 and 47% polystyrene and increased the grafting efficiency from 41 to 54%. However, grafting was absent at 0.25 moles styrene/liter. Grafting efficiencies in these two experiments may not be strictly comparable since styrene conversion was 61% in expt. I-2 and only 22% in expt. I-3.

TABLE VII

EFFECT OF STYRENE CONCENTRATION ON GRAFTING AT -50°C USING Et_2AlCl

Experiment No.	Styrene (m/1)	Styrene Conversion Time (min.)	(%)	(g/min.)	Grafting Efficiency (%)	Styrene in Graft (% wt.)
Series A[a] [13.3 g/1 Cl-EPM (1.6% wt. Cl), 25% vol. MeCl]						
I-1	0.25	45	0	0	-	-
I-2	0.50	45	61	0.21	41	44
I-3	1.0	45	22	0.16	54	47
Series B[a] [13.3 g/1 Cl-EPM (3.0% wt. Cl), 16% vol. MeCl]						
I-4	0.50	90	17	0.03	36	16
I-5	1.0	45	11	0.07	38	21
I-6	1.5	20	9	0.23	47	31
I-7	2.5	10	3	0.21	36	13

[a] $Et_2AlCl = 2 \times 10^{-2}$M, cosolvent = hexane, total volume = 300 ml.

Increasing the styrene concentration from 0.5 to 1.5 m/l in Series B gave increased grafting efficiencies and styrene conversions. Increasing the styrene concentration beyond 1.5 m/l gave no further rate increase and indeed grafting efficiency appeared to decrease after this point (expt. I-7). However, the trend toward higher grafting efficiency with increased styrene concentration may be related to conversions since this parameter decreases with increasing styrene concentration. The effect of styrene concentration on grafting should be studied at equivalent styrene conversions.

Influence of chlorine content of Cl-EPM on grafting. The effect of changing the chlorine content from 1.0 to 3.0% in the Cl-EPM backbone on grafting efficiency was examined. In this range, the chlorine content appeared to have no effect on the grafting efficiency.

Influence of temperature on grafting. Graft polymerizations were carried out in the range from -5 to -80°C using n-hexane/methyl chloride or ethyl chloride mixed solvents. The results are shown in Table VIII.

In Series A, with 3% chlorine in the EPM and 16% methyl chloride, styrene conversion decreased in the -30 to -80°C range. This effect is likely due to a decrease in the rate of initiation with decreasing temperature. At -80°C, there appears to be essentially no initiation. These observations are in agreement with the data reported by Kennedy[8] for the polymerization of styrene using Et_2AlCl/t-Butyl chloride in solvents of low polarity. Grafting efficiency increases from -30 to -50°C which is probably due to increasing branch lengths, e.g., reduced transfer at decreasing temperatures.

Similar trends were observed in Series B in the range from -5 to -50°C using 20% ethyl chloride and a Cl-EPM backbone containing 1.6% chlorine.

A decrease in temperature increases the dielectric constant of the medium which also leads to increased

TABLE VIII

EFFECT OF TEMPERATURE ON GRAFTING USING Et_2AlCl

Experiment No.	Temp. °C	Solvent Type/vol.%	Time (min.)	Styrene Conv. (%)	Grafting Efficiency %	Styrene in Graft (% wt.)
Series A[a] [13.3 g/l Cl-EPM(3.0% wt. Cl)]						
II-1	-30	MeCl/16	7	15	29	30
II-2	-50	"	45	11	38	21
II-3	-80	"	90	<5	-	0
Series B[a] [13.3 g/l Cl-EPM(1.6% wt. Cl)]						
II-4	-5	EtCl/20	10	21	33	36
II-5	-20	"	30	12	43	28
II-6	-50	"	90	<2	-	0
II-7	-50	MeCl/20	60	<5	-	~5

[a] $Et_2AlCl = 2 \times 10^{-2}$ M, cosolvent = hexane, total volume = 300 ml.

grafting efficiency (see below).

Influence of the polarity of the reaction medium. The influence of the polarity of the medium on the grafting reaction was examined by adding varying amounts of methyl chloride to the hexane solution of Cl-EPM up to the incipient precipitation point of the EPM (~45% methyl chloride at -50^0C). The styrene concentration and the total volume of the system were kept constant by decreasing the hexane concentration to compensate for the increased amount of methyl chloride. Results are shown in Figure 9.

Minoura and coworkers[5] also noted a linear increase in grafting efficiency with increasing polar solvent concentrations for the chlorobutyl/$SnCl_4$/styrene/nitrobenzene + cyclohexane system. This increase in grafting efficiency with increasing polar solvent concentration is likely related to the generally observed increase in degree of polymerization (DP) with increased dielectric constant of the medium in cationic polymerization. This trend in DP was also observed for the styrene/Et_2AlCl/t-BuCl model system in the present study. Both the increase in DP and grafting efficiency with increased

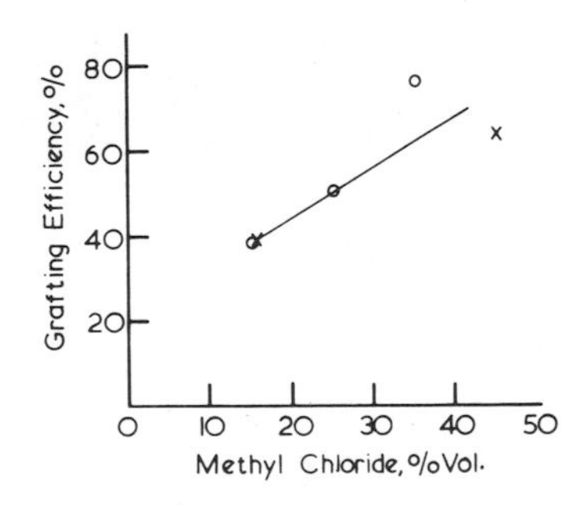

Figure 9. Grafting efficiency vs. methyl chloride concentration.

polarity are due to reduced rates of transfer relative to propagation. A probable explanation is that the counterion in the proximity of the growing carbenium ion impedes propagation to a greater extent than the competing transfer process[22]. Addition of polar solvent increases ion pair separation and leads to increased molecular weights. Several workers [12,28,29] have shown that free carbenium ion polymerizations (i.e., radiation-induced polymerization) give polymers of significantly higher molecular weight than those synthesized by conventional ion-pair propagation.

Grafting did not occur in the absence of methyl chloride. In contrast, chlorobutyl rubber was readily grafted in hydrocarbon solvent alone[5,30]. Perhaps the tertiary chlorine in Cl-EPM is a less active initiating site than the allylic chlorine in chlorobutyl rubber.

Influence of the nature of alkylaluminum coinitiator. Experimental conditions and results are summarized in Table IX. With Et_2AlCl, an increase in polar solvent concentration from 30 to 50% and a decrease in temperature from -10 to -50°C increases grafting efficiency. These trends are in agreement with the data previously discussed concerning the effects of temperature and solvent polarity on grafting efficiency. Me_3Al or Et_3Al gave substantially higher grafting efficiencies than Et_2AlCl. In view of previous publications on the polymerization of isobutylene and styrene with the RCl/Et_2AlCl and RCl/R'_3Al systems[8,9], these findings were unexpected. Highest grafting efficiency was obtained with Et_3Al at the lowest temperature.

To gain further insight into this interesting finding, several grafting experiments have been carried out in which the reaction was followed by withdrawing samples of the charge at suitable intervals. The samples were immediately quenched and analyzed as to styrene conversion, graft molecular weight and graft composition.

TABLE IX

EFFECT OF THE NATURE OF THE ALKYLALUMINUM COINITIATOR ON GRAFTING EFFICIENCY[a]

Expt. Nos.	Temp./EtCl °C/% vol.	Grafting Efficiency, (%)		
		Et_2AlCl	Me_3Al	Et_3Al
III-1 to III-3	-10/50	24	49	78
III-4 to III-6	-35/30	35	69	87
III-7 to III-9	-35/50	61	67	87
III-10 to III-12	-50/50	69	78	88

[a] 20 g/l Cl-EPM (3.0% wt. Cl), cosolvent = hexane, [styrene] = 1.5 M, styrene conversions 10 to 35%, $[Et_2Al] = 2 \times 10^{-3}$ M, $[Et_3Al] = 4 \times 10^{-2}$ M, $[Me_3Al] = 1 \times 10^{-2}$ to 4×10^{-2} M; total volume = 300 ml.

Figure 10 is a plot of styrene conversion versus time using 30% ethyl chloride at -35°C with the three coinitiators. The number average molecular weights and the polystyrene contents of the graft copolymers are also given. Significantly, with Et_3Al, the final styrene conversion level (~5%) is reached in 30 min. and remains <u>un</u>changed thereafter. The polystyrene content and M_n of the copolymer also remain essentially unchanged after the first 30 minutes. In contrast, with Et_2AlCl or Me_3Al, styrene conversion, graft M_n and polystyrene content increase with time. The relative rates of polymerization, i.e., $Et_2AlCl > Me_3Al > Et_3Al$, appear to follow the Lewis acidity of the coinitiator used.

Figure 11 shows styrene incorporation versus styrene conversion. Evidently, most efficient grafting is obtained with Et_3Al, followed by Me_3Al and Et_2AlCl. While Et_3Al induces only a limited amount of polystyrene incorporation (~25%) and the reaction stops

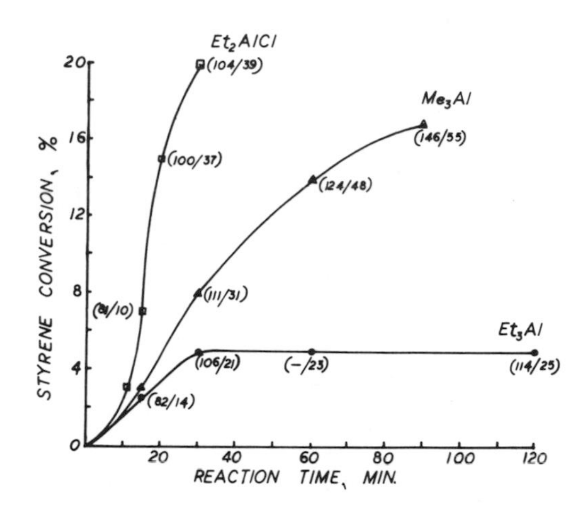

Figure 10. Conversion-time curves for graft reactions using various alkylaluminum coinitiators.

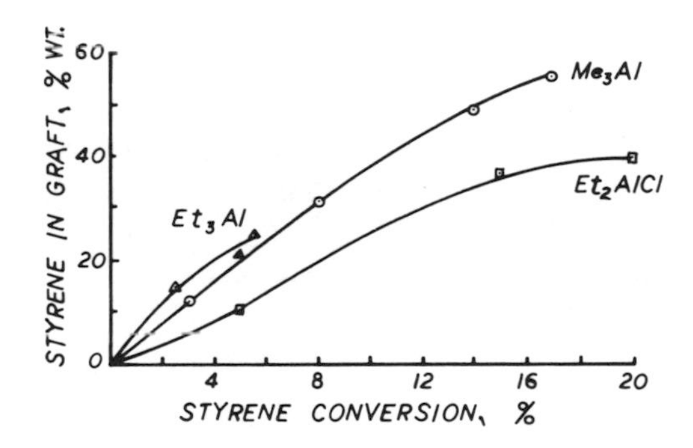

Figure 11. Styrene incorporation-conversion curves for graft reactions using various alkylaluminum coinitiators.

at low conversion (~5%), Et_2AlCl and Me_3Al produce 40-60% incorporation and the level increases with styrene conversion up to 16-20%.

The polystyrene content of the grafts prepared with Et_3Al can be controlled by a judicious choice of temperature and polar solvent concentration as shown in Figure 12. Thus, by increasing the ethyl chloride concentration from 30 to 50% and decreasing the reaction temperature from -35 to -50°C, the polystyrene content can be increased from 25 to 56%. However, regardless of conditions, the final polystyrene content is reached after the first 20-30 minutes.

Grafting by Et_3Al terminates at very low styrene conversions (<25%) while much higher conversions can be attained with Me_3Al and Et_2AlCl. This is difficult to explain on the basis of published information. The high grafting efficiencies (~90%) obtained with Et_3Al (Table IX) and polymerization cessation at low styrene conversions (Figure 10) strongly suggest rapid termination and the virtual absence of chain transfer. Evidence for fast termination by hydridation operating with the Et_3Al coinitiator even at lowest temperatures has recently been obtained in our laboratories[31].

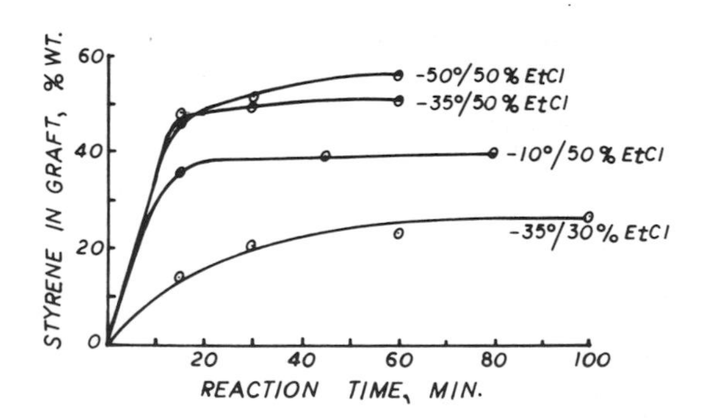

Figure 12. Styrene incorporation-time curves for graft reactions with Et_3Al at various reaction conditions.

The influence of the three alkylaluminum coinitiators on grafting appears to be best rationalized on the basis of relative rates of termination versus transfer. Highest grafting efficiency was obtained by using coinitiators with the greatest tendency toward termination, i.e., $Et_3Al > Me_3Al > Et_2AlCl$[31].

In sum, temperature, medium polarity and the nature of the alkylaluminum coinitiator strongly affect grafting efficiency. With Et_2AlCl, high grafting efficiency was obtained only at low temperature and/or high polar solvent concentration. Styrene conversion and polystyrene content of the graft copolymer increased with reaction time. Et_3Al gave high grafting efficiencies at every temperature and medium polarity, however, styrene conversion and the polystyrene content do not increase substantially with time. Grafting with Me_3Al resulted in efficiencies intermediate between those of Et_3Al and Et_2AlCl. The unique position of Et_3Al has been explained by postulating rapid termination by hydridation and virtually no transfer.

Characterization of EPM-g-Polystyrene Copolymers

Characterization of the EPM-g-polystyrene system involved: (a) separation and purification, (b) composition analysis and (c) molecular weight determination. Separation by selective extraction with acetone lead to two fractions, homopolystyrene and a fraction containing the Cl-EPM-g-polystyrene copolymer and presumably some ungrafted Cl-EPM backbone. Attempts to separate pure Cl-EPM from the latter fraction by selective extraction or selective precipitation were unsuccessful due to the similar solubilities of the Cl-EPM and graft copolymers and/or the relatively minor amount of pure Cl-EPM remaining after grafting. The fraction obtained after removal of homopolystyrene was considered to be EPM-g-polystyrene plus small amounts of ungrafted Cl-EPM contaminants and used for property studies.

Graft homogeneity. A typical graft reaction mixture was analyzed by GPC. The usual work-up procedure of the graft mixture (i.e. precipitation into excess

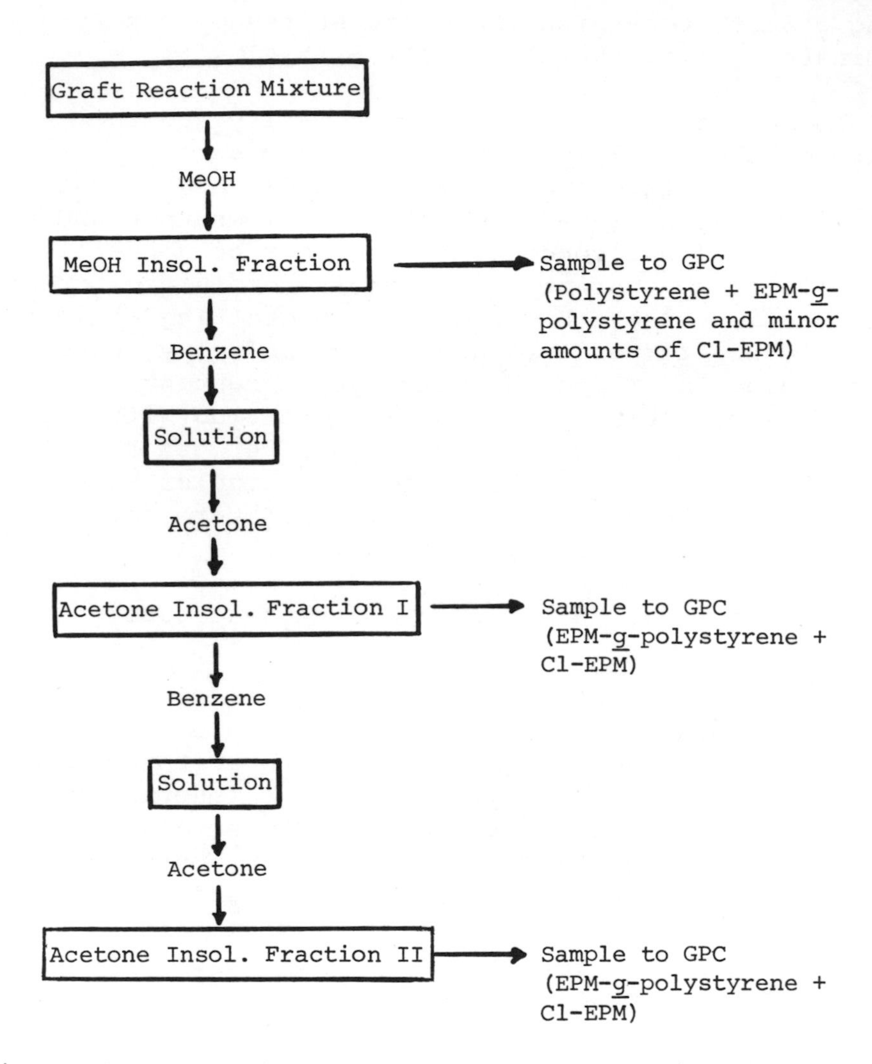

Figure 13. Fractionation scheme used to investigate the reliability of the acetone separation technique.

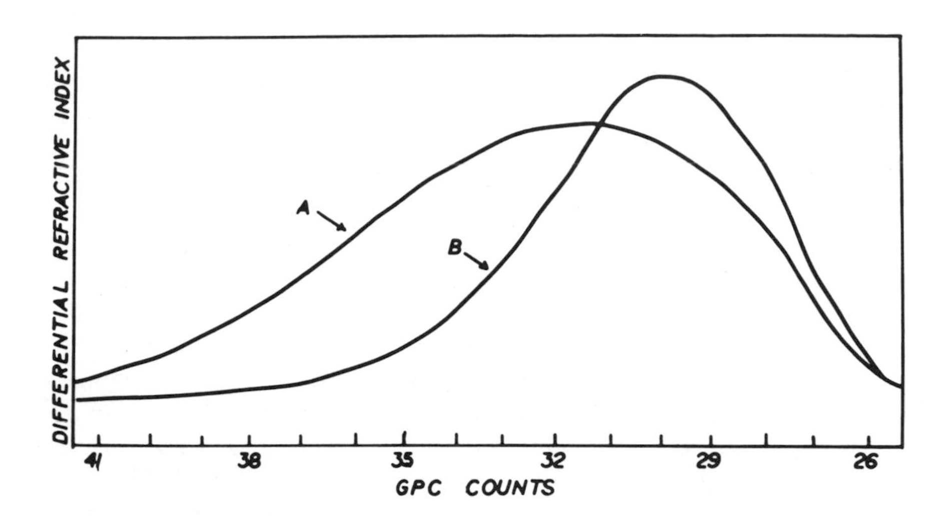

Figure 14. GPC curves of precipitated graft reaction mixture from exp. II-1. A. Methanol insoluble fraction, B. Acetone insoluble fractions.

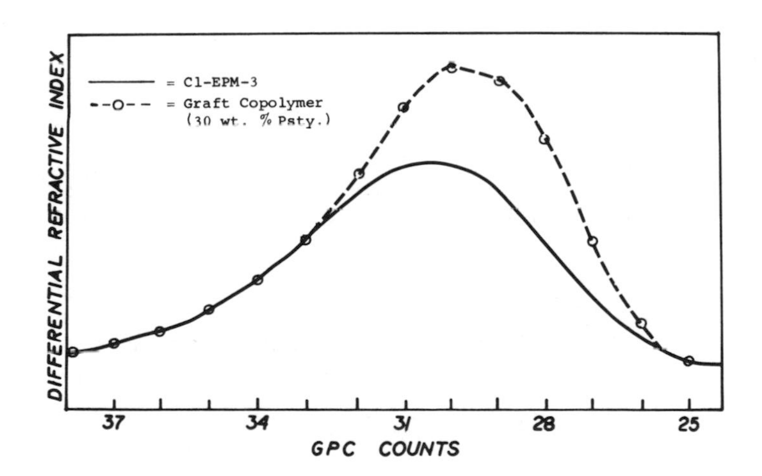

Figure 15. GPC curves of Cl-EPM-3 and purified graft copolymer II-1 after second acetone precipitation.

acetone) was modified in this experiment as shown in Figure 13.

The insoluble fractions were analyzed by GPC Figure 14. The GPC traces of both acetone insoluble fractions were superimposable so that only one curve is shown in Figure 14. Negligible quantities of the crude reaction mixture were soluble in methanol so that curve A (the methanol insoluble fraction) represents the crude mixture (i.e., homopolystyrene, EPM-g-polystyrene, and minor amounts of Cl-EPM). Nearly 30% wt. of the methanol insoluble fraction, however, was soluble in acetone. The GPC traces and compositions (by infrared analysis) of acetone insoluble fractions I and II were essentially identical. This study, along with the previously discussed blend separations indicate that acetone extraction is effective in removing homopolystyrene of molecular weight up to 100,000 from the crude mixtures.

In Figure 15, the GPC traces of the starting Cl-EPM backbone and the purified graft copolymer (II-1) are compared. The chromatogram of the graft is shifted slightly toward lower elution counts compared to the Cl-EPM as would be expected for the higher molecular weight grafted species. In addition, the greater refractive index response of the graft compared to Cl-EPM at equivalent sample concentrations is a qualitative indication of the presence of polystyrene in the graft since polystyrene has a higher refractive index than Cl-EPM.

Graft Composition. The polystyrene contents of the grafts have been determined by infrared spectroscopy (cf. Experimental). To verify the accuracy of our infrared analyses, the compositions of three graft copolymers have also been determined by NMR spectroscopy. The NMR and infrared data are compared in Table X.

TABLE X

COMPARISON OF OVERALL GRAFT COMPOSITIONS DETERMINED BY INFRARED AND NMR ANALYSES

Copolymer	Polystyrene Content of Graft, % wt. Infrared	NMR[a]
A	32	30
B	47	46
C	33	30

[a]Converted from mole % to weight %.

The agreement of the composition data from the two independent methods is within the limit of error of either method. Thus, the infrared method (i.e. calibration curve, Figure 3) is an accurate as well as rapid procedure for determining the polystyrene content of EPM-g-polystyrene copolymers.

Molecular weight and molecular weight distribution of graft branches. The approach used to estimate MW and MWD of graft branches was to characterize the homopolymer extracted from the crude grafting mixture. The adducted (grafted) branches were assumed to have the same MW and MWD as the homopolymer. This approach is permissible in cationic grafting since molecular weights are usually determined by chain transfer.

Characterization data for homopolystyrenes extracted from several graft reaction mixtures are given in Table XI.

TABLE XI

CHARACTERIZATION OF HOMOPOLYSTYRENE EXTRACTED FROM GRAFTING REACTIONS

	GRAFT REACTION CONDITIONS[a]			HOMOPOLYSTYRENE[b]		
Experiment No.	Alkylaluminum Initiator	Temperature (^{0}C)	EtCl (% v.)	$\bar{M}_n$ x 10^{-3}	$\bar{M}_w$ x 10^{-3}	$\bar{M}_w/M_n$
III-1	Et_2AlCl	-10	50	32	81	2.5
III-10	Et_2AlCl	-50	50	77	140	1.8
c	Et_2AlCl	-50	30	40	90	2.3
III-3	Et_3Al	-10	50	38	81	2.2
III-9	Et_3Al	-35	50	37	72	2.0
III-12	Et_3Al	-50	50	33	69	2.1

[a] Synthesis details in Table IX

[b] Molecular weights calculated from GPC chromatograms

[c] Prepared from Cl-EPM-5 (13.3 g/l), $[Et_2AlCl] = 2 \times 10^{-2}$ M.

To support the validity of the MW and MWD results, model experiments were carried out in which styrene was polymerized under simulated grafting conditions by replacing the macromolecular initiator (Cl-EPM) with t-butyl chloride. Polystyrenes produced with the t-BuCl/R_3Al system can be directly characterized and should be similar in MW and MWD to the grafted polystyrene branches obtained with Cl-EPM/R_3Al. The results are summarized in Table XII.

The good agreement in the molecular weight trends of the polystyrenes produced with the t-BuCl/RCl system and the polystyrenes extracted from grafting reactions indicate that the MW and MWD of the polystyrene branches are similar to those of the respective extracts. Also, the effect of temperature and EtCl concentration on branch length and grafting efficiency are similar with Et_2AlCl and Me_3Al (Tables IX and XII). This may indicate that improved grafting efficiencies are primarily due to increased branch lengths with Et_2AlCl and Me_3Al. Using Et_3Al, branch lengths and grafting efficiency are essentially independent of temperature and EtCl concentration.

Number of branches in the graft. To estimate the number of branches per Cl-EPM molecule, the residual chlorine content of several graft copolymers has been determined. The number of chlorine atoms per molecule of graft copolymer was calculated using the following equation:

$$\text{Chlorine atoms per polymer molecule} = \frac{\bar{M}_n \times Z}{100 \times 34.5}$$

where: $\bar{M}_n$ is the number average molecular weight of the graft

Z is the wt. % chlorine of the graft

The difference in chlorine contents of the graft and the Cl-EPM backbone should correspond to the number of branch sites. Data from these calculations are given in Table XIII. The backbone used (Cl-EPM-4) had 63±6 atoms of chlorine per molecule. The calculated chlorine contents of the grafts ranged from 38 to 92 atoms of

TABLE XII

CHARACTERIZATION OF POLYSTYRENE PREPARED WITH THE t-BuCl/ALKYLALUMINUM SYSTEM

	Polymerization Parameters			Polystyrene[a]		
Experiment No.	Alkylaluminum Initiator	Temp. (^{0}C)	EtCl (% v.)	$\overline{M}_n$ x 10^{-3}	$\overline{M}_w$ x 10^{-3}	$\overline{M}_w/\overline{M}_n$
1	Et_2AlCl	-30	15	20	40	2.0
2	Et_2AlCl	-30	45	92	179	1.9
3	Et_2AlCl	-30	90	151	294	1.9
4	Me_3Al	-30	25	47	116	2.5
5	Me_3Al	-30	50	79	200	2.5
6	Me_3Al	-30	80	188	343	1.8
7		0	50	27	72	2.7
8		-30	50	79	200	2.5
9		-50	50	80	181	2.3
10	Et_3Al	-30	30	30	65	2.2
11	Et_3Al	-30	80	46	67	1.5
12	Et_3Al	-50	30	41	99	2.4
13	Et_3Al	-50	50	43	115	2.5

[a] Molecular weights calculated from GPC chromatograms.

TABLE XIII

CHLORINE CONTENTS OF GRAFT COPOLYMERS

Experiment No.[a]	Graft Characterization			
	Polystyrene Content (% wt.)	Chlorine[b] Content (% wt.)	M_n x 10^{-3}	Atoms Chlorine / Mole EPM
Cl-EPM-4	0	3.00[c]	74.5	63±6
III-4	39	2.24	104	66
III-5	55	1.42	146	59
III-6	25	1.68	114	55
III-1	29	2.09	99	59
III-2	58	1.62	140	65
III-3[d]	38	1.38, 1.50	96	38,41
III-10	64	1.83	130	68
III-11[d]	57	1.31,1.54	210	79,92
III-12[d]	56	1.12,1.48	179	57,76
III-8	55	1.65	143	67
III-9	51	1.57	140	63

[a] Refer to Table IX for synthesis details.

[b] All chlorine analyses by Galbraith Laboratories.

[c] Same value obtained on two different samples.

[d] Two samples submitted for chlorine analysis.

chlorine per polymer molecule. Of the eleven graft copolymers examined, however, nine had 55 to 68 atoms of chlorine per polymer-molecule. Thus, considering the combined error limits ($\pm$ 5-10%) of the chlorine analyses and the molecular weight determinations, the Cl-EPM used for grafting contained 63$\pm$6 chlorine atoms per molecule. Hence, relatively few branches are formed on these copolymers and this method of estimating branch density is not sensitive to distinguish between small variations in branch density.

An alternate method of estimating the number of branches per molecule was to calculate the total $\bar{M}_n$ of polystyrene in the graft ($\bar{M}n_g$ - $\bar{M}n_{Cl-EPM}$) and divide this by the average $\bar{M}_n$ of homopolystyrenes extracted from the grafting experiment.

No. of Branches = $\frac{\bar{M}n_g - \bar{M}n_{Cl-EPM}}{\bar{M}n_{PS}}$, where $\bar{M}n_g$ = number average molecular weight of graft, $\bar{M}n_{Cl-EPM}$ = number average molecular weight of backbone prior to grafting, $\bar{M}n_{PS}$ = number average molecular weight of polystyrene extracted from crude reaction mixture. The assumption implied in this approach is that the $\bar{M}_n$ of the extracted homopolystyrene is identical to the average $\bar{M}_n$ of the branches on the graft. The results of these calculations are given in Table XIV.

The fact that only a small number of branches (1-3) per graft are calculated by this method agrees with the data obtained by chlorine analysis. Interestingly, the graft copolymers having more than one branch per molecule (III-9 and III-12) were prepared with Et_3Al. This fact will be of importance when considering copolymer properties.

Glass transition temperature. The glass transition temperatures (Tg) of two grafts with different polystyrene contents (31 and 47 wt. %) were determined by DTA. The Tg's corresponding to the EPM portions of the copolymers were independent of the polystyrene content and were identical to the Tg determined for

TABLE XIV

CALCULATION OF NUMBER OF GRAFT BRANCHES FROM ESTIMATED BRANCH LENGTHS

Experiment No.[a]	$\overline{M}_n$ of Graft Copolymer ($\times 10^{-3}$)	$\overline{M}_n$ of Grafted Polystyrene ($\times 10^{-3}$)[b]	M_n of Graft Branch ($\times 10^{-3}$)[c]	Number of Branches Per Molecule[d]
III-1	99	25	32	~1
III-3	96	24	38	~1
III-9	140	66	37	1.8
III-10	130	56	77	~1
III-12	179	105	33	3.2

[a] Refer to Table IX for synthesis details.

[b] $\overline{M}_n$ (Graft Copolymer) - $\overline{M}_n$ (Cl-EPM-4)

[c] From extracted homopolystyrene characterization (Table XI)

[d] $[\overline{M}_n$ (Graft Copolymer) - $\overline{M}_n$ (Cl-EPM-4)$]/\overline{M}_n$ (Graft Branch)

Cl-EPM (-61°C). A transition temperature corresponding to the glassy polystyrene branches of the grafts was not detectable by DTA.

In sum, our characterization results demonstrate the reliability of both acetone extraction for separating graft copolymer from the crude reaction mixture and the infrared method for determining polystyrene content of the grafts. According to model experiments, branch lengths of about 30,000 to 80,000 M_n are probably present when using Et_2AlCl or Me_3Al at various temperatures and EtCl concentrations whereas the branch lengths are probably ~35,000 M_n with Et_3Al under all reaction conditions examined. The number of branches per graft molecule was estimated to be very small, probably 1 to 3, with copolymers prepared using Et_3Al having the highest branch densities. The Tg associated with the EPM phase (~-61°C) of the graft copolymers did not vary with graft composition in the 31 to 47 wt. % polystyrene range.

Physical Properties of EPM-g-Polystyrene Copolymers

Effect of graft composition on physical properties. As shown in Table XV and Figure 16, the tensile strengths and moduli of solution cast EPM-g-polystyrene films increase with polystyrene content. In some cases, films of identical composition exhibited vastly different moduli and tensile strengths. The reason for this apparent discrepancy is believed to be related to the variation in the number of branches per molecule.

Copolymers giving opaque films exhibited poor tensile properties compared to copolymers of similar composition but giving clear films.

The appreciable tensile strengths of the films indicate the presence of networks similar to thermoplastic elastomers. Solution blends of EPM with 30 and 50 wt. % polystyrene resulted in incoherent films with negligible strength.

TABLE XV

PHYSICAL PROPERTIES OF EPM-g-POLYSTYRENE FILMS CAST FROM BENZENE SOLUTION

Experiment No.	Polystyrene Content (wt. %)	Modulus 300% (kg/cm^2)	Tensile Strength (kg/cm^2)	Elongation (%)	Film Clarity
III-6	28	10	40	1300	clear
III-1	29	-	10	250	opaque
I-6	31	15	59	1100	clear
III-4	33	4	4	700	turbid
II-4	36	33	93	1000	clear
III-3	38	36	103	1000	clear
I-2	44	72	93	600	turbid
I-3	47	56	65	400	opaque
III-9	51	101	237	650	clear
III-5	52	128	283	650	clear
III-8	53	89	95	400	turbid
III-12	56	126	288	650	clear
III-11	57	113	136	500	turbid
III-2	58	111	176	600	clear

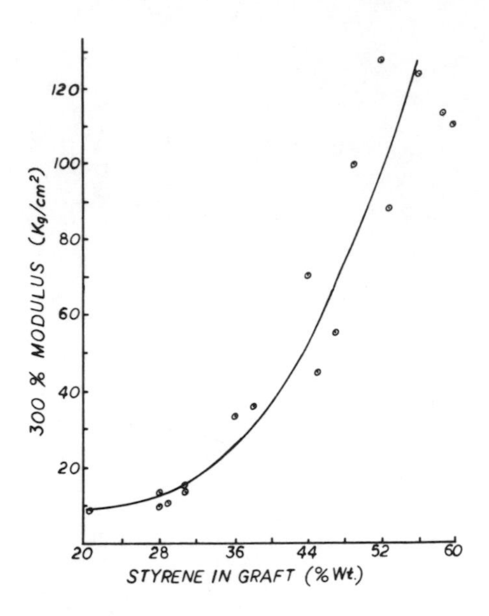

Figure 16. Effect of polystyrene content of the EPM-g-polystyrene system on the 300% modulus.

The effect of polystyrene content on the stress-strain properties of a series of grafts prepared with Et_3Al is shown in Figure 17. The stress-strain behavior changes from elastomeric to tough plastic with increasing polystyrene content.

The physical properties of SBS triblock copolymers were also found to be dependent on composition[32]. Moduli of the block polymers increased with polystyrene content in the 20 to 40% polystyrene range. Tensile strengths increased from about 100 to 250 kg/cm^2 by increasing the polystyrene content from 20 to 30%, and a further increase to 40% resulted in a slight decrease in tensile strength (220 kg/cm^2). In comparison, tensile strengths in the 100 to 300 kg/cm^2 range were attained with the graft copolymers only at significantly higher (i.e. 40-50%) polystyrene contents. The higher polystyrene levels required for high strength in the graft copolymers suggests that a portion of the grafted polystyrene is not useful in network formation. This would be the case if grafts containing only one branch per molecule were present. Such grafts would be analogous to diblock polymers which have a deleterious effect on the tensile strengths of the styrene-diene triblock

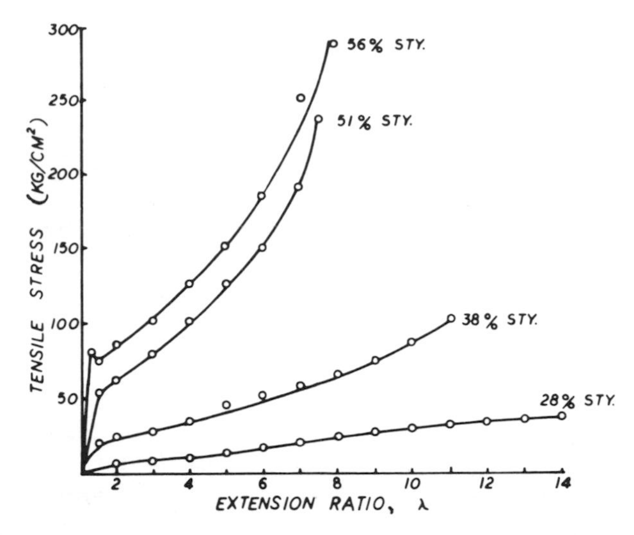

Figure 17. Stress-strain curves of EPM-g-polystyrene copolymers prepared with Et_3Al.

polymers. Single-branched structures are probably present in the grafts since the number of branches per chain is quite low. A second factor that could contribute to the high polystyrene content requirement of the graft copolymers is the broad molecular weight distribution of the polystyrene branches. The M_w/M_n of these branches was estimated to be at least 1.5 to 2.5 (Tables XI and XII). Conceivably, branches that are too short would not participate in domain formation. If a major portion of the branches of a graft were unable to participate in network formation, diminished tensile properties and macrophase separation would result.

Influence of synthesis variables on graft properties. It was previously noted that certain grafts of the same composition had vastly different tensile properties even though they had the same work-up history. An explanation is that copolymers prepared under different conditions may differ substantially in terms of branch length and branch density. The effect of branch length and/or branch density on graft properties can be inferred by examining the effect of synthesis variables on properties of copolymers similar overall composition.

In Table XVI, the effect of temperature and solvent polarity on graft properties is examined by comparing three series of experiments using either Me_3Al or Et_2AlCl coinitiator. In each series, the overall composition of the two copolymers is essentially the same. Decreasing the EtCl concentration (Series A) and increasing the temperature (Series B and C) produce an increase in tensile strengths. Previous studies with Me_3Al and Et_2AlCl have shown (Table XII) that increasing the temperature or decreasing the EtCl concentration decreases branch length. Two copolymers of similar polystyrene content but different branch lengths must have different branch densities. The data in Table XVI indicate that copolymers with highest tensile strengths are prepared with Et_2AlCl and Me_3Al by employing conditions favoring optimum short branch lengths. Optimum properties are likely obtained when the proper balance between branch length and branch density are achieved.

The effect of the nature of the alkylaluminum coinitiator on graft properties was investigated by comparing the properties of copolymers of similar overall composition.

The data in Table XVII indicate that tensile strengths of copolymers prepared with Et_3Al are substantially higher than those for copolymers prepared with Me_3Al or Et_2AlCl. Also, Me_3Al gives higher tensile strengths than Et_2AlCl (Series C).

As shown in Tables XI and XII, the molecular weights of polystyrenes prepared with Et_3Al are relatively low ($\sim$35,000 M_n) and do not vary substantially with reaction conditions. In contrast, the molecular weights of polystyrenes prepared with Me_3Al and Et_2AlCl did show a significant dependence on conditions (30,000 to 80,000 M_n). As a consequence, grafts prepared with Et_3Al would be expected to have shorter branches than those prepared with Me_3Al or Et_2AlCl. Evidently, branch length and density are very important in determining graft properties.

TABLE XVI

EFFECT OF REACTION TEMPERATURE AND POLARITY OF REACTION MEDIUM ON THE TENSILE STRENGTH OF GRAFTS

Experiment No.	Initiator Type	Temp. (°C)	EtCl (% vol.)	Grafted Styrene (% wt.)	Tensile Strength (kg/cm^2)
Series A					
III-8	Me_3Al	-35	50	53	95
III-5	Me_3Al	-35	30	52	283
Series B					
III-11	Me_3Al	-50	50	57	136
III-2	Me_3Al	-10	50	58	176
Series C					
-	Et_2AlCl	-50	25	31	13
II-4	Et_2AlCl	-5	20	36	93

TABLE XVII

EFFECT OF THE NATURE OF THE ALKYLALUMINUM COINITIATOR ON THE TENSILE STRENGTH OF GRAFTS

Experiment No.	Coinitiator	Grafted Styrene (% wt.)	Tensile Strength (kg/cm^2)
Series A			
III-8	Me_3Al	53	95
III-9	Et_3Al	51	237
Series B			
III-4	Et_2AlCl	33	4
III-6	Et_3Al	28	40
Series C			
III-10	Et_2AlCl	61	55
III-11	Me_3Al	57	136
III-12	Et_3Al	56	288

Finally, the effect of the chlorine content of the Cl-EPM backbone on the properties of the resulting copolymers was examined using Et_2AlCl coinitiator (Table XVIII). It appears that increasing the chlorine content of the Cl-EPM from 1.0 to 3.0% (Series A) or from 1.6 to 3.0% (Series B) gives substantially increased tensile strengths. Increasing the chlorine content of the Cl-EPM should result in more sites for branch initiation. Thus, the improvement in tensile properties with the apparent increase in the number of branches is consistent with the previously established trends.

The low tensile properties of many of our materials are probably due to the presence of grafts with only one branch on the backbone. The chlorine content of the Cl-EPM should be increased to obtain more branches per backbone and consequently better tensile strength.

Effect of added homopolystyrene on graft properties. The stress-strain properties of polystyrene/EPM-*g*-polystyrene blends have been examined to determine the

TABLE XVIII

EFFECT OF THE CHLORINE CONTENT OF Cl-EPM ON THE TENSILE STRENGTH OF GRAFTS

	Chlorine Content (% wt.)	Grafted Styrene (% wt.)	Tensile Strength (kg/cm^2)
Series A			
	1.0	28	14
	3.0	31	59
Series B			
	1.6	31	13
	3.0	29	32

effect of homopolystyrene on tensile properties of our graft system. Homopolystyrenes of two molecular weights, M_n = 40,000 (III-9 extract, Table IX) and M_n = 80,000 (Exp. 9, Table XII) were used. The polystyrene were combined with a graft containing 28 wt. % polystyrene (III-6) to obtain blends containing a total of 38 wt. % polystyrene. Cationically prepared polystyrene of 40,000 M_n was used for blending to achieve perfect mixing of this polymer with the graft whose polystyrene branches were also estimated to have an M_n of ~40,000. The polystyrene of 80,000 M_n was not expected to mix well with the graft branches.

The stress-strain curves of the two blends are shown in Figure 18. For comparison, stress-strain curves of pure graft copolymers containing 28 and 38 wt. % polystyrene are included. Both blends exhibited increased moduli but decreased tensile strengths and elongations compared to the pure graft copolymer containing 28% polystyrene. Furthermore, the blends had poorer tensile properties in all respects as compared to a pure graft containing 38% polystyrene. The blend prepared with the lower molecular weight homopolystyrene had only slightly higher tensile strength and ultimate elongation than that obtained from the higher molecular weight polystyrene. Thus, the modulus of the EPM-g-polystyrene system is somewhat increased by added

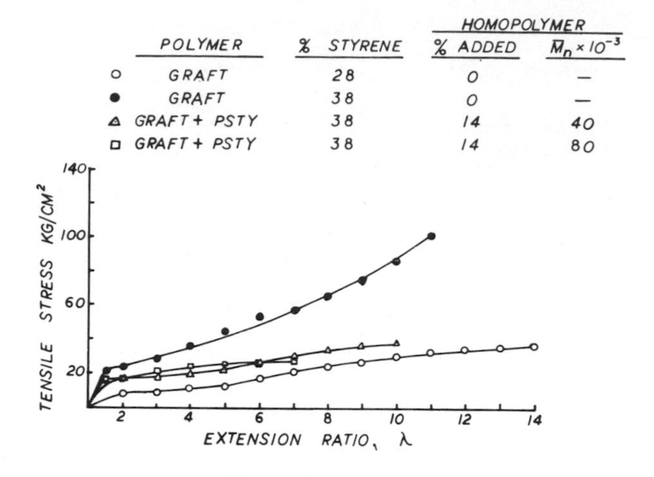

Figure 18. Effect of homopolymer on stress-strain curves of graft copolymer.

homopolystyrene but the tensile strength is diminished.

In conclusion, the appreciable tensile strengths exhibited by our graft copolymers indicate that these materials are able to form networks in the absence of chemical crosslinks. Copolymers prepared with Et_3Al have higher tensile strengths than copolymers of equivalent overall composition prepared with Et_2AlCl or Me_3Al. The low tensile properties of some of our copolymers was explained on the basis that these materials contained grafts with only one branch per molecule.

CONCLUSION

Chlorinated EPM copolymers containing 1.1 to 5.1 wt. % chlorine were prepared by photo-chlorination of EPM. Preliminary experiments of both a chemical and physical nature indicated that graft copolymers could indeed be prepared with the Cl-EPM/styrene/R_3Al system with the complete absence of backbone degradation. A systematic examination of the effect of reaction variables on grafting parameters such as grafting efficiency, graft composition and graft structure indicated that temperature, solvent polarity and nature of the alkylaluminum coinitiator were all important variables. Grafting efficiencies of nearly 90% were obtained with Et_3Al.

Membrane osmometry, GPC, NMR, IR and extraction techniques were utilized in determining graft homogeneity, graft composition and graft molecular weight. Preliminary experiments were also carried out to estimate the branch lengths and branch densities of the grafts. Branch lengths of about 30,000 to 80,000 M_n were estimated for grafting with Et_2AlCl and Me_3Al at various temperatures and EtCl concentrations. With Et_3Al, branch lengths of about 35,000 M_n were obtained irrespective of reaction conditions. The number of branches per molecule was estimated to be quite small, probably 1 to 3. Copolymers prepared with Et_3Al appeared to have the highest branch densities.

The tensile properties of solution cast films of the grafts indicated appreciable tensile strengths (288 kg/cm^2 maximum) in the absence of chemical crosslinking and hence thermoplastic behavior. The stress-strain behavior of the grafts ranged from elastomeric to tough plastic depending on their polystyrene content. Compared to the poly(styrene-b-butadiene-b-styrene) triblock copolymers, a higher weight percent of polystyrene must be incorporated into the graft copolymers to achieve comparable strengths.

REFERENCES

1. H. C. Haas, P. M. Kamath and N. W. Schuler, J. Polymer Sci., 24, 85 (1957).
2. C. G. Overberger and C. M. Burns, J. Polymer Sci., A-1, 7, 333 (1969).
3. V. Jaacks and W. Kern, Makromol. Chem., 83, 71 (1965).
4. G. Kockelberg and G. Smets, J. Polymer Sci., 33, 227 (1958).
5. Y. Minoura, T. Hanada, T. Kasaba and Y. Ueno, J. Polymer Sci., A-4, 4, 1665 (1966).
6. P. H. Plesch, Chem. Ind. (London), 954 (1958).
7. O. F. Solomon, M. Dimonie and C. Ciuciu, J. Polymer Sci., A-1, 8, 777 (1970).

8. J. P. Kennedy, J. Macromol. Sci. Chem., A3 (5), 861 (1969).
9. J. P. Kennedy, J. Polymer Sci., A-1, 6, 3139 (1968).
10. N. G. Gaylord and A. Takahashi, J. Polymer Sci., B8, 361 (1970).
11. N. G. Thame, R. D. Lundberg and J. P. Kennedy, J. Polymer Sci., A-1, 10, 2507 (1972).
12. J. P. Kennedy, XXIII International Congress of Pure and Applied Chemistry, Marcomol. Preprint, Vol. 1, 105 Boston, 1971.
13. G. Natta, F. Severini, M. Pegoraro and A. Crugnola, Chim. Ind. (Milan), 47, 1176 (1965).
14. G. Natta, M. Pegoraro, F. Severini and S. Dabhade, Chim. Ind. (Milan), 47, 384 (1965).
15. G. Natta, M. Pegoraro, F. Severini and G. Aurello, Chim. Ind. (Milan), 50, 18 (1968).
16. A. A. Buniyat-Zade and A. E. Portyanskii, Plast. Massy, 7, 6 (1967).
17. G. Natta, F. Severini, M. Pegoraro, E. Beati, G. Aurello and S. Toffano, Chim. Ind. (Milan), 47, 960 (1965).
18. Y. Ueno, T. Kasabo, H. Hironaka and Y. Minoura, Nippon Gomu Kyokaishi, 41, 10 (1968).
19. A. G. Sirota, B. G. Fedotov, E. P. Ryabokov, P. A. Il'chenko, A. L. Gol'denberg, L. I. Zyuzena and E. E. Manusevich, Plast. Massy. 7, 10 (1967).
20. J. Pellon and K. I. Valan, J. Appl. Polym. Sci., 9, 2955 (1965).
21. P. Hamed and E. T. McDonel, U.S. Patent 3,622,652, Nov. 23, 1971.
22. J. P. Kennedy, A. Shinkawa and F. Williams, J. Polymer Sci., A-1, 9, 1551 (1971).
23. J. P. Kennedy and R. M. Thomas, Adv. in Chem. Ser., 34, 111 (1962).
24. H. S. Makowski, W. P. Cain, and P. E. Wei, Ind. Eng. Chem., Prod. Res. Develop., 3, 282 (1964).
25. F. L. Ramp, J. Polymer Sci., C24, 247 (1968).
26. D. P. Tate, A. F. Halasa, F. J. Webb, R. W. Koch and A. E. Oberster, J. Polymer Sci., A-1, 9, 139 (1971).
27. J. P. Kennedy and R. R. Phillips, J. Macromol. Sci., Chem. A4, 1759 (1970).

28. D. J. Metz and D. S. Ballantine, Ann. N. Y. Acad. Sci., 155, (2), 468 (1969).
29. K. Hayashi, K. Hayashi, and S. Okamura, Nippon Genshiryoku Kenkyusho Nempo, JAERI 5026, 169 (1970); Chem. Abstr. 74, 126147Z (1971).
30. J. P. Kennedy, British Patent, 1,174,323, July 5, 1967.
31. J. P. Kennedy and S. Rengachary, Adv. Polymer Sci., in press (1974).
32. P. C. Juliano, Ph.D. Thesis, University of Akron (1968).

Acknowledgement - Financial support received in the form of NSF grant GH-37985 is gratefully acknowledged.

THE CAUSES OF PITTING AND HAZE ON MOLDED ABS PLASTIC SURFACES

T. F. Reed,* H. E. Bair, and R. G. Vadimsky

Bell Telephone Laboratories, Incorporated
600 Mountain Avenue, Murray Hill, New Jersey 07974
*General Tire and Rubber Co.
2990 Gilchrist Rd., Akron, Ohio 44329

INTRODUCTION

A widely used plastic resin made from acrylonitrile, butadiene, and styrene (ABS) has been characterized as a two-phase system.[1] The polybutadiene (BD) constitutes a rubbery phase and exists as particles in a glassy matrix of poly(styrene-co-acrylonitrile), i.e. SAN. In addition, SAN is usually grafted to the BD balls to increase compatibility between the two phases.

The use of ABS for appearance parts requires that processing conditions be selected to give acceptable surfaces as well as adequate physical properties. One kind of common visual blemish called haze appears as a whitening of the plastic surface. It has been reported[2] that the degree of haze on injection molded parts varies inversely as injection pressure and mold temperature and directly with injection time; in short, selecting processing variables which produce low cavity pressure during the injection part of the molding cycle will also tend to produce a hazy molded surface.

During investigations preliminary to the present studies, we found that haze (i.e. diminuation of gloss) is caused by a microscopic surface roughness. This is shown in Figure 1 where photomicrographs (180X) of sample surfaces injection molded from a black ABS compound using 15 different molding variables are compared to the gloss measurements of the sample surfaces made with a

Hunter gloss meter (60° angle). The sequence of photomicrographs in Figure 1 was made by subjectively arranging them according to their degree of surface roughness without knowing the sample identification, and we see from the comparison with the gloss values that a good correlation exists.

The present studies are directed toward revealing the origin of haze and why it depends on the cavity pressure during molding. We also investigated a second adverse surface condition found on injection molded parts called pitting; this appears as discrete, pinhole-size blemishes on the molded surfaces. The two independent variables in this study were the BD content and the concentration of material that volatilizes during molding.

EXPERIMENTAL

Three BD contents were obtained by selecting a commercial black ABS, a clear commercial SAN to which carbon black was added, and a 50/50 blend of the two. The SAN was chosen as having approximately the same styrene/acrylonitrile ratio as the ABS. The concentrations of volatile materials in these three samples were controlled by processing them in a vacuum-vented, 0.8-inch, counter-rotating, twin-screw extruder. The barrel temperature in the vented section was maintained at 500°F, and the temperature of the extrudate was 430°F; the residence time was estimated to be three minutes. Most of the volatiles were collected in a dry ice cold trap located between the extruder vents and the vacuum pump. Volatiles from the two samples containing ABS were identified by infrared and calorimetric analyses as primarily styrene and water; at the interface between the two was an unidentified waxy substance with a melting range slightly above room temperature and presumed to be a fatty acid. The SAN volatile material was primarily styrene. During the drying of the ABS sample which was passed six times through the extruder, it was found that most of the volatile material was removed during the first pass; therefore, the SAN and blend were extruded only once.

Test plaques 2.25 by 2.25 by 0.1 inch were molded from the original ABS and these three dried samples using a 2-ounce capacity screw injection molding machine. The important variables are given in Table I. It has been found previously[2] that "Condition 1" produces glossy test plaques using the black ABS compound, whereas "Condition 2" causes haze to form during molding.

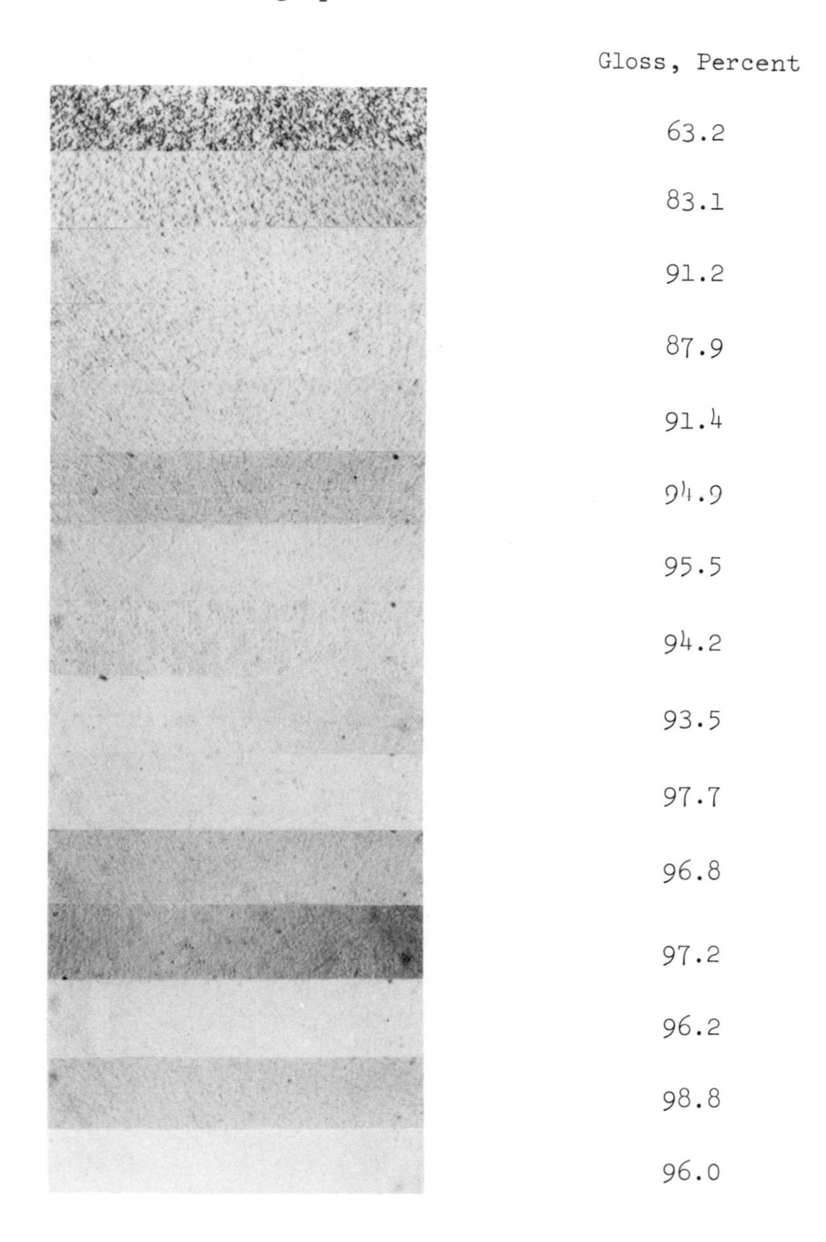

Figure 1

Comparison of microscopic surface roughness and gloss of ABS.

100μ

The amount of volatile material present in the ABS and SAN samples was measured by thermogravimetric analysis (TGA) using a Perkin-Elmer Model No. TGS-1. The instrument was operated isothermally at 240°C using a 160°C-per-minute temperature rise from ambient; all tests were conducted in nitrogen. The volatiles as reported here are the total weight losses from heating samples three minutes at 240°C. In order to assess the residual material after heating, one sample was cycled four times through the three-minute volatile test using a 320°C-per-minute programmed cooling rate between cycles (Figure 2). Carbon black contents of two samples are reported as the residues from heating the plastic samples at 550°C under nitrogen to constant weight; this required approximately five to eight minutes at the test temperature.

Glass transition temperatures (T_g) and concentrations of the SAN and BD in the ABS and SAN samples were measured by differential scanning calorimetry (DSC) using a Perkin-Elmer DSC-1 instrument.[3] Also the concentrations of a molding lubricant and a suspected fatty acid were measured from their heats of fusion in the plastic samples compared to those measured of the free materials. The heating rate in all cases was 40°C per minute.

Residual water content in the plastic samples was measured with a duPont Model No. 26-321A moisture analyzer. The temperature of the analyzer was adjusted to 160-170°C, and sample weights of approximately 90 mg were used. Styrene contents were measured by solid injection of the plastic samples into a Varian 1700 gas chromatograph (GC) equipped with a flame ionization detector. A 5-foot by 1/8-inch-diameter Porapak R column with a helium flow rate of 22.4 ml per minute was used; the column and injection port temperatures were 225°C and 260°C, respectively. A standard of styrene in acetone was used for calibration of the GC. Besides the styrene peak at four minutes retention time, there were observed six other unidentified peaks - two before and four after the styrene. Assuming the same detector response as styrene, these extra six peaks were approximately 14 percent of the concentration of styrene.

Molecular weights of the soluble portions of the ABS and SAN samples were measured by gel permeation chromatography (GPC).[4] The insoluble portions of the plastic samples that could not be analyzed were those bound to the carbon black and crosslinked BD particles.

TABLE I

MOLDING CONDITIONS

	Condition 1	Condition 2
Mold Heater (°F)	140	140
Fill Pressure (psi)	6,500	4,550
Injection Time (Seconds)	0.5	5

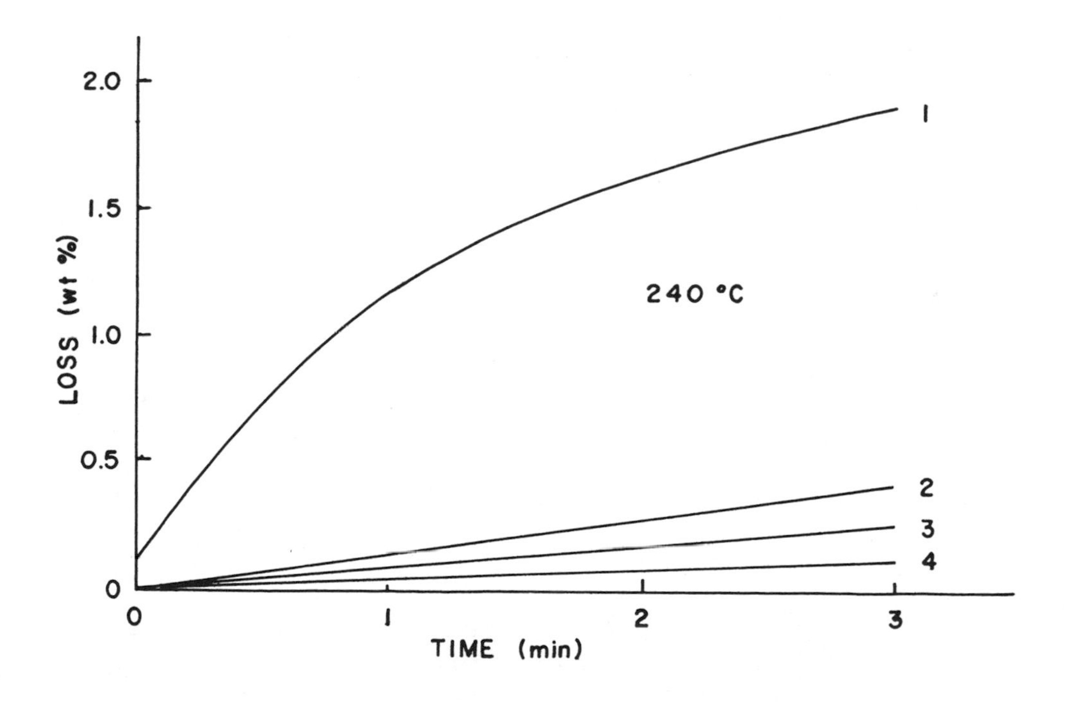

Figure 2

Isothermal TGA weight loss curves for four consecutive heating cycles of ABS.

Gloss of the molded plaques was measured with a Hunter glossmeter according to ASTM D-523 using a 60° angle. The plaque surfaces were also examined with a Reichert light microscope (180X) using dark-field illumination. With dark-field, only light reflected from surface irregularities (concave or convex) enters the microscope objective. A perfectly smooth surface appears completely dark, whereas a rough surface produces bright spots in a dark background. In addition, preliminary surface and fracture morphological studies were carried out by direct observation in a scanning electron microscope. Subsequent clarification of these results was obtained by examination of replicas in a transmission electron microscope. A two-stage replication procedure was employed with poly(acrylic acid) used to produce the first replica, and evaporated carbon, the second. Replicas were shadowed with carbon-platinum at arc tan 1/2.

RESULTS AND DISCUSSION

The data given in Table II and Figures 2 and 3 summarize the numerical results from the ABS, SAN, and 50/50 blend. The carbon black contents reveal that the concentration mixed into the SAN sample is approximately that supplied in the commercial ABS; no effects from small differences in the amount of carbon black were observed in other test results.

Most of the volatile material removed from the ABS during the first pass through the extruder came from the SAN portion of the plastic; this is concluded from examination of the T_g results in Table II. It is well known that the presence of soluble low molecular weight materials in a polymer lowers its glass transition temperature.[5] In the ABS sample all of the increase in the T_g of the SAN resulted from the first pass through the extruder; conversely, the major increase in T_g of the BD portion occurred between the first and fourth passes. This indicates that the most easily removed volatiles are those contained in the SAN and that those in the BD require much more work to expel. This result was confirmed by Blyler,[6] who found that all of the viscosity change measured by a capillary rheometer as a function of number of passes through the extruder occurred during the first pass. The viscosity change was attributed entirely to the loss of volatile components.

TABLE II

RESULTS FROM THE ABS, SAN, AND BLEND

	ABS	ABS	ABS/SAN	SAN
Extruder Passes	0	6	1	1
Carbon Black (Weight Percent)	1.8	-	-	2.4
Volatiles (Weight Percent)				
As Received	1.90	-	-	1.40
Plus Carbon Black	-	-	-	1.00
Extruded 1	1.05	-	0.37	0.25
Extruded 4	0.45	-	-	-
Extruded 6	0.34	0.34	-	-
Water (Weight Percent)	0.22	0.08	0.14	0.15
Styrene (Weight Percent)				
Pellets				
As Received	1.01	-	-	0.36
Extruded 1	0.09	-	-	0.02
Extruded 4	0.003	-	-	-
Extruded 6	0.003	0.003	-	-
Condition 1 Plaque	1.08	0.005	0.07	0.03
Condition 2 Plaque	0.94	0.002	0.02	0.04
Glass Transition of SAN and <u>BD</u> (°C)				
As Received	93, <u>-91</u>	-	-	94
Plus Carbon Black	-	-	-	99
Extruded 1	100, <u>-88</u>	-	-	105
Extruded 4	99, <u>-81</u>	-	-	-
Extruded 6	100, <u>-81</u>	-	-	-
Molecular Weight				
$\overline{M}_n \times 10^{-3}$	51	47	-	81
$M_w \times 10^{-3}$	127	139	-	189
Gloss (Percent)				
Condition 1 Plaque	97.8	96.1	97.0	100.0
Condition 2 Plaque	82.8	69.9	90.6	94.5

It should be evident that the capillary rheometer results also show that there was negligible degradation of the SAN during the extruder drying process. This conclusion is supported by the molecular weight data given in Table II for the original and six-pass ABS samples. Although these results measured by GPC apply only to the soluble portions of the ABS, any effects of processing on the molecular weight should be reflected in these numbers. The small differences in the values of $\overline{M}_n$ and $\overline{M}_w$ between the original and six-pass samples in Table II can be ascribed to experimental error. The absence of degradation at the temperatures of extruder drying has also been reported for pure SAN copolymers of varying compositions.[7]

The thermal analysis of undried ABS revealed not only the expected second order transitions at -91°C and +93°C but also two first order transitions at 70° and 150° (see Figure 3). The discontinuous increase in specific heat, ΔC_p, at -91° is due to the BD phase of the composite. (Note the inset in Figure 3.) With ΔC_p equal to 0.013 cal°C^{-1}g.$^{-1}$, the amount of BD is estimated to be 14 weight percent. The T_g at 93°C and ΔC_p = 0.070 cal°C^{-1} g.$^{-1}$ is due to 76 weight percent SAN, the matrix in the ABS resin. This polymeric composition agrees reasonably well with that measured for similar resins by chemical techniques.[8]

Above the T_g of SAN a fusion peak was observed with ΔH_f = 0.85 cal g.$^{-1}$. This melting, which terminated at 150°C, is attributed to 3.60 weight percent of the mold lubricant added to the ABS. A second small endotherm (ΔH_f = 0.11 cal g.$^{-1}$) occurred near 70° and is suspected to be the melting of crystals of stearic and palmitic acids. Assuming ΔH_f = 45 cal g.$^{-1}$, the average of the heats of fusion of stearic and palmitic acids, these fatty acids were calculated to be 0.24 weight percent in the undried ABS sample. At 240°C and atmospheric pressure the mold lubricant and fatty acids are relatively nonvolatile. However, under the conditions of reduced pressure and elevated temperature which existed in the vacuum extruder, the level of mold lubricant in the ABS was decreased during the six passes from the original 3.60 to 1.10 percent. No melting of fatty acids was detected in the six-pass extrudate (<0.05 percent limit of sensitivity), which indicates that they have been removed by the extrusion process.

Comparison of the volatile contents with the gloss measurements in Table II and the visual appearances of the plaques in Table III with the same gloss measurements yields five useful

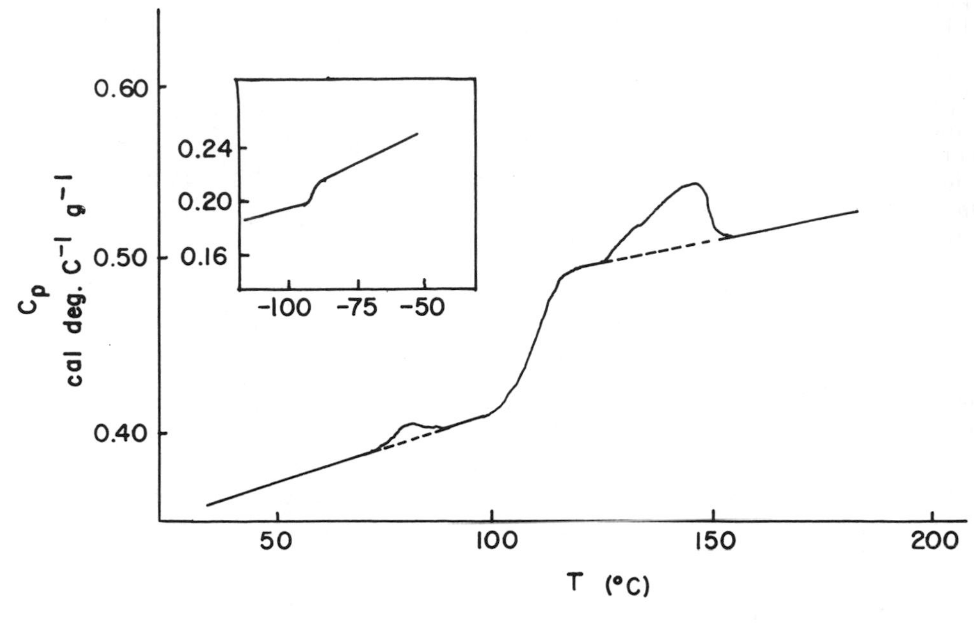

Figure 3

Specific heat curve for ABS showing the T_g of SAN and of BD (inset) and two first order transitions from additives.

TABLE III

VISUAL OBSERVATIONS OF THE TEST PLAQUES

	ABS	ABS	ABS/SAN	SAN
Extruder Passes	0	6	1	1
Condition 1				
Pitting	None	None	None	None
Haze	None	None	None	None
Condition 2				
Pitting	Some	None	Some	Much
Haze	Some	Much	Little	None

conclusions about surface roughness. First, Condition 1 molding variables produce only glossy plaques from all four samples; this expected result is shown by the gloss measurements, visual examination of the plaques, and the light photomicrographs in Figure 4. Secondly, Condition 2 molding variables produce plaques with unacceptable surfaces. There are, however, two kinds of surface roughness responsible for the poor appearance - a macroscopic, pinhole-size pitting and the microscopic unevenness normally called haze. The relation in size between these two kinds of roughness is seen in Figure 4 for the zero-pass ABS molded according to Condition 2 variables; the large blemishes in this photograph are the pits, and the uniform unevenness throughout the picture is the haze. The third conclusion is that pitting does not markedly affect the gloss measurements; e.g. the SAN sample molded with Condition 2 variables and observed to have a severely pitted surface registered a 94.5 percent gloss (Table III). The gloss measurements, however, are severely affected by a hazy surface without any pits; an example of this is the six-pass ABS sample (Condition 2) shown in Table III to have no pitting and much haze. The result is a low gloss reading. The fourth conclusion is that the major cause of pitting is from the volatile components in the plastic samples. We see this by comparison of the visual observations in Table III of plaques made from the zero-pass and six-pass ABS using Condition 2 variables. The zero-pass sample, containing much more volatile material, was found to be severely pitted, whereas no pits were detected visually on the six-pass sample surface. The final conclusion is that the volatile content is not responsible for haze. This is shown by comparing the gloss measurements of the undried and six-pass ABS (Condition 2) samples with their volatile contents given in Table II. The gloss of the six-pass ABS plaque is lower than that of the undried sample; one would expect just the opposite if haze resulted from the presence of volatiles in the plastic. The lower gloss of the plaque molded from the six-pass ABS as compared to that molded from the undried material is attributed to the higher viscosity of the dried resin; during molding this higher viscosity causes a greater pressure drop between the screw and the mold. This, in turn, reduces the packing pressure within the cavity and results in a hazier surface.

The dark-field photomicrographs in Figure 4 show the microscopic surface roughness responsible for haze. The correlation between the microscopic surface roughness in these eight photomicrographs and the gloss measurements in Table II agrees well with that found previously (Figure 1). A direct correlation also exists between haze and the BD content of the samples shown in Figure 4. Recall that the BD content of the ABS was found to be 14 weight percent; the 50/50 blend and SAN, therefore, have 7 and 0 percent BD, respectively. A comparison of these values with the

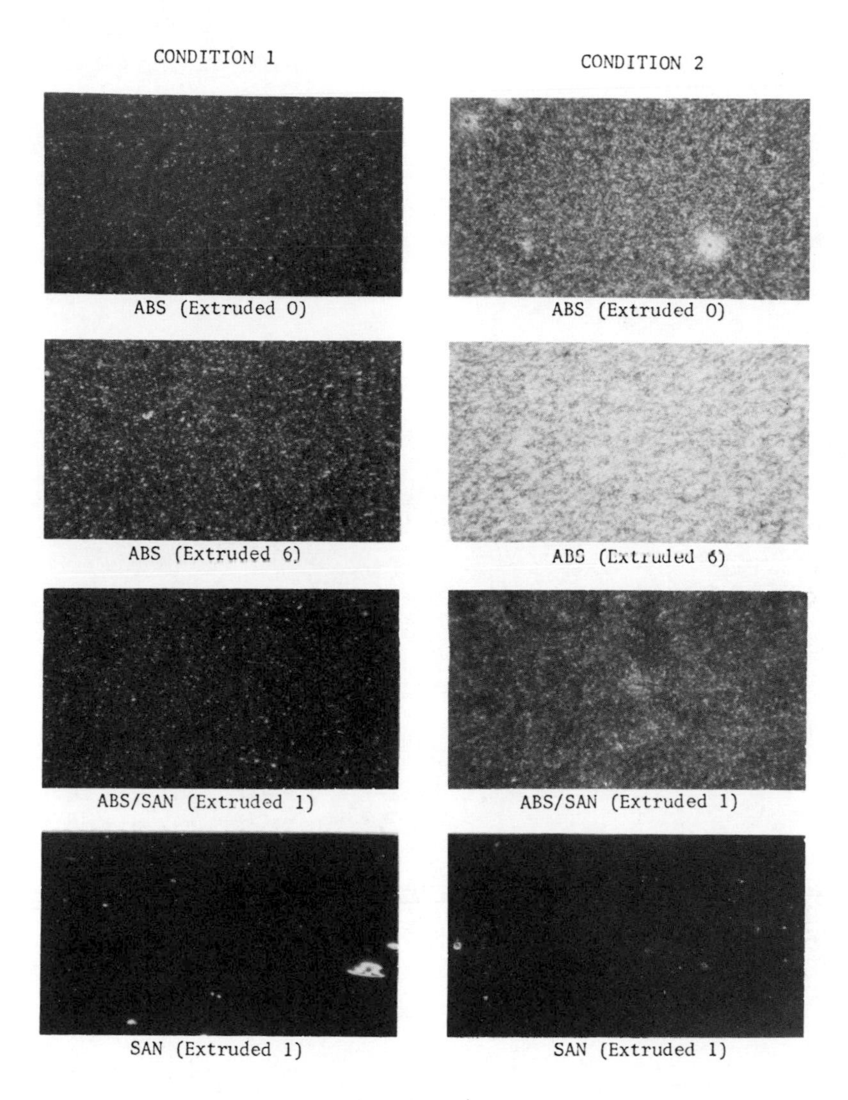

Figure 4

Dark-field photomicrographs of test plaques molded from four compounds showing the degree of surface roughness. ⊢—⊣ 100μ

gloss measurements (Table II) or microscopic surface roughness (Figure 4) shows that haze increases with increasing BD content; i.e. haze on the molded ABS surface results from the presence of the BD particles.

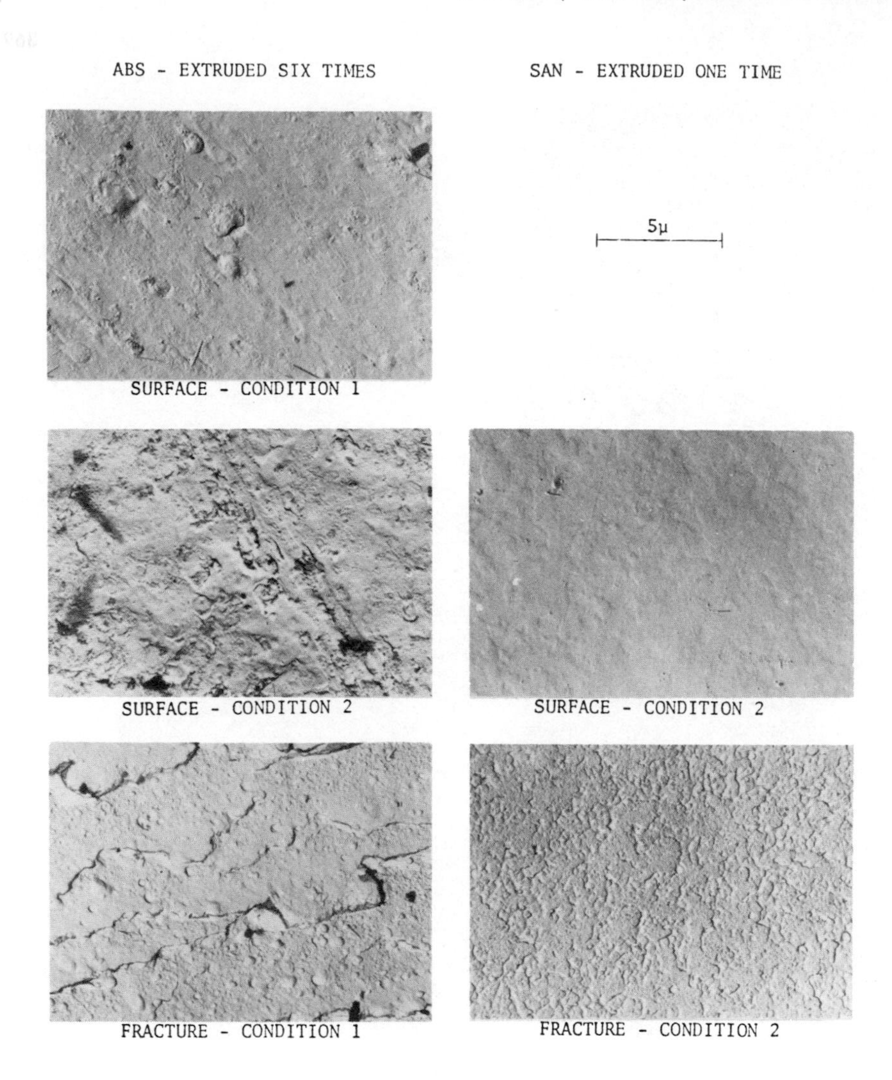

Figure 5

Transmission electron micrographs from replicas of the test plaque surfaces molded from two compounds.

At the higher magnification of the electron micrographs shown in Figure 5, the same dependence of surface roughness on BD content is shown. Here the SAN, containing no BD, produces practically a featureless molded surface even using Condition 2 variables; the only features on the surface created by fracturing the SAN sample at liquid nitrogen temperature are the result of

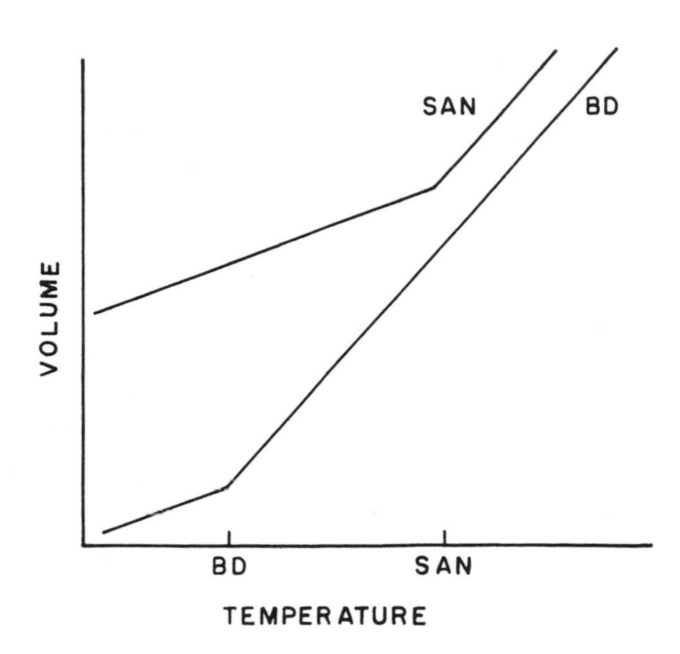

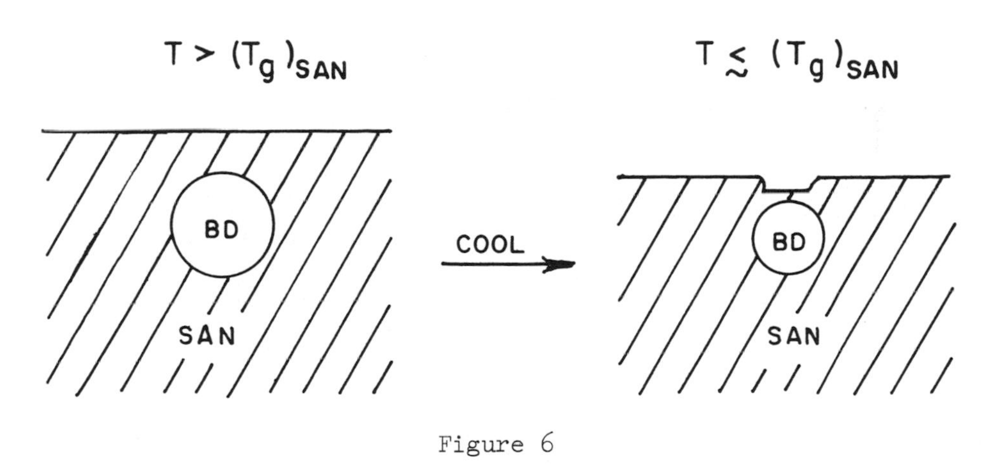

Figure 6

Model of ABS composite explaining the role of BD in causing surface roughness.

ductile failure of the SAN during fracture. The six-pass ABS, using high-pressure Condition 1 molding variables, however, has some molded-in surface roughness. The molding variables of Condition 2 increase the number of surface blemishes on the ABS but do not appreciably affect their size. Furthermore, the size distribution of surface blemishes of the Condition 2 ABS plaque is similar to that of the BD particles observed on the fracture surface of the same compound. We conclude, therefore, that the haze is caused by the BD particles present in the ABS.

One simple model explaining the observed effects of BD on the hazy type of surface roughness is shown in Figure 6. For temperatures above T_g of the SAN, the thermal expansion coefficients of the SAN and BD are approximately the same.[9,10] Cooling of the composite in this temperature range produces a uniform shrinkage of the surface. For temperatures below T_g of the SAN, the coefficient of thermal expansion (volume) of SAN is 2.9×10^{-4}°C^{-1},[9] and that of BD is 7.8×10^{-4}°C^{-1};[10] the large decrease of the thermal expansion of SAN at T_g causes a differential thermal contraction between the SAN glass and the BD rubber. Two possible extremes that can occur as a result of this difference in volume change are a cavitation either within the BD or between the BD and SAN or a deformation of the SAN surrounding the BD particle. Near the sample surface deformation of the SAN is more likely than cavitation, and a microscopic blemish similar in size to the BD ball will be observed. In the limit of zero SAN thickness, this effect will be maximum.

SUMMARY

Two kinds of surface roughness have been observed on samples molded from ABS. The first of these, normally called haze, appears as a whitening of the surface and can be reduced to an acceptable level by selecting molding conditions which provide adequate pressure within the cavity. Microscopic examination of molded ABS samples shows that the difference between hazy and acceptable surfaces is the degree of surface roughness; i.e. even acceptable surfaces display some evidence of roughness.

The cause of the microscopic surface roughness responsible for haze is the presence of the BD particles in the ABS. The role of adequate cavity pressure in minimizing haze is to compress the ABS within the cavity and cause the plastic to replicate the smooth mold surface. A simple model based on the differential thermal contraction between the BD and SAN explains these results.

The other surface blemish, commonly called pitting, appears as pinhole-size defects on the molded ABS samples. Pitting is caused by the escape of volatiles from the plastic surface during molding and can be eliminated by removal of the volatile components from the ABS. It is more economical, however, to correct pitting conditions by providing adequate cavity pressure during molding. Pressure increases the solubility of volatile components and also prevents their escape by packing the plastic tightly against the mold.

ACKNOWLEDGMENTS

The authors thank W. N. Nissle of Welding Engineers, Incorporated for help with extruder drying, and P. C. Warren and F. J. Padden for advice and help with experimental techniques. We have also had the benefit of several discussions with our colleagues at the Western Electric Engineering Research Center, who have also been working in this field.

REFERENCES

1. K. Kato, Polymer 8, 33 (1966); ibid 9, 225 (1966); M. Matsuo, C. Nozaki, and Y. Jyo, Polymer Eng. and Sci. 9, 197 (1969).

2. P. Hubbauer, Society of Plastics Engineers Technical Papers, Volume XIX, 523 (1973).

3. H. E. Bair in "Analytical Calorimetry," R. S. Porter and J. F. Johnson, editors, Plenum Press, New York (1970), Page 51.

4. E. P. Otocka, private communication.

5. J. D. Ferry, "Viscoelastic Properties of Polymers," Wiley, New York (1961), Chapter 16.

6. L. L. Blyler, Jr., private communication.

7. N. Grassie and D. R. Bain, J. Polymer Sci. Part A-1, 8, 2653 (1970); ibid 2665 and 2679.

8. B. D. Gesner, J. Polymer Sci. A3, 3825 (1965).

9. E. Scalco, T. W. Huseby, and L. L. Blyler, Jr., J. Appl. Polymer Sci. 12, 1343 (1968).

10. F. Bueche, "Physical Properties of Polymers," Interscience, New York (1962), Page 110.

POLY(BUTADIENE-CO-STYRENE)/POLYSTYRENE IPN'S, SEMI-IPN'S AND GRAFT COPOLYMERS: STAINING BEHAVIOR AND MORPHOLOGY

A. A. Donatelli, D. A. Thomas, and L. H. Sperling

Lehigh University

Bethlehem, Pennsylvania 18015

INTRODUCTION

Since the synthesis of interpenetrating polymer networks (IPN's) in this laboratory in 1967,[1] the existence and importance of two-phase morphologies have become increasingly apparent.[2,3] A close relationship to graft copolymers has also been recognized. The research reported here is the first systematic study of the morphological relationships between IPN's and graft copolymers of the solution graft type. Because of the critical role of staining in revealing morphologies on the 100 Å scale, we have studied the kinetics of OsO_4 staining as well.

IPN's are a unique type of polyblend, synthesized by swelling a crosslinked polymer (I) with a second monomer (II), plus crosslinking and activating agents, and polymerizing monomer II *in situ*.[1] The term IPN was adopted because, in the limiting case of high compatibility between crosslinked polymers I and II, both networks could be visualized as being interpenetrating and continuous throughout the entire macroscopic sample. However, if components I and II consist of chemically distinct polymers, incompatibility and some degree of phase separation usually occurs.[1-5] Even under these

The authors wish to thank the National Science Foundation for support under Grant GK-13355, Amendment No. 1, and Mr. Richard Korastinsky for his expert assistance with the electron microscopy.

conditions, the two components remain intimately mixed, the phase domain dimensions being on the order of hundreds of Ångstroms.

Electron microscopy studies on the poly(ethyl acrylate)/polystyrene (PEA/PS)* IPN system indicated a cellular structure of about 1000 Å in diameter, and a fine structure of about 100 Å.[2] When poly(methyl methacrylate) replaced the polystyrene as polymer II, the cell size decreased and the fine structure predominated. The fine structure is thought to arise because of a secondary phase separation late in the polymerization of monomer II.

A comparison of the phase separation modes of graft-type polymer blends with IPN's led us to predict[2] a fine structure within the cell walls of the graft copolymers as well. In the PEA/PS IPN system, osmium tetroxide staining was rather light, facilitating the original discovery of the fine structure. Staining in the polybutadiene/polystyrene IPN's[3], graft copolymers,[6-8] and block copolymers[9-11] is much deeper because of the more numerous double bonds. It was apparent that detection of any fine structure in the butadiene-containing graft copolymers would require special attention.

Relationship of IPN's to Graft Copolymers

Shall the IPN's be considered a new class of polymer mixture parallel to blends, blocks, and grafts? The synthetic method clearly places the IPN's as a subclass of the graft copolymers. Grafting in the normally defined sense does indeed take place. However, the crosslinking of both polymer I and polymer II introduces a new structure-influencing element. In most IPN's of the type discussed here, the deliberate crosslinks outnumber the accidental grafts. Thus new morphologies and properties arise because of the controlling effects of the more numerous crosslinks. As we will show below, the ordinary usage of the term graft copolymer becomes more suitable as the number of crosslinks is reduced. As commonly employed, the term "graft copolymer" means the polymerization of monomer II in the presence of polymer I, regardless of the true extent of chemical graft formation. The relationships among the blends, grafts, and IPN's are considered in detail elsewhere in this symposium volume.[12]

---

* The polymer written first indicates polymer I, and the one written second, polymer II, synthesized in that order.

At this point, it is useful to recognize the concept of a semi-IPN. A semi-IPN may be defined as a graft copolymer in which one of the polymers is crosslinked, and the other is linear. Two semi-IPN's may be distinguished. A semi-IPN of the first kind has polymer I in network form, and polymer II is linear. A semi-IPN of the second kind is reversed, polymer I being linear and polymer II being crosslinked. As further discussed in reference 12, there are many related systems containing one or both polymers in network form.

We will first report a series of staining experiments, followed by a study of the morphology of IPN's, semi-IPN's, and graft copolymers.

STAINING BEHAVIOR

Since Kato's[7,13] use of OsO_4 to harden and stain unsaturated polymers, electron microscopy of ultramicrotomed thin sections has played a major role in elucidating the morphology of polymer blends. Yet, little or no information has become available concerning the actual process of staining. Swelling, diffusion, and rate of reaction data have rarely been reported. In this section we present recent results on impact resistant polystyrene (HiPS) and pure polystyrene (PS) which show that staining proceeds by sorption (solution and diffusion) of OsO_4 into the PS phase, followed by reaction with the unsaturated rubber.

Experimental

In order to understand the staining process of OsO_4 in polymer blends, two diffusion experiments were performed.

1. Pure commercial PS and HiPS[14] samples, about 2 x 2 x 20 mm, were enclosed in a glass container in the presence of vapor from solid OsO_4. Periodically the samples were weighed in order to determine the amount of OsO_4 absorbed.
2. A second set of PS and HiPS samples was exposed to OsO_4 vapor. At various time intervals during the course of the experiment, a cross sectional segment was sliced from the samples and viewed under a microscope with a calibrated eyepiece, in order to follow the penetration of OsO_4.

A separate experiment was performed to substantiate the observed plasticizing effect of OsO_4 on PS and HiPS. Samples of PS and HiPS of approximately 2 x 6 x 30 mm in size were exposed to OsO_4, and periodically, over a 264 hour time span, the shear modulus of the samples was measured at room temperature using a Gehman-type torsion apparatus.

Results and Discussion

The percentage weight gain of PS and HiPS and the penetration depth of OsO_4 as a function of the square root of time are presented in Figs. 1 and 2. The penetration of OsO_4 into the samples was easily followed since its diffusion was characterized by a dark, sharp, moving boundary separating the outer stained region from the inner unstained region. This phenomenon is illustrated in Fig. 3 for samples of PS and HiPS stained for 12 hours. The weight gain of PS and HiPS and the penetration of OsO_4 into these materials follows a $t^{\frac{1}{2}}$ relationship, suggesting Fickian sorption. A similar behavior for diffusion in polymers was observed by Kwei and coworkers (15,16). Using an analogous mathematical treatment, estimates of the diffusion coefficients for OsO_4 diffusing into pure PS and HiPS were obtained:

$$D_{OsO_4\text{-}PS} = 4 \times 10^{-11} \text{ cm}^2/\text{sec} \tag{1a}$$

$$D_{OsO_4\text{-}HiPS} = 3 \times 10^{-11} \text{ cm}^2/\text{sec} \tag{1b}$$

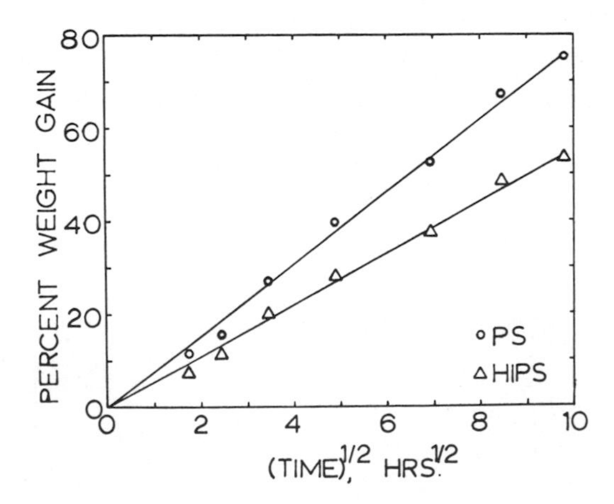

Fig. 1. Percentage weight gain of samples as a function of the square root of time in OsO_4 vapor.

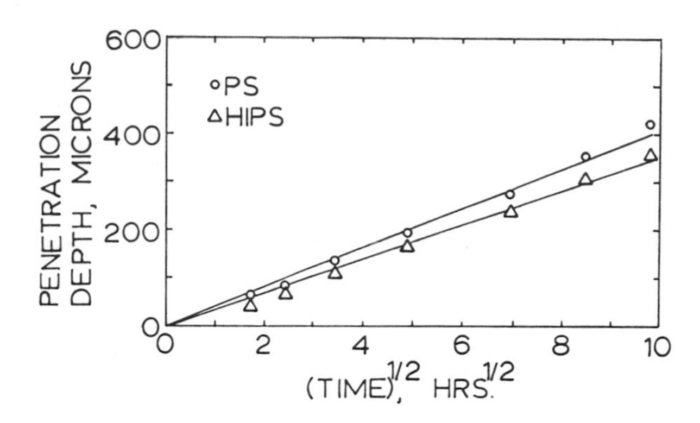

Fig. 2. Penetration of OsO_4 into samples as a function of the square root of time.

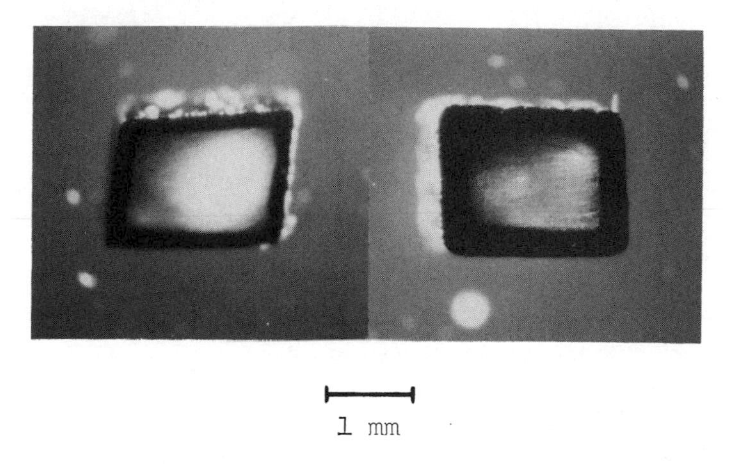

Fig. 3. Sharp diffusion boundary of OsO_4 in HiPS (left) and PS (right) after staining for 12 hours. The core of the HiPS sample is white while the core of PS is clear.

In the second case the diffusion analysis is not strictly applicable because of the reaction of OsO_4 with the polybutadiene in HiPS, but the similar values of D are expected since diffusion proceeds mainly through the continuous PS matrix.

The values of the diffusion coefficients and the data of Figs. 1 and 2 indicate that OsO_4 diffuses through pure PS at a faster rate than through HiPS. If HiPS is assumed to be a polymer with a randomly dispersed rigid, spherical, impenetrable filler, a tortuosity factor, T, can be calculated for the diffusion of OsO_4 through the PS matrix using the equation of Fricke[17]:

$$T = 1 + v_f/\chi \qquad (2)$$

where v_f = volume fraction of filler
χ = shape factor with a value of 2 for spheres

Using low magnification electron micrographs, the volume fraction of dispersed rubber particles in HiPS was estimated to be

$$v_f = 0.57$$

This yields a tortuosity factor of

$$T = 1.29$$

An experimental tortuosity factor can be calculated:[17]

$$T_{exp} = \frac{D_{OsO_4\text{-}PS}}{D_{OsO_4\text{-}HiPS}} = 1.33 \qquad (3)$$

The calculated and experimental tortuosity factors are in close agreement, suggesting that only a very small fraction of the OsO_4 which enters the HiPS reacts with the dispersed rubber phase. After the rubber particles have been stained and hardened they apparently act as an impenetrable filler and provide a more tortuous path to the diffusing OsO_4 molecules.

With regards to the plasticizing effect of OsO_4 on both PS and HiPS the results listed in Table 1 show a steady decrease in 3G(10) with increasing exposure time. Three times the shear modulus at ten seconds, 3G(10), is approximately equal to Young's modulus. The outer stained zone is soft to the touch, but the specimens remain fairly stiff because of the unplasticized core.

The above experiments suggest that a twofold effect is present in the staining process of blends with polystyrene-type continuous phases. First, OsO_4 diffuses into and plasticizes the matrix material through its glass transition. The penetration of OsO_4 follows Fickian diffusion characterized by a sharp advancing boundary. On the swollen side, the polymer is softened by the presence of massive amounts of swellent and diffusion is rapid. Across a rather sharp boundary, the

Table 1

Modulus of PS and HiPS Exposed to OsO_4 Vapor

Time (hrs)	$3G(10) \times 10^{-10}$ (dynes/cm^2) PS	HiPS
0	3.2	2.0
30	2.4	1.7
60	1.9	1.5
144	1.6	1.4
264	0.9	0.9

diffusivity is rather low, and little swellent penetrates. The boundary itself is the region of critical swelling, wherein the diffusivity jumps remarkably as the polymer is transformed from a stiff to a soft material by the plasticizing action. Second, OsO_4 stains and hardens the unsaturated rubbery phase by means of a crosslinking reaction. These effects play an important role in the successful ultramicrotoming of the specimens for electron microscopy. Sufficient OsO_4 must be introduced into the specimens and allowed to react with the rubbery phase; however, excess OsO_4 must be allowed to diffuse out of the sample once staining has been completed to permit the matrix material to regain its rigidity so that microtoming can be performed.* Although not studied, it may be speculated that diffusion of OsO_4 out of the specimens will involve normal Fickian modes.

MORPHOLOGY

Group Theory Applications

The number of ways of synthesizing polymer blends, grafts, and IPN's is large. In reference 12, a quantitative representation of their molecular architecture was developed. It was

*For the samples used in the electron microscopy experiments in the following section, excess OsO_4 was allowed to diffuse out for approximately 36-48 hours.

pointed out that each change in the synthetic steps, no matter how slight, may lead to important differences in morphology and properties. The present syntheses are a case in point. The relatively fine nuances of structural differences among the several materials result in significant changes in morphology. At this time, let us apply the concepts of group theory to these materials, to better understand their structure.

The relationship between the graft copolymers, semi-IPN's, and full IPN's is schematically depicted in Figure 4. A simple graft copolymerization may be expressed as:[12]

$$M_1P_1M_2(P_2G_{12}) \qquad (4)$$

where the symbols represent different operations. M_1 and M_2 represent monomer I and II additions, respectively, P_1 and P_2 the polymerization operations, and G_{12} symbolizes the grafting of polymer II side chains to polymer I backbone chains. In the above case, grafting is simultaneous with P_2, as indicated by including the two operations in parentheses. Otherwise, the operations are sequential in time. Expression (4) thus reads: add monomer I, polymerize it, add monomer II, and polymerize it at the same time permitting grafting.

In the preparation of network polymers such as IPN's, crosslinking operations, C, also take place. The two semi-IPN's prepared in the present study are expressed as:

$$M_1P_1C_1M_2(P_2G_{12}) \qquad (5)$$

$$M_1P_1M_2(P_2C_2G_{12}) \qquad (6)$$

where (5) is a semi-IPN of the first kind and (6) is a semi-IPN of the second kind.

The full IPN of the present study may be written:

$$M_1P_1C_1M_2(P_2C_2G_{12}) \qquad (7)$$

The group theory representation makes it evident that many other IPN's and semi-IPN's could be prepared, by different sequences of crosslinking the polymers and by inverting the order of polymerization of I and II. In all cases considered here, polymer I is poly(butadiene-co-styrene) and polymer II is polystyrene. All of the materials considered were prepared by bulk polymerization processes.

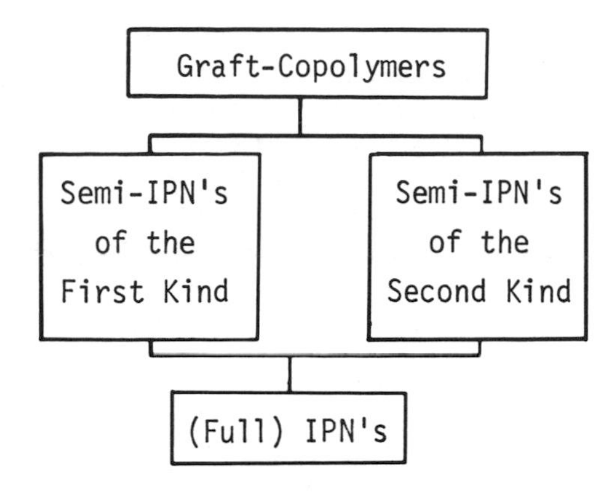

Fig. 4. Organization chart for IPN's and related materials.

Experimental

Semi-IPN's of the first kind were prepared by crosslinking linear poly(butadiene-co-styrene) (SBR)[18] with either 0.1% or 0.2% dicumyl peroxide (DiCup) at 165°C. A styrene (S) monomer solution containing 0.4% lauroyl peroxide (LP) initiator was swollen into the SBR network, and polymerization was conducted thermally at 50°C. The final product contained 80% PS by weight.

A semi-IPN of the second kind was prepared by dissolving 20% linear SBR by weight in a monomer solution containing 2% divinyl benzene (DVB) crosslinker and 0.4% LP and subsequently thermally polymerizing the solution at 50°C without stirring.

Full IPN's were prepared by crosslinking linear SBR with either 0.1% or 0.2% DiCup at 165°C. A S monomer solution containing 2% DVB and 0.4% LP was swollen into the SBR network, and polymerization was conducted thermally at 50°C. The final product contained 80% PS by weight.

A solution graft copolymer was prepared by dissolving 20% linear SBR by weight in a S monomer solution containing 0.4% LP then polymerizing the solution at 50°C without stirring.

Samples from the above materials were exposed to OsO_4 vapor for approximately 50 hours prior to ultramicrotoming.

Microscopy was performed on a Philips 300 electron microscope equipped with a high resolution stage.

Results and Discussion

The morphology of semi-IPN's and IPN's is shown in Figs. 5-9. Figs. 5 and 6 show the morphology of semi-IPN's of the

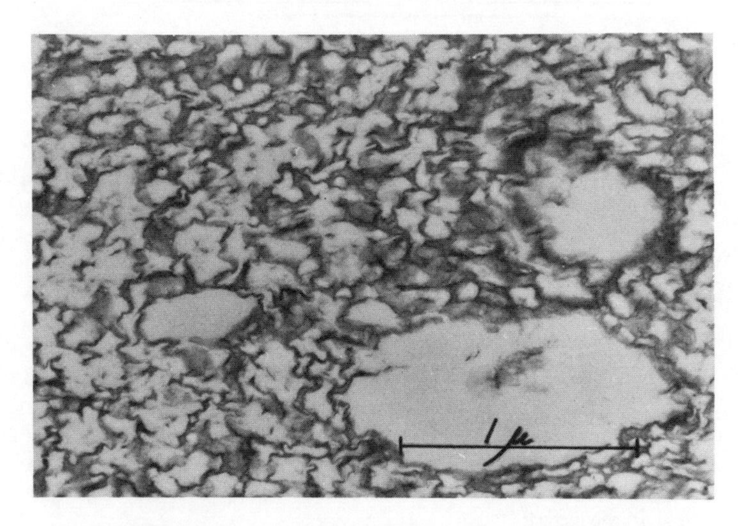

Fig. 5. Semi-IPN of the first kind. SBR is crosslinked with 0.1% DiCup, and PS is linear.

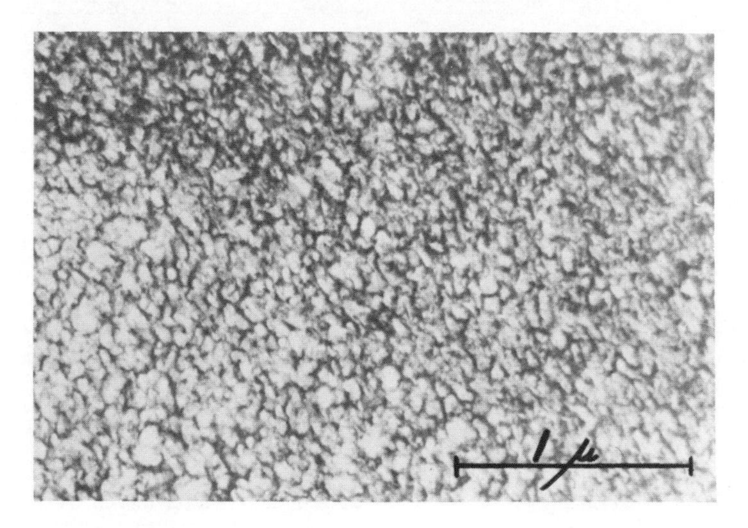

Fig. 6. Semi-IPN of the first kind. SBR is crosslinked with 0.2% DiCup, and PS is linear.

first kind. The SBR (dark areas) appears to be the more continuous phase, forming a network or cellular structure with the size of this structure governed by the degree of crosslinking of the SBR. In Fig. 5, where the SBR is crosslinked with 0.1% DiCup, the cellular structure is approximately 2000 Å in size while in Fig. 6, where the SBR is crosslinked with 0.2% DiCup, the cellular structure is about 1000 Å in size. Therefore, it is apparent that increasing the crosslink density of the SBR polymer I phase results in a much finer cellular structure.

The morphology of a semi-IPN of the second kind is shown in Fig. 7. This material also has a continuous rubbery phase, but the morphology is of a much coarser nature with PS domains varying in size from about 1 μ to less than 0.01 μ (100 Å).

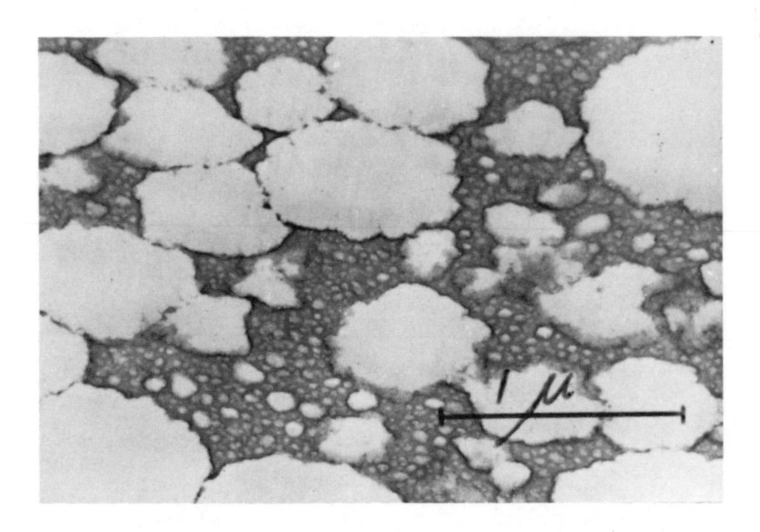

Fig. 7. Semi-IPN of the second kind. SBR is linear, and PS is crosslinked with 2% DVB.

Figures 8 and 9 show the morphology of full IPN's. As with the semi-IPN's of the first kind, the SBR appears to be the more continuous phase, forming a network structure whose size is governed by the degree of crosslinking of the SBR. For SBR crosslinked with 0.1% DiCup, Fig. 8, the cellular structure is about 1600-1800 Å in size whereas for SBR crosslinked with 0.2% DiCup, Fig. 9, the structure is on the order of 800-1000 Å.

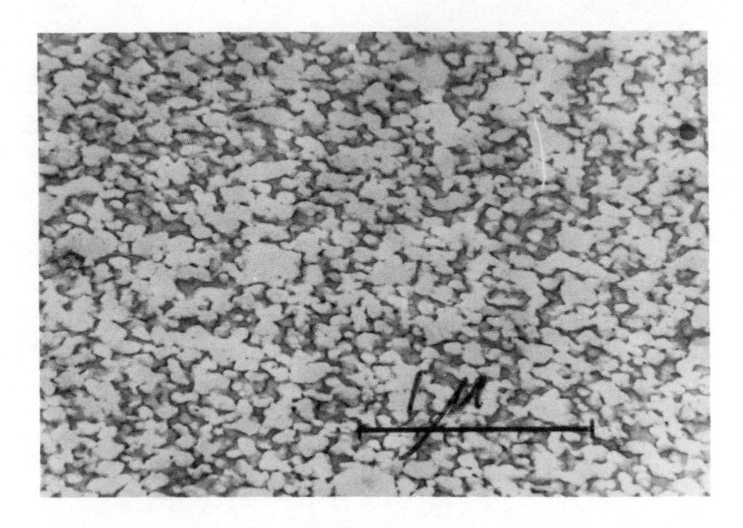

Fig. 8. Full IPN. SBR is crosslinked with 0.1% DiCup, and PS is crosslinked with 2% DVB.

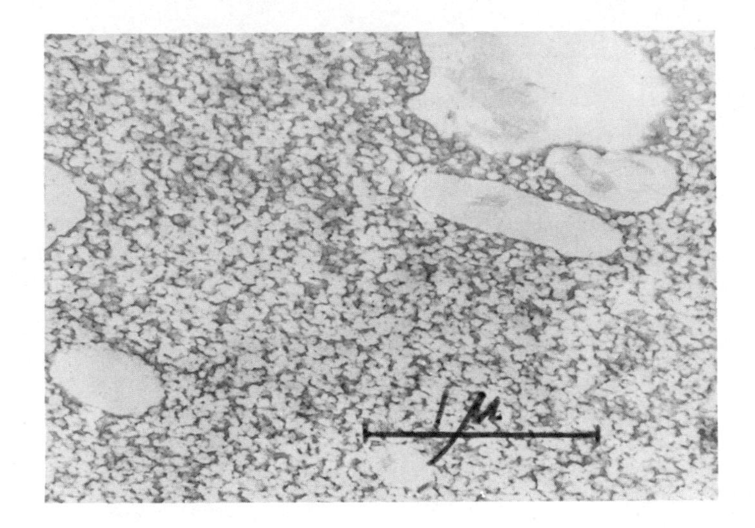

Fig. 9. Full IPN. SBR is crosslinked with 0.2% DiCup, and PS is crosslinked with 2% DVB.

In Fig. 10 the morphology of the solution graft copolymer, synthesized without stirring, is shown. The structure of this material resembles that of a semi-IPN of the second kind where the rubber component forms the continuous phase. Similarly, the structure is of a coarse nature with PS domains varying

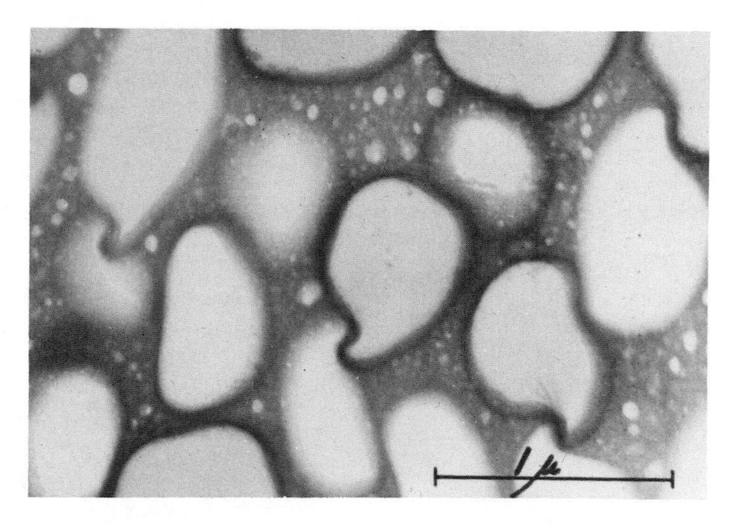

Fig. 10. Solution graft copolymer polymerized without stirring.

from 1 μ to less than 0.01 μ (100 Å) in the cell wall.

Finally, the morphology of the more common type of solution graft copolymer is shown in Fig. 11. This is a commercial HiPS material in which the polymerization reaction of the S monomer is conducted with stirring so that a phase inversion can occur to permit the PS phase, present in a greater amount, to become the more continuous phase[19]. The

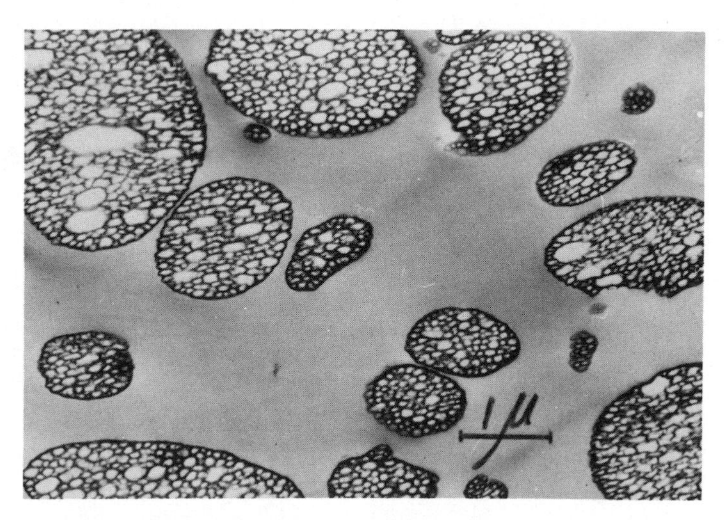

Fig. 11. Commercial HiPS graft copolymer.

rubber particles, containing PS occlusions, vary in size from about 2.5 μ to 0.35 μ. In each particle of the HiPS, the rubber network has a similar appearance to the morphology of the IPN's.

From the preceding discussion it is apparent that the degree of crosslinking in the first polymer component, SBR in our case, is the controlling factor in determining the morphology of the system, while the crosslinking of the second polymer plays a minor role. This effect is exhibited by considering the semi-IPN's of the first kind and the full IPN's. With equal degrees of crosslinking in the SBR phase the morphology of these two types of materials is similar, although the structure of the full IPN is slightly finer. Now comparing the semi-IPN of the second kind and the solution graft copolymer prepared without stirring, a similarity in morphology is noted between these materials because of an analogous cellular structure and PS domains of the same size distribution. These two comparisons should clearly elucidate the factors governing the morphology of these materials. The cellular structures are the same as the type II occlusions described by Molau and Keskkula.(19)

The preceding electron micrographs were all obtained at fairly low magnifications and serve to illustrate the cellular structures. At higher magnifications, the fine structure becomes visible, as shown in Figs. 12-16. These micrographs show the appearance of fine structures in semi-IPN's, full IPN's, and graft copolymers. The morphology in Fig. 16 will be examined first because its importance as a commercial material lends credence to the structures observed. In the top center portion of the figure, a stained cell wall appears to be parallel to the ultramicrotomed section and there appears to be a submerged cellular structure such that the stained diene portion is rather thin. The schematic diagram in Fig. 17 shows that such a region is particularly favorable for observing structures on the 100 Å scale. A fine structure of about 100 Å diameter then becomes more easily seen. The various sized structures in Fig. 16, and particularly in Fig. 14, suggest that phase separation is occuring continuously throughout the polymerization rather than in two discrete steps.

The fine structure was predicted by Huelck, Thomas, and Sperling,(2) after their discovery of it in lightly stained IPN's. The basis for the generalization to semi-IPN's and graft copolymers lies in the fact that after the initial cellular formation polymer I is still highly swollen with monomer II. Upon continued polymerization, further phase separation is inevitable because the two polymers remain incompatible.

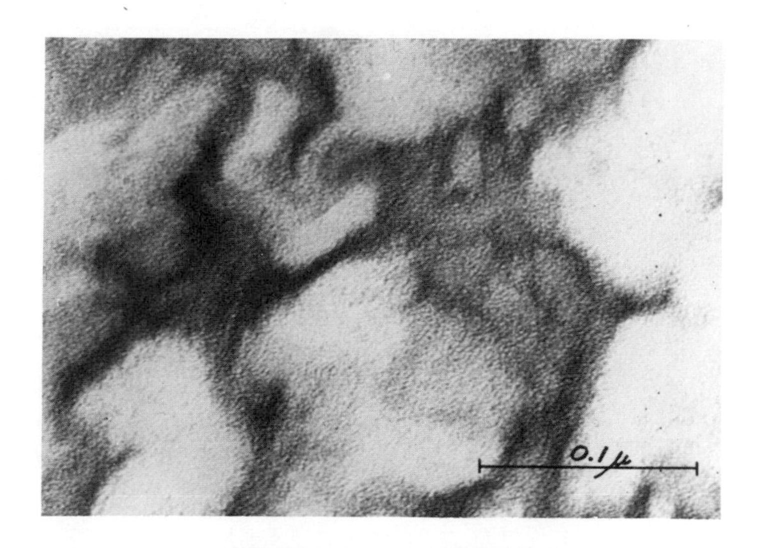

Fig. 12. High magnification electron micrograph of semi-IPN of the first kind. SBR is crosslinked with 0.1% DiCup, and PS is linear.

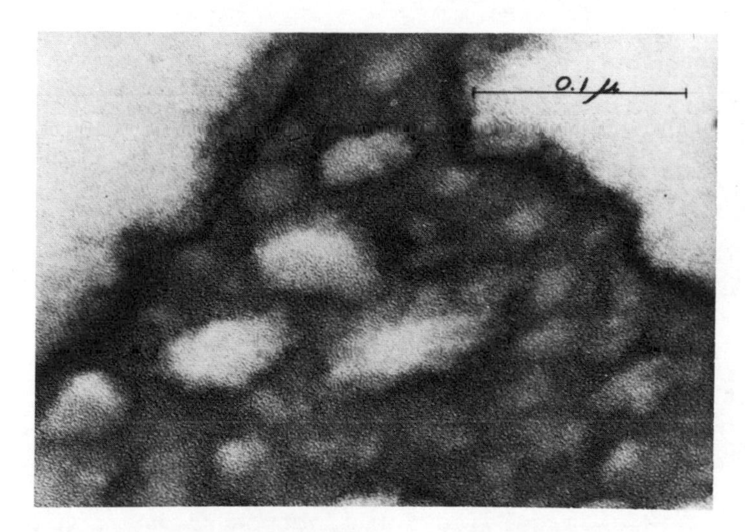

Fig. 13. High magnification electron micrograph of semi-IPN of the second kind. SBR is linear, and PS is crosslinked with 2% DVB.

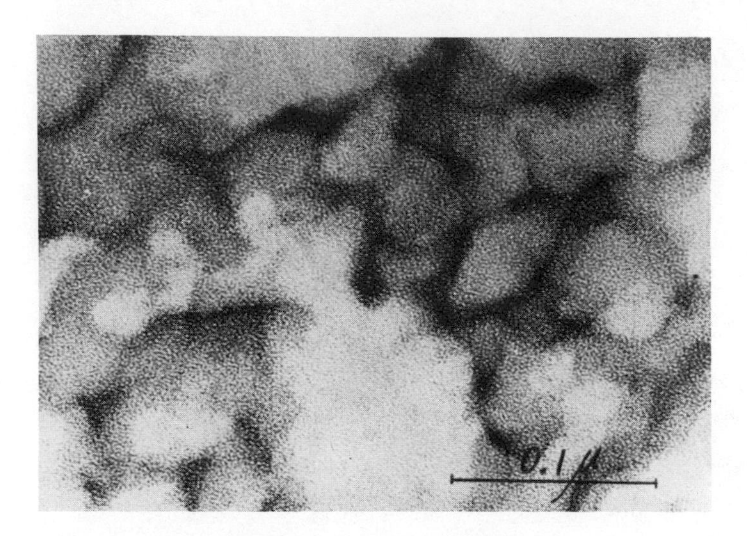

Fig. 14. High magnification electron micrograph of full IPN. SBR is crosslinked with 0.1% DiCup, and PS is crosslinked with 2% DVB.

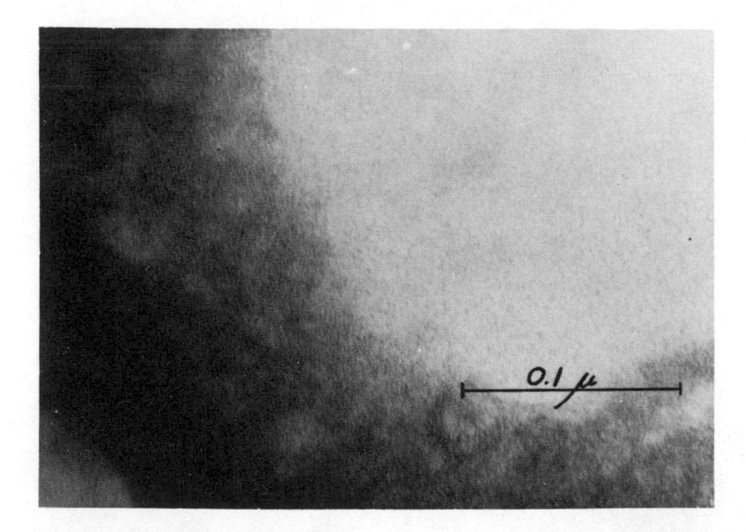

Fig. 15. High magnification electron micrograph of solution graft copolymer prepared without stirring.

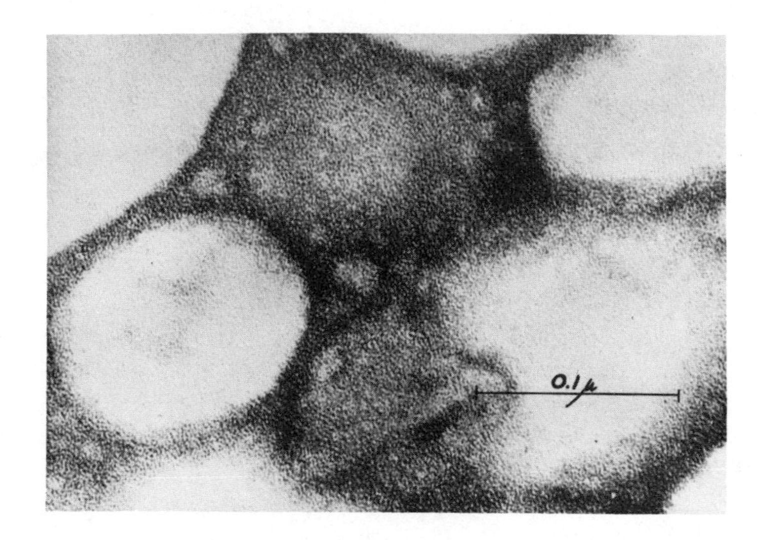

Fig. 16. High magnification electron micrograph within a rubber particle of commercial HiPS.

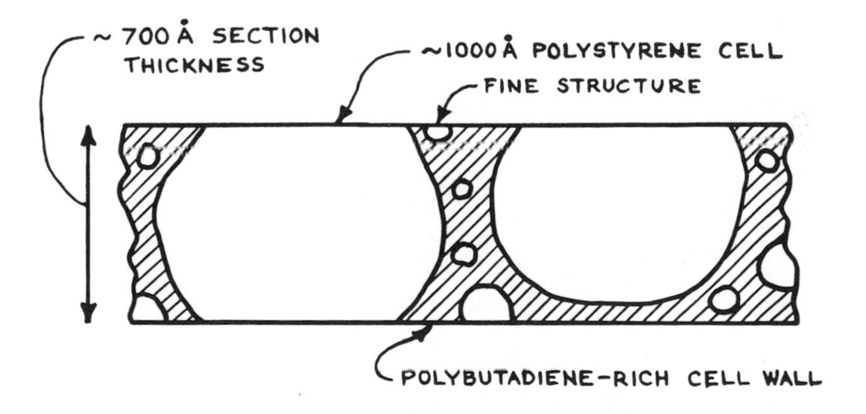

Fig. 17. Schematic edge view of ultramicrotome section through "rubber" particle in HiPS. Thin areas of PB-rich cell walls, as at right, enable 100 to 200 Å PS fine structure and approximately 20 Å ultrafine structure to be observed more clearly.

However, the high viscosity of the components at later stages of the polymerization hinders diffusion, forcing subsequent domain formations to be smaller and smaller.

An ultrafine structure also pervades Figs. 12-16, of the order of 15-25 Å in size. Structures similar in appearance have been reported by Yeh[20,21] and Geil.[22,23] Yeh has interpreted his structures as an ordered chain packing associated with amorphous polymers. It may be that either the two component nature of our materials or the use of OsO_4, or both, has brought out this ultrafine structure more clearly than previously possible.

We can conclude that the IPN's, semi-IPN's, and graft copolymers have cellular and fine structures of a similar nature. While the cellular mode of phase separation is known to occur in both IPN's and graft copolymers, it is interesting to note that the crosslinking in polymer I, much more than in polymer II, controls the phase domain size. The decrease in cell size with increasing crosslinking density suggests that the cell size is controlled by the swellability of the network. Qualitatively, the diameter of the cells, of the order of 1000 Å, must be related to the distances between neighboring crosslink sites. However, a good quantitative theory is still lacking and would be useful at this point.

REFERENCES

1. L. H. Sperling and D. W. Friedman, J. Polym. Sci. A-2, 7, 425 (1969).

2. V. Huelck, D. A. Thomas, and L. H. Sperling, Macromolecules, 5, 340, 348 (1972).

3. A. J. Curtius, M. J. Covitch, D. A. Thomas, and L. H. Sperling, Poly. Eng. and Sci., 12, 101 (1972).

4. L. H. Sperling, Tai-Woo Chiu, C. P. Hartman, and D. A. Thomas, Intern. J. Polym. Mater., 1, 331 (1972).

5. L. H. Sperling, Tai-Woo Chiu, and D. A. Thomas, J. Appl. Poly. Sci., 17, 2443 (1973).

6. M. Matsuo, Japan Plastics, 2, 6 (July, 1968).

7. K. Kato, Japan Plastics, 2, 6 (April, 1968).

8. J. A. Manson and L. H. Sperling, "Polymer Blends and Composites, Broadly Defined", to be published by Plenum Press, 1974.

9. G. E. Molau, Ed., "Colloidal and Morphological Behavior of Block and Graft Copolymers", Plenum, 1971.

10. S. L. Aggarwal, Ed., "Block Polymers", Plenum, 1970.

11. M. Szwarc, Poly. Eng. and Sci., 13, 1 (1973).

12. L. H. Sperling, "On the Generation of Novel Polymer Blend, Graft, and IPN Structures through the Application of Group Theory Concepts". Chapter elsewhere in this volume.

13. K. Kato, Poly. Letters, 4, 35 (1966).

14. Commercial materials kindly supplied by Monsanto Chemical Co.: Hi-Flow 55 general purpose polystyrene and Hi-Test 88 high-impact polystyrene.

15. T. K. Kwei and H. M. Zupko, J. Poly. Sci., A-2, 7, 867 (1969).

16. H. L. Frisch, T. T. Wang, and T. K. Kwei, J. Poly. Sci., A-2, 7, 879 (1969).

17. J. Crank and G. S. Park, "Diffusion In Polymers", Academic Press, New York, Ch. 6, 1968.

18. The SBR, containing 23.5% styrene by weight, was kindly supplied by General Tire and Rubber Co..

19. G. E. Molau and H. Keskkula, J. Poly. Sci., A-1, 4, 1595 (1966).

20. S. L. Lambert and G. S. Y. Yeh, Bull. Am. Phys. Soc., 14, 423 (March, 1969).

21. G. S. Y. Yeh, Poly. Preprints, 14, 718 (1973).

22. D. M. Gezovich and P. H. Geil, Intern. J. Poly. Mater., 1, 3 (1971).

23. P. K. C. Tsou and P. H. Geil, Intern. J. Poly. Mater., 1, 223 (1972).

Topologically Interpenetrating Polymer Networks

K. C. Frisch, D. Klempner; H. L. Frisch
H. Ghiradella

U.D.; State U. New York
Detroit, Mich.; Albany, New York

Ever since the concept of chemical topology was introduced,[1] a number of investigators have prepared molecules showing topological isomerism,[2-16] most of which were catenanes, i.e., interlocking rings with no chemical bonds between them, shown below.

These materials are topological isomers of their non-threaded counterparts. Construction of space-filling models of such catenanes has demonstrated that the minimum ring size (if a carbon sp^3 ring) is twenty atoms.

In recent years, the attention has turned to interpenetrating polymer networks (IPN).[7-16] One method of producing these materials is by mixing two initially linear polymers in the liquid state (dispersion, solution, or bulk liquid prepolymers), fabricating them into the desired form (i.e. casting films, molding sheets, etc.) evaporating the vehicle (if any) and then crosslinking them *in situ*. Permanent entanglements will then occur, depending on the relative cohesive energy densities of the two materials. If they differ too greatly, total phase separation will result. This technique results in Simultaneous Interpenetrating Networks (SIN). In selecting the polymers, it must be borne in mind that little or no reaction between the differing polymers must occur during cure, otherwise true chemical topology will not result (analogous to covalent bonds between the rings above.) Previous studies in this laboratory[7-10] have indicated that it is desirable to select the polymers such that one is a rubbery material and the other is glassy. In this manner, additional reinforcement of the composite structures is achieved, and a morphological analysis by glass transition behavior is facilitated.

Another method of producing IPN's is the swelling method employed by Sperling et al.[11-13] In this technique, a crosslinked film of one polymer is allowed to imbibe monomer and crosslinking agent of the second polymer. Subsequent polymerization and crosslinking of the second polymer in situ results in IPN's, in this case Sequential Interpenetrating Polymer Networks (SIPN). These SIPN's, in all cases, exhibited phase separation, as evidenced by electron microscopy and glass transition studies. The reinforcement effects seen by these authors[12] were explained by the mutual glass-rubber reinforcement effect (depending on which is the continuous phase).

Latex IEN's, (interpenetrating elastomeric networks), produced in a previous study[7-10] by combining latexes of linear polymers, together with crosslinking agents and stabilizers, casting films and curing them, thereby fusing the particles and crosslinking them in situ, resulted in a similar morphology. However, enhancement in tensile strength was too great to be explained by a mere glass-rubber mutual reinforcement. A maximum in crosslink density at about 70% polyurethane and 30% polyacrylate, corresponding to the tensile strength maximum suggested that interpenetration also must play a part in this enhancement.

In the present study, a number of polymer combinations have been produced by mixing in bulk or in solution the linear prepolymers, together with crosslinking agents, and curing the combinations in situ, thereby hypothetically producing IPN's. The combinations were selected such that reactions between the constituent networks during crosslinking would be minimized, thereby preserving chemical topology. The SIN's thus produced all appeared to possess a single phase morphology. The enhancement in mechanical properties is explained by an increase in crosslink density due to the apparent extensive interpenetration. The possibility of intermolecular crosslinking between the constituents to produce a better cured, tougher system is, however, not entirely ruled out.

EXPERIMENTAL

A. Materials: The raw materials used and their descriptions are listed in Table I.

All polyols were dried at 80°C for five hours under a vacuum of 2.0 mm Hg. 2-Butanone-oxime was dried by refluxing under a vacuum of 1.5mm Hg for six hours. Solvents used were reagent grade and dried over molecular sieves. All other materials were used without further purification.

B. Preparation

1) Polyurethanes (PU)

A number of different polyurethanes were synthesized in order to better determine structure-property relationships in the IPN's.

a) Prepolymer preparation

Eight different isocyanate-terminated prepolymers (NCO/OH=2) (both polyesters and polyethers) were prepared: Poly(tetramethylene glycol), M.W.=661 (PM 660) + tolylene diisocyanate (TDI); a poly(oxypropylene) adduct of trimethylolpropane, M.W.=420)(TP 440) + TDI; PM 1000 + TDI; PM 600 + 4, 4'-methylene bis (cyclohexyl isocyanate) ($H_{12}MDI$); PM 1000 + $H_{12}MDI$; PM 660 + 4, 4'-diphenylmethane diisocyanate (MDI); PM 660 + xylylene diisocyanate(XDI); and trimethylolpropane (TMP) + $H_{12}MDI$. Table II shows the composition and designation of these prepolymers. A resin kettle, equipped with a nitrogen inlet, stirrer, thermometer, and reflux condenser was charged with two equivalents of isocyanate. To this was added slowly with stirring one equivalent of polyol. The reactions were carried out under nitrogen at 60°C (for the MDI prepolymer) and 80°C (for the TDI, XDI and $H_{12}MDI$ prepolymers) until the theoretical isocyanate contents (as determined by the di-n-butylamine method[17]) were reached.

b) Blocking

Since combinations were made between some of the above polyurethanes and a melamine cured polyacrylate which crosslinks via pendant hydroxyl groups, it was necessary to block the isocyanates of these polyurethanes to prevent inter-reaction between the polyacrylate and the isocyanate. Thus, 50% solutions in cellosolve acetate of each of the following prepolymers were made: PM 660 + $H_{12}MDI$, PM 1000 + $H_{12}MDI$, PM 660 + MDI, and TMP + $H_{12}MDI$. A slight equivalent excess of 2-butanone oxime and 0.2% by weight of dibutyltin

Table I Materials

Designation	Description	Source
Polymeg 660 (PM 660)	Poly(1,4-oxybutylene)glycol [Poly(tetramethylene glycol)] M.W.=661, hydroxyl no. 169.8	Quaker Oats Co.
Polymeg 1000 (PM 1000)	Poly(1,4-oxybutylene)glycol[Poly(tetramethylene) glycol] M.W.=1004, hydroxyl no. 111.8	Quaker Oats. Co.
TMP	Trimethylolpropane	Celanese Chem. Corp.
2-But. Ox.	2-Butanone oxime	Matheson Coleman&Bell
Pluracol TP 440	Poly(oxypropylene) adduct of trimethylol propane, M.W.=420, hydroxyl no. 410	BASF Wyandotte
$H_{12}MDI$	4,4'-Methylene bis (cyclohexyl isocyanate)	Allied Chem. Co.
MDI	4,4'-Diphenylmethane diisocyanate	Mobay Chem. Co.
XDI	Xylylene diisocyanate; 70/30 mixture of meta and para isomers NCO=94.1	Takeda Chem. Co.
TDI	Tolylene diisocyanate; 80/20 mixture of 2,4 and 2,6 isomers; NCO = 87.0	BASF Wyandotte
T-12	Dibutyltin dilaurate	M & T Chem. Co.
T-9	Stannous octoate	M & T Chem. Co.
Acrylic 342-CD 725	Random copolymer of butyl acrylate,methacrylic acid,styrene	

Table I (continued)

Designation	Description	Source
	and hydroxyethyl methacrylate; 50% solution in xylene: cellosolve acetate (1:1); hydroxyl no. 60; acid no. 13.5	Inmont Corp.
Melamine	Butylated melamine formaldehyde resin; 60% solution in xylene: cellosolve acetate (1:1)	Inmont Corp.
Silicone L-522	Poly(dimethyl siloxane)-poly (oxyalkylene) copolymer	Union Carbide Corp.
CAB	Cellulose acetate butyrate EAB-381-2;ASTM viscosity 15	Eastman Chem. Corp.
BD	1,4-Butanediol	GAF Corp.
Elastonol JX 2057	Hydroxyl-terminated polyester of 1,4-butanediol and adipic acid, M.W.=2036, hydroxyl no. 55.11	North American Urethanes
Epon 828	Bisphenol A-epichlorohydrin resin; Epoxy 189	Shell Chem. Co.
Epon 152	Novolac-epichlorohydrin resin; Epoxy 175	Shell Chem. Co.
DMP-30	2,4,6-tris(Dimethylaminomethyl) phenol	Rohm & Haas Co.
Polyester P-373	Unsaturated polyester; dipropylene glycol maleate	Marco Division W.R. Grace & Co.

Table I (continued)

Designation	Description	Source
Styrene		Dow Chem. Co.
Adipic Acid		Monsanto Chem. Co.
Maleic Anhydride		Monsanto Chem. Co.
DEG	Diethylene glycol	Dow Chem. Co.
DPG	Dipropylene glycol	Dow Chem. Co.

dilaurate (catalyst)(T-12) were added to each of the prepolymer solutions in a 3-necked flask equipped with a stirrer, reflux condenser, thermometer, and nitrogen inlet. The reactions were carried out under nitrogen at 80°C until the isocyanate content reached zero (complete blocking). Subsequent chain extension and crosslinking (see below) can only occur at 150°C, and proceeds slowly since deblocking must first occur. Since the melamine cure of the polyacrylate proceeds quite readily at 100°C and very rapidly at 150°C, it is expected that this reaction will take place before the isocyanate is deblocked, thus reducing the possibility of grafting occurring between the two polymers. However, the possibility of reaction with excess polyacrylate hydroxyls does exist.

c) Chain extension and Curing

An equivalent weight of the curing agent (50% solution in cellosolve acetate) and 0.1% by weight of stannous octoate (catalyst) (T-9) were added to the prepolymer solution. Films 0.002"-0.003" in thickness were cast on glass with a doctor blade and oven cured. The curing conditions are shown in Table II as well as the composition and designation of the fully cured polyurethanes.

2. Polyepoxides (E)

Two epoxy resins were employed, Epon 828 (bisphenol A-epichlorohydrin resin) (EI) and Epon 152 (Novolac-epichlorohydrin resin) (E2). Films of both were cast and cured with 2,4,6 tris (dimethylaminomethyl) phenol (0.5%) (DMP-30). Epon 828 contains a small amount of free hydroxyls which may possibly react with the isocyanate during IPN formation. Epon 152 has none.

Table II
Polyurethane Curing Conditions

Description	Curing Agent	Prepolymer Composition	Temp. (°C)	Time (Hrs.)
PU 1	TP 440	PM660+TDI	100	16
PU 2	Elasto-nol JX2057	TP440+TDI	85 130	16 2
PU 3	BD+TMP (1:1)	PM1000+TDI	85 130	16 2
PU 4	TMP	PM660+H_{12}MDI	150	4
PU 5	TMP	PM1000+H_{12}MDI	150	4
PU 6	TMP	PM660+MDI	150	4
PU 7	TMP	PM660+XDI	150	4
PU 8	TMP	TMP+H_{12}MDI	150	4

TP 440-Poly(oxypropylene) adduct of trimethylolpropane

PM 660-Poly(tetramethylene) glycol

TDI- Tolylene diisocyanate

Elastonol JX 2057- Polyacrylate copolymer

BD-Butanediol

TMP-Trimethylolpropane

H_{12}MDI-4,4'-methylene bis (cyclohexyl isocyanate)

MDI- 4,4'-diphenylmethane diisocyanate

XDI- Xylylene diisocyanate

3. Polyesters (PE)

Two unsaturated polyesters were used, one of which was highly unsaturated (dipropylene glycol maleate) (PE-2) while the other one was more flexible, containing adipic acid in addition to maleate units in the polyester moiety (PE-1). Both were crosslinked by means of syrene monomer.

a) PE 1: a mixture of 196g (0.2 mole) maleic anhydride, 116.8g (0.8 mole) adipic acid, 26.8g (0.2 mole) dipropylene glycol, and 95.4g (0.9 mole) diethylene glycol was reacted at 150°C under nitrogen for 24 hours. To 100g of this unsaturated prepolymer were added 30g of styrene and 1.3g (1%) benzoyl peroxide. The mixture was stirred and castings made between glass plates sealed with rubber gaskets (to prevent monomer evaporation) and cured at 85°C for 16 hours.

b) PE 2: This unsaturated polyester (dipropylene glycol maleate) (P-373) was cured with styrene and films made as above.

4. Polyacrylate (PA)

The polyacrylate employed was a 50% solution in cellosolve acetate and xylene. It consisted of a random copolymer of butyl acrylate, methacrylic acid, styrene, and hydroxyethyl methacrylate and had a hydroxyl number of 60. The styrene and butylacrylate were present in equal concentration. Only a small amount (~1%) of methacrylic acid was present and served as catalyst for the melamine cure. Crosslinking occurred by reaction of the pendant hydroxyl groups on the chain with a butylated melamine formaldehyde resin in 60% solution in xylene and cellosolve acetate. To 149g of the acrylic solution were stirred in 5g of the modified melamine resin solution. Films were cast on glass using a doctor blade and cured at 150°C for 4 hours.

5. Interpenetrating polymer networks (IPN's)

The IPN's were made by combining the linear prepolymers, together with crosslinking agents and catalysts, casting films, and subsequently curing them. All IPN's made were binary systems composed of one of the polyurethanes and one of the other components. Table III shows the designation, composition, and curing conditions.

C. Measurements

1. Stress-strain

The tensile strengths and elongations at break were measured on an Instron Tensile Tester at a crosshead speed of 2in/min.

Table III

IPN Composition and Curing Conditions

Description	Composition	Temp (°C)	Time (Hrs)
IPN 1 (Urethane-polyester)	PU 1 + PE 1	110	16
IPN 2 (Urethane-polyester)	PU 1 + PE 2	110	16
IPN 3 (Urethane-epoxy)	PU 2 + E 1	85 130	16 2
IPN 4 (Urethane-epoxy)	PU 3 + E 1	85 130	16 2
IPN 5 (Urethane-epoxy)	PU 3 + E 2	85 130	16 2
IPN 6 (Urethane-polyacrylate)	PU 4 + PA	150	4
IPN 7 (Urethane-polyacrylate)	PU 5 + PA	150	4
IPN 8 (Urethane-polyacrylate)	PU 6 + PA	150	4
IPN 9 (Urethane-polyacrylate)	PU 7 + PA	150	4
IPN 10 (Urethane-polyacrylate)	PU 8 + PA	150	4

PU- Polyurethane E- Epoxy
PE- Polyester PA- Polyacrylate

2. Calorimetric measurements

The glass transitions (Tg) were determined on a Perkin-Elmer differential scanning calorimeter, DSC-1b. Measurements were carried out from -100°C to +150°C under nitrogen at a scanning rate of 10°C/min.

3. Infrared analysis

Infrared spectra were obtained on a Perkin-Elmer Model 457 infrared spectrophotometer at room temperature.

4. Electron microscopy

Samples were stained in osmium tetroxide for two weeks, after which they were embedded in epoxy resin[18], sectioned on an LKB Ultratome III and observed with an AEI 6B and a Phillips 300 electron microscope.

Results and Discussion

A. Stress-strain Properties

The tensile strengths and breaking elongations as a function of urethane content in the IPN's are shown in Figures 1 and 2, respectively.

Maxima in tensile strength which were significantly higher than the tensile strenths of the components occurred in all cases except for IPN 10 . This behavior is typical of IPN's made previously in this laboratory[7-10]. The maximum may be attributed to an increase in the crosslink density due to interpenetration, since it is well known that the tensile strength of a crosslinked rubber goes through a maximum as the crosslink density increases. Previous studies showed there to be a maximum in crosslink density corresponding to the maximum in tensile strength. The position of the maximum, however, varied depending upon the materials employed. IPN 10, made from a very highly crosslinked polyurethane (trifunctional prepolymer as well as a trifunctional chain extender) showed no such enhancement in tensile strength, as would be expected. This high degree of crosslinking lowers the statistical probability of threading, thereby precluding the reinforcing effect of interpenetration. IPN's 4,5,6, and 7 showed minima at 75% polyurethane, and IPN 8 showed no minimum. The minima may be attributed to initial weakening of hydrogen bonding (or some other such strong polar intermolecular force) at small values of interpenetration (more of one component than the other). After about 25% of one component, the crosslink density effect takes over, thus raising the tensile strength

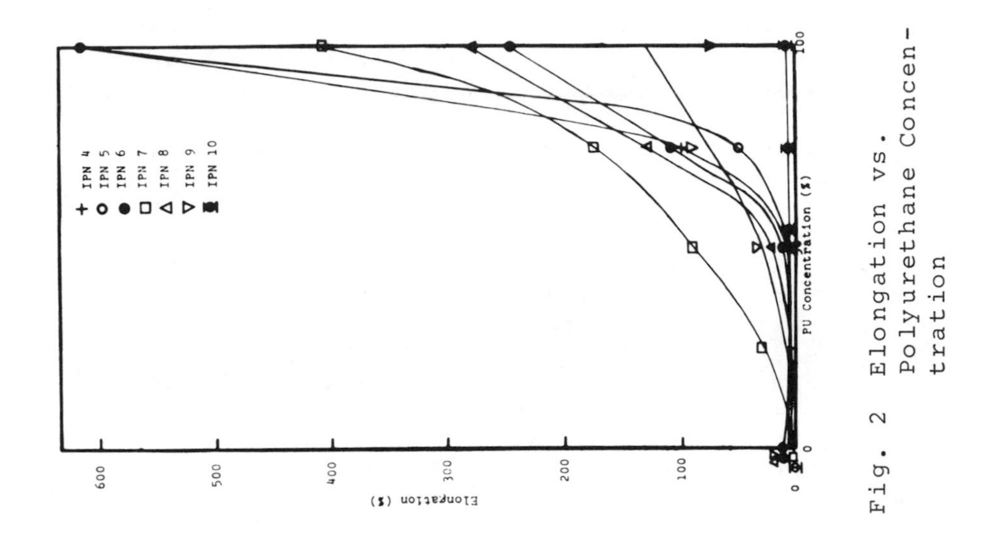

Fig. 2 Elongation vs. Polyurethane Concentration

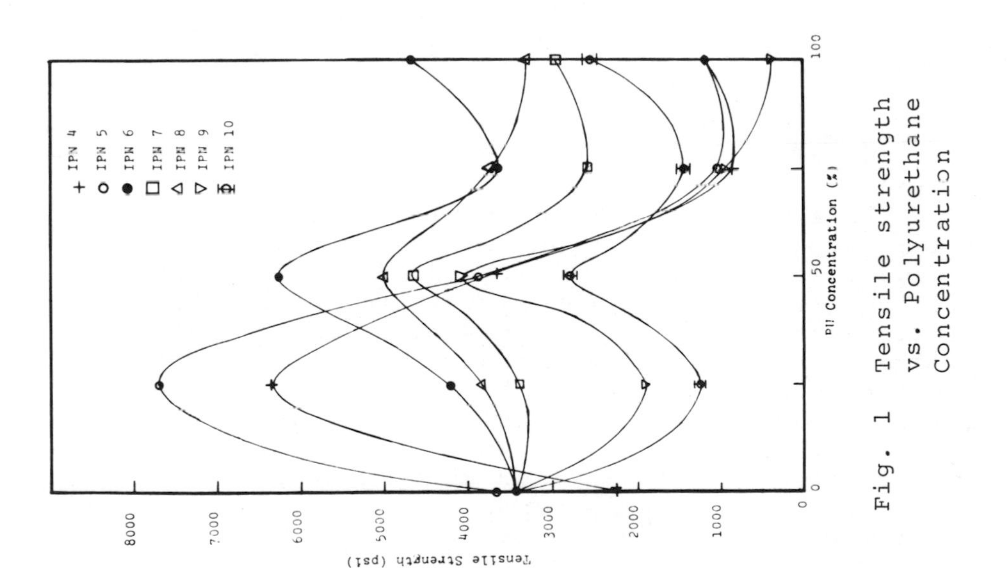

Fig. 1 Tensile strength vs. Polyurethane Concentration

to a maximum. IPN's 6 and 7 (minima at 75% polyurethane) are made from polyurethanes with H_{12}MDI (cycloaliphatic) as the isocyanate, while IPN 9 used XDI (an aralkyl) and IPN 8, MDI (an aromatic). Thus, it appears from this work that aliphatic polyurethanes yield IPN's with tensile strength minima near the pure polyurethane, while aromatic (and aralkyl) polymers result in either no minima or minima near the pure polyacrylate. IPN's 4 and 5 (made from an aromatic polyurethane and polyepoxides, a relatively non-polar polymer) showed minima near the pure polyurethane. Thus, a plausible explanation for the positions of the minima would be that aliphatic polyurethanes have more hydrogen bonding than aromatic polyurethanes. This might be expected since aliphatic groups are electron releasing and aromatics electron withdrawing. Thus, with aliphatic polymers, the nitrogen atom in the urethane linkage would be more electronegative and therefore, more available for hydrogen bonding. Thus, the plasticizing effect of the second component (the non-urethane) on aliphatic polyurethanes is greater and minima occur near the polyurethane. When the polyurethane is aromatic, the plasticizing effect is less and minima occur either near the second component or not at all, unless the second component is even less polar than the aromatic polyurethane, as with polyepoxides (IPN's 4 and 5) in which case the minima occur near the polyurethane.

Another possible reason for the maximum in tensile strength is intermolecular crosslinking occurring between the two networks to result in a better cured system. Specifically, Epon 828 does contain a small amount of free hydroxyls which could react with the isocyanate terminated prepolymer. However, Epon 152 contains none. Since the behavior of IPN's 4 and 5 were quite similar, we may assume either that this reaction does not occur or, that if it does, it has a minimal effect on the properties of the IPN. Also, the isocyanate terminated prepolymer may possibly react with excess hydroxyl on the polyacrylate backbone, or with a small amount of amine hydrogen possibly present on the melamine. However, infrared analysis of IPN's 6-10 (below) suggests these side reactions to be minimal.

Except for IPN 10 (made from the highly crosslinked rigid polyurethane) the elongations behave basically the same. They all decrease rapidly as the polyurethane content decreases until about 50% at which point they approach the value of the pure polyacrylate or polyepoxide.

B. Infrared Analysis:

Infrared spectra of IPN's 6-10 showed all the bands of the constituent networks with no new ones appearing. The position of all the bands in the IPN's were the same as in the components. This is an indication that within the limitations of this technique, little interreaction between component polymers has taken place during cure. This measurement would be expected to detect at least 2 grafts per hundred monomer units. The degree of interpenetration would be expected to be much greater than this, probably on the level of segmental mixing due to the single phase nature of these systems. Thus, even if occasional grafts did occur at a level too low to be seen, the degree of interpenetration would probably be greater than the grafting.

C. Calorimetric Measurements

A common technique in determining the morphology of a blend of two polymers is measurement of the glass transition temperatures (Tg's) by differential scanning calorimetry. If complete phase separation has occurred, two distinct Tg's, corresponding to the transitions of the constituent polymers, will occur in the blend. If the situation is single phase, one sharp intermediate Tg would occur. Any inward shifting of the Tg's is indicative of some phase mixing. The IEN's discussed earlier showed one broad glass transition. The Tg of the rigid component (a polyacrylate) ran into that of the rubbery network (a polyurethane) indicating that some interpenetration had occurred. Table IV shows the Tg's of IPN's 1,2,3, and 6 and their component networks. Also shown are the arithmetic means of the Tg's of the components. In all cases, the IPN's showed one Tg intermediate in temperature to the Tg's of the components and as sharp as the Tg's of the components. This suggests a situation in which at least extensive phase mixing occurs. In fact, a single phase morphology may be the case. Thus, interpenetration of at least a similar extent as that found for the latex IPN's[7-10] made previously by us must occur. In fact, interpenetration is probably more extensive now since the Tg's are much sharper than those of the latex IEN's in which the transition of one component ran into that of the other.[10] Also, the enhancement in tensile strength of these SIN's was much greater than that of the IEN's.

Table IV

Glass Transition Temperatures

Sample	Composition	Tg,°K (found)	Tg,°K (Ave.)*	θ**
IPN 1	50% PU1/50% PE1	252	257	0.020
IPN 2	50% PU1/50% PE2	306	330.5	0.080
IPN 3	50% PU2/50% E1	283	302.5	0.069
	25% PU2/75% E1	304	336.75	0.108
IPN 6	75% PU4/25% PA	246	249	0.012
	50% PU4/50% PA	274	288	0.051
	25% PU4/75% PA	321	327	0.019
PU 1	100%	266		
PU 2	100%	234		
PU 4	100%	209		
PE 1	100%	248		
PE 2	100%	395		
E	100%	371		
PA	100%	367		

PU- Polyurethane
PE- Polyester
E- Epoxy
PA- Polyacrylate

*Average: $Tg = W_1 Tg_1 + W_2 Tg_2$

**$\frac{Tg - Tg\ (Ave.)}{Tg\ (Ave.)} = \frac{-\theta}{1+\theta}$

In examining the Tg's of the IPNs (Table IV) we see immediately that the Tg of the IPN is always lower than the Tg (av) defined by:

$$Tg\ (av) = W_1 Tg\ (1) + (1 - W_1)\ Tg\ (2) \qquad (1)$$

where W_1 is the weight fraction of component 1 and Tg (1) and Tg (2) are the Tg's of components 1 and 2 respectively.

This observation may be consistent with a modification of a theoretical equation of DiBenedetto (unpublished result quoted in reference 19) relating the shift in glass transition temperature to degree of crosslinking. Ordinarily, chemical crosslinking in conventional polymers raises the Tg. If Tg is the glass temperature of the crosslinked polymer, Tg,o the glass temperature of the uncrosslinked polymer, X_c the mole fraction of monomer units which are crosslinked in the polymer, $\varepsilon_x/\varepsilon_m$ the ratio of the lattice energies for crosslinked and uncrosslinked polymer and F_x/F_m the ratio of segmental mobilities for the same two polymers, then the DiBenedetto equation reads:

$$\frac{Tg - Tg,o}{Tg,o} = \frac{\left(\frac{\varepsilon_x}{\varepsilon_m} - \frac{F_x}{F_m}\right) X_c}{1 - \left(1 - \frac{F_x}{F_m}\right) X_c} \qquad (2)$$

Any copolymer effect due to crosslinking is to be accounted for by modifying Tg,o. For chemically crosslinked polymers $\varepsilon_x/\varepsilon_m \neq 1$ (DiBenedetto estimates this to be about 1.2) and the mobility of a chemically crosslinked segment $F_x \ll F_m$ so that $F_x/F_m \simeq 0$. Hence (2) simplifies in first approximation to:

$$\frac{Tg - Tg,o}{Tg,o} \simeq \frac{1.2\, X_c}{1 - X_c} \qquad (3)$$

which exhibits the often experimentally observed increase of Tg with X_c.

In the case of an IPN we must modify (2) by replacing Tg,o with Tg (av). This should account in first approximation for the copolymer effect which is obviously present with the IPN's. Next we note that $\varepsilon_x/\varepsilon_m = 1$ since the monomer units of both networks are not chemically modified on forming an IPN as a result of forming permanent entanglements by topological interpenetration. Thus, (2) now becomes:

$$\frac{Tg - Tg\ (av)}{Tg\ (av)} = \frac{[1 - F_x/F_m]\ X_c}{1 - (1 - F_x/F_m)\ X_c} \qquad (4)$$

with X_c the entanglement mole fraction. In general, secondary intra-molecular bonding of a network, Van der Waals or hydrogen bonding (if present), are reduced by the permanent entanglement of portions of two different networks; hence the mobilities of the segments of an

IPN, F_x, are larger than in the non-interpenetrating separate network, F_m, i.e. $F_x/F_m > 1$. Setting,

$$\theta = [F_x/F_m - 1] \quad X_c \geq 0 \; , \; 1 \geq X_c \geq 0 \qquad (5)$$

we can rewrite (4) as:

$$\frac{Tg - Tg\ (av)}{Tg\ (av)} = \frac{-\theta}{1 + \theta}$$

which would predict that the Tg of an IPN would be less than or equal to Tg (av), the relative negative shift being given quantitatively by $\theta/1+\theta$ which increases monotonically from zero to $(F_x-F_m)/F$ as X_c goes from zero to one. θ values for the IPNs are listed in Table IV.

The values of F_x/F_m and X_c depend on the chemical nature, crosslink densities and weight fraction W, of the constituent networks of the IPN. For a series of IPN's of differing W made from the same two constituent networks θ is expected to reach a maximum not far from the maximum of X_c as a function of W.

Inspection of Table IV shows that for the PU/PA IPN series, θ reaches a maximum at 50% PU. This IPN also exhibits a maximum in tensile strength (Figure 1) at 50% PU. Since this has already been separately theorized to be due to a maximum in crosslink density, the two theories are self-consistent. We thus have strong support for the occurrence of extensive interpenetration.

D. Electron Microscopy

The electron micrographs of IPN 6 are shown in Figures 3 and 4. There appears no evidence of phase separation as evidenced by the absence of any microstructure. At higher magnifications, some granularity is visible in the high-polyurethane IPN's, but this is present in highest quantity in the pure polyurethane sample and so is probably not a characteristic of the IPN's themselves.

This single phase situation that occurs between normally incompatible polymers is most probably due to topological interpenetration, which does not allow

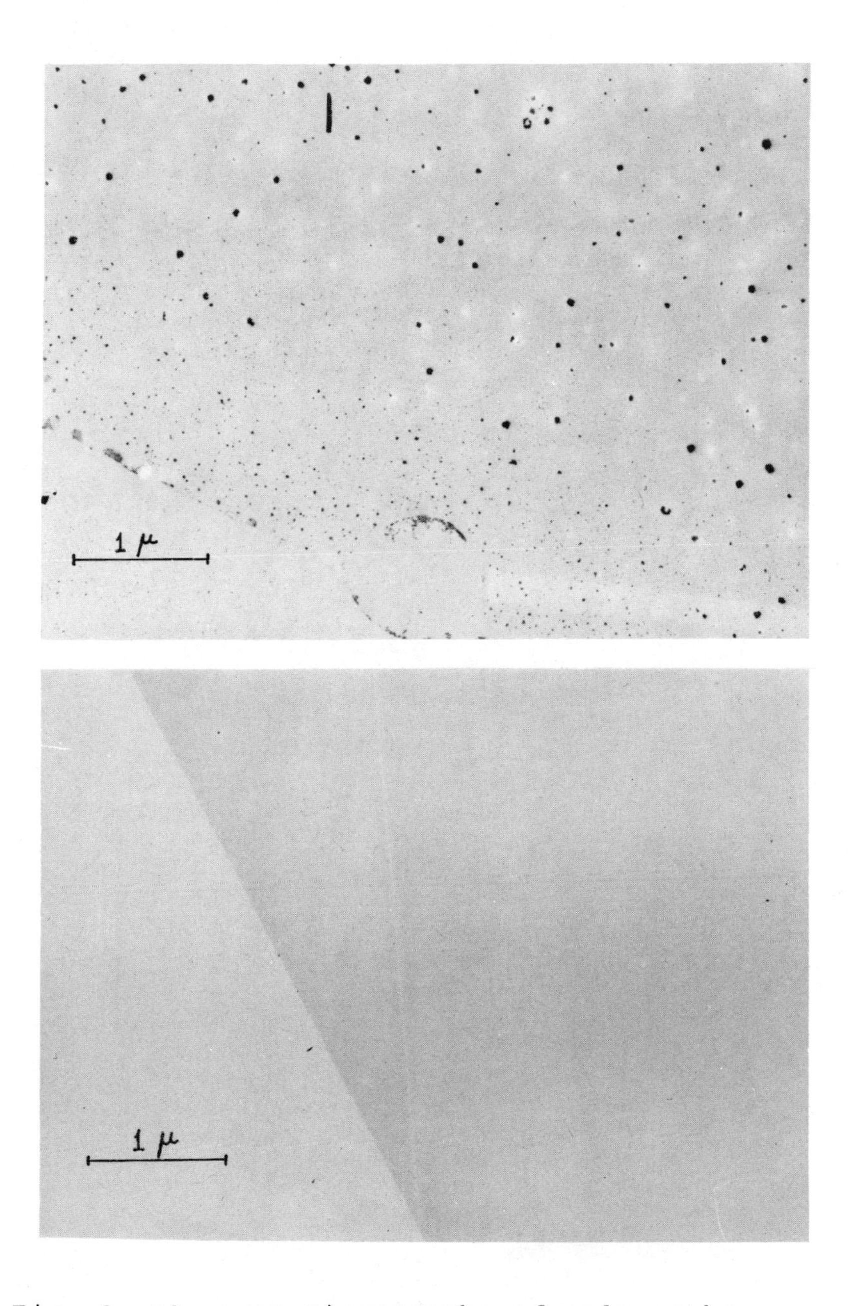

Fig. 3 Electron micrographs of polyurethane (PU4)(top) and the polyacrylate (bottom) (11,000x).

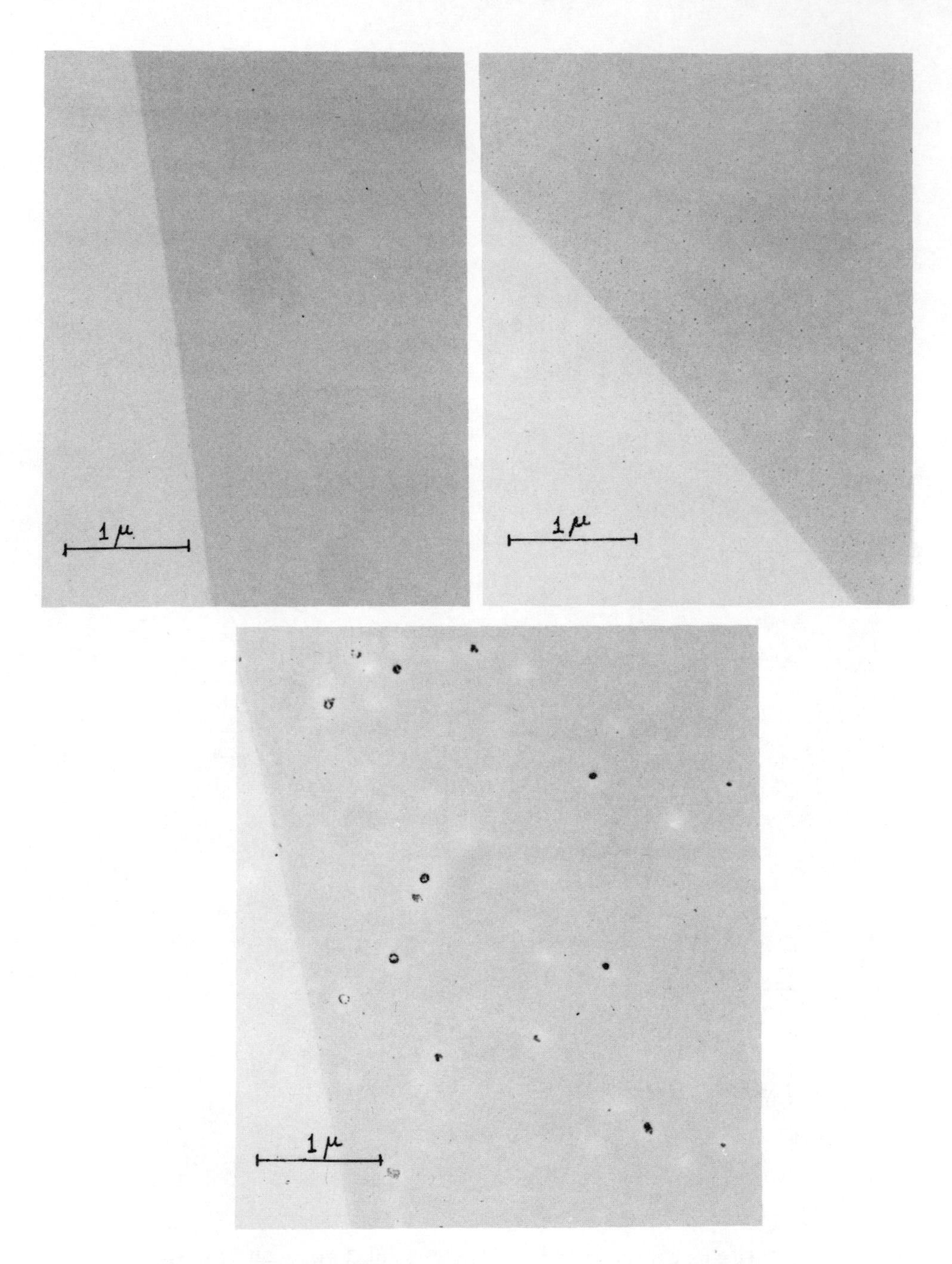

Fig. 4 Electron micrographs of IPN 6. PU4/PA=25/75 (left, above) 50/50 (right, above) 75/25 (below)(11,000x).

thermodynamic equilibrium (phase separation) to be achieved. There is the possibility, of course, that intermolecular crosslinking between the two networks is the reason for the appearance of a single phase. However, as stated earlier, this is expected to be minimal.

Conclusions

We may conclude that at least partially (in some cases totally) interpenetrating polymer networks were produced with possibly a small degree of reaction occurring between the component networks. The existence of interpenetration was evidenced by the appearance of one glass transition in the cases measured. In one case, electron microscopy revealed no evidence of phase separation. These SIN's exhibited enhancement in mechanical properties which may be explained by the phenomenon of interpenetration.

Bibliography

1. H. L. Frisch and E. Wasserman, J. Am. Chem. Soc., 83, 3789 (1961)

2. E. Wasserman, Sci. American, 207 (5), 94 (1962)

3. A. Luttringhaus and G. Isele, Angew. Chem., 6, 956 (1967)

4. G. Schill, Chem. Ber., 100 (6), 2021 (1967)

5. E. Wasserman, D. A. Ben-Efraim and R. Wolovsky, J. Am. Chem. Soc. 90, 3286 (1968).

6. J. Wang and H. Schwartz, Biopolymers, 5, 953 (1967)

7. H. L. Frisch, D. Klempner, and K. C. Frisch, J. Polymer Sci. (B) 7, 775 (1969).

8. D. Klempner, H. L. Frisch, and K. C. Frisch, J. Polymer Sci. (A-2) 8, 921 (1970).

9. M. Matsuo, T. K. Kwei, D. Klempner and H. L. Frisch, Polymer Eng. and Sci. 10 (6), 327 (1970)

10. D. Klempner and H. L. Frisch, J. Polymer Sci. (B), 8, 525 (1970).

11. L. H. Sperling and D. W. Friedman, J. Polymer Sci. (A-2) 7 425 (1969).

12. A. J. Curtius, M. J. Covitch, D. A. Thomas and L. H. Sperling, Polymer Eng. and Sci., 12, 101 (1972).

13. V. Huelck, D. A. Thomas and L. H. Sperling, Macromolecules, 5 340 (1972).

14. K. C. Frisch, D. Klempner, S. K. Mukherjee, and H. L. Frisch, J. Appl. Polmer Sci. (in press)

15. K. C. Frisch, D. Klempner, T. Antczak, S. K. Mukherjee and H. L. Frisch, J. Appl. Polymer Sci. (in press)

16. K. C. Frisch, D. Klempner, S. Midgal, T. Antczak and H. L. Frisch, J. Polymer Sci. (in press).

17. E. J. Malec and D. J. David, Analytical Chemistry of Polyurethanes, edited by D. J. David and H. B. Staley, John Wiley, New York, 87 (1969).

18. A. R. Spurr, J. Ultrastr. Res., 26, 31 (1969).

19. L. E. Nielson, Macromol. Chem., 4, 76 (1970), ed by G. B. Butler and K. J. Driscoll.

TIME-TEMPERATURE SUPERPOSITION IN TWO-PHASE POLYBLENDS*

D. Kaplan and N. W. Tschoegl

Division of Chemistry and Chemical Engineering
California Institute of Technology
Pasadena, California 91109

ABSTRACT

The temperature dependence of the mechanical properties of a 50/50 blend of PVAC and lightly crosslinked PMMA has been examined using the data of Kawai et al. The shift distances, log a_T, were generated by bringing the experimental data into coincidence on master curves calculated from a Takayanagi model whose parameters were varied in different regions of temperature. This method allows one to construct a master curve for a thermorheologically complex two-phase material if the model and the mechanical properties of the constituent phases and their temperature dependence is known. The shift distances then provide insight into the intricate relations between the time and temperature dependence of the mechanical properties of the composite.

INTRODUCTION

In order to characterize the mechanical properties of a polymer completely, measurements are generally conducted over a limited time or frequency scale and at a series of temperatures. The response curves are then shifted empirically along the logarithmic time or frequency axis to form a master curve extending over 10 to 20 decades [1]. This procedure is based on the principle of equivalence of time and temperature effects and is valid only when all relaxation mechanisms are affected equally by a change in temperature.

*Reprinted from the January, 1974, issue of Polymer Engineering and Science.

For a thermorheologically simple polymeric material, the amount of shift, $\log a_T$, needed to translate a response at temperature T to coincide with a response at a reference temperature T_r is the same for all values of time. All relaxation times, $\tau_i(T)$, at temperature T are related to the same relaxation times, $\tau_i(T_r)$, at temperature T_r by a constant ratio, a_T, i.e.

$$\frac{\tau_i(T)}{\tau_i(T_r)} = a_T \tag{1}$$

For a thermorheologically complex material this condition is not met. For example, the relaxation times for a two-phase material are not equally affected by a change in temperature, and simple time-temperature superposition would not be admissible unless the behavior were completely dominated by one phase or the other. Thus, a two-phase material is inherently thermorheologically complex. In such a material the shift factors depend on time as well as on temperature. A general treatment of time-temperature superposition in such materials has been presented in a previous publication [2]. It was shown that, for such a material,

$$\left(\frac{\partial \log D(t)}{\partial \log t}\right)_T = \left(\frac{\partial \log D[t/a_T(t)]}{\partial \log t/a_T(t)}\right)_{T_r} \left[1- \left(\frac{\partial \log a_T(t)}{\partial \log t}\right)_T\right] \tag{2}$$

i.e. in logarithmic coordinates the slope of the compliance with respect to time at a given temperature equals the slope of the compliance with respect to time at the reference temperature multiplied by a factor representing the effect of time on the shift factor. For a thermorheologically simple material this factor is unity because the shift factor does not depend on time.

Unfortunately, theoretical equations such as Eq. (2), although useful for the insight they provide, cannot be integrated to allow their application to the shifting of isothermal data obtained on thermorheologically complex materials. As pointed out earlier, [2, 3] it is necessary to use a model from which the response of the composite material can be constructed as a function of the properties of the pure constituent phases. In the previous publications from our laboratory [2,3] a simple model based on the additivity of the compliances of the two pure phases was proposed specifically for use on polystyrene/1,4-polybutadiene/polystyrene (SBS) block copolymers of moderate (about 25%) polystyrene content, for which the assumption of additivity appears justified [4]. However, as explained elsewhere [3], in the aforementioned block copolymers lack of superposition is never clearly apparent because of the wide separation of the two transitions and the relative smallness of the experimental window (generally four logarithmic decades of time at best).

This paper treats published data on a polyblend in which lack of superposition is clearly evident. The method of generating the time-temperature shift factors is extended from the additive compliance model to a more sophisticated one. Instead of generating shift factors from synthetic curves, the original data were shifted to the master curve constructed on the basis of the model. This method is preferable when the mechanical properties of the pure phases are known.

EXPERIMENTAL DATA

Measurements of the relaxation moduli of a 50/50 (by volume) blend of lightly crosslinked poly(methyl methacrylate), PMMA(cr), and of poly(vinyl acetate), PVAC, obtained by Kawai et al. [5] will be used here. The preparation of the samples and the experimental procedures are described in detail elsewhere [5, 6]. The data were obtained at 19 temperatures between 29.6° and 163.2°C and are reproduced in Fig. 1. Lack of superposition is clearly apparent in the PMMA(cr) dispersion region, particularly between 103.6° and 128.5°C. Figure 2 shows an enlarged portion of Fig. 1 in which superposition is attempted but clearly not possible.

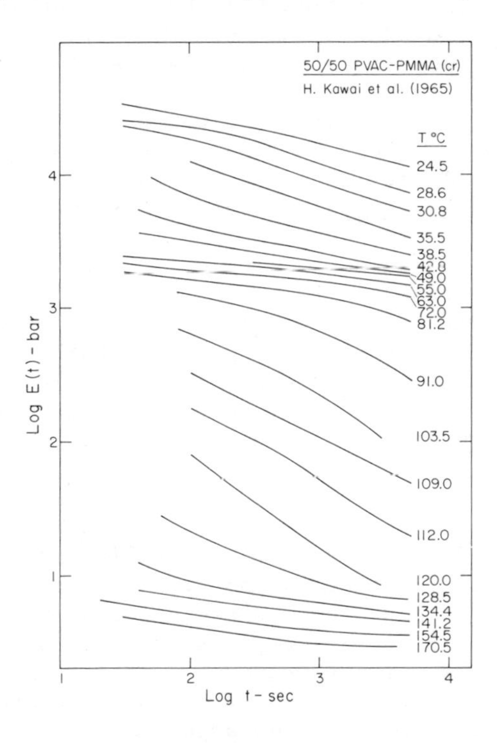

Fig. 1. Relaxation moduli for a 50/50 blend of PVAC and PMMA(cr) as function of time at different temperatures. Replotted from reference [5].

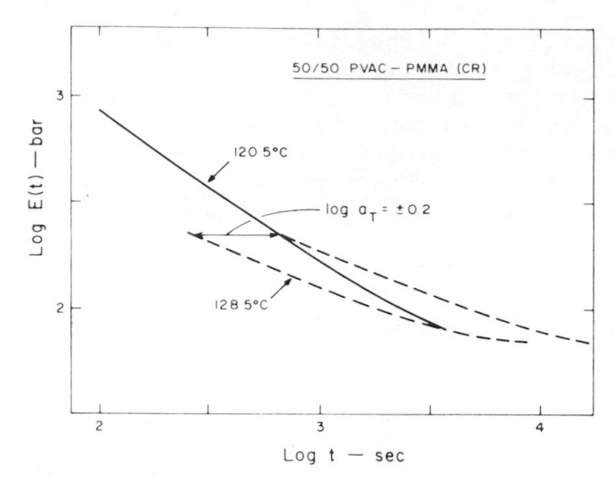

Fig. 2. Enlarged portion of Fig. 1 showing lack of superposition.

MODEL

That the lack of superposition becomes apparent in this blend through an experimental window of no more than 2.5 logarithmic decades of time is probably due to the relative closeness of the glass transition temperatures of the two phases [about 35°C for PVAC, and about 110°C for PMMA(cr)] and to the presence of the two phases in equal amounts. Under these circumstances additivity of the compliances, as used previously [2, 3] for the SBS block copolymer, cannot be valid. We therefore attempted to represent the data by the more complex Takayanagi model [7] of which the additive compliance model is a limiting case [2]. The model has been proposed in two variants. Both assume that the two phases are connected partly in parallel and partly in series. The two variants are shown in the Appendix to be equivalent. Both predict the same results, provided the appropriate values are chosen for the model parameters. Our calculations in this paper will be based on the variant which we call Model I in the Appendix.

Figure 3 shows the isochronal (500 second) relaxation moduli of the polyblend as calculated from the model letting the model parameter λ vary from 0.500 to 1.000. The solid curves apply when PMMA(cr) is taken as the continuous phase; the dotted lines were obtained for PVAC as the continuous phase. It can be seen that the prediction of the model is relatively insensitive to the choice of the parameter λ when the correct phase is chosen as the continuous one. It is possible however, to choose the model parameters so as to predict the same value of the modulus over a fair span of temperatures in the intertransition region. This is exemplified by the dotted line for $\lambda = 0.5005$ and the solid line for $\lambda = 0.850$.

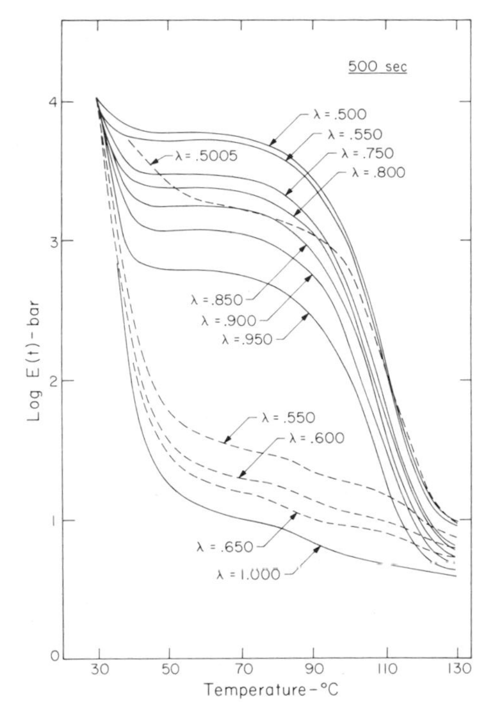

Fig. 3. Predictions of the Takayanagi model for the blend using different values of the model parameters. Calculations were based on the isochronal (500 sec) properties of the constituent phases. Solid and dotted lines represent calculations with PMMA(cr) or PVAC as continuous phase.

Figure 4 shows the 500 second relaxation moduli of the pure phases and the polyblend as a function of temperature. The curve for the polyblend was obtained by crossplotting from Fig. 1. Those for the pure phases were crossplotted from the data in the original reference [5]. Lacking electronmicroscopic or other suitable evidence, it is not clear which of the two phases is the continuous one. Indeed, in a 50/50 blend both phases may be continuous. We were not able to fit the experimental data satisfactorily with a single value of λ whether PVAC or PMMA(cr) was taken as the continuous phase. It is reasonable to assume, however, that the degree of coupling, i.e. the parameters λ and ϕ, are themselves functions of the temperature. As a practical expedient in the absence of information on this temperature we divided the temperature response into three regions, the two transition regions and the intertransition region. Regarding PVAC as the continuous phase, we then fitted the data with three values of λ and ϕ in these three regions. The values and temperature regions are shown below.

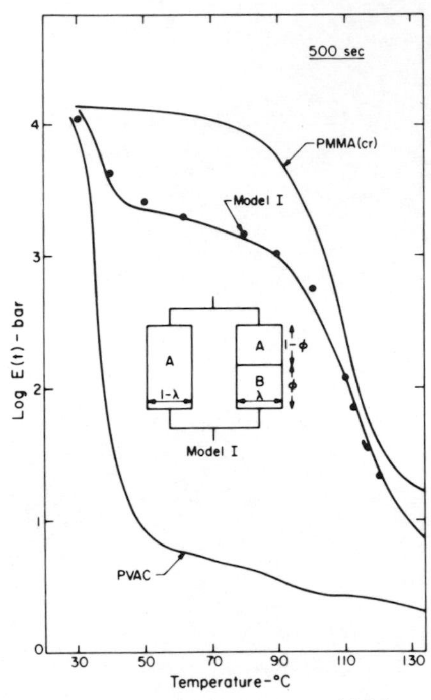

Fig. 4. Fit of Takayanagi model using different values of the parameters in different regions of temperature.

λ	φ	Temperature (°C, 500 sec)	Log Time (sec, 112.0°c)	Log Time (sec, 40.8°C)
0.5010	0.9980	$T < 40$	$t < -4.6$	$t < 1.7$
0.5005	0.9990	$40° \leq T \quad 110$	$4.6 \leq t < 0.0$	$1.7 \leq t < 10.4$
0.5000	1.000	$T \geq 110$	$t \geq 0.0$	$t \geq 10.4$

It should be noted that a choice of λ automatically determines the value of φ. Another choice of the isochronal time (500 seconds in Fig. 4) does not change the values of λ but will change the constraint on the temperature axis from what is shown above. The full circles in Fig. 4 represent the prediction of the model with these parameters and with PVAC as the continuous phase. No satisfactory fit could be obtained with PMMA(cr) as the continuous phase

TEMPERATURE DEPENDENCE

Figure 5 shows a master curve, at 112.0°C, of the relaxation moduli for the two phases obtained in the usual way by empirical

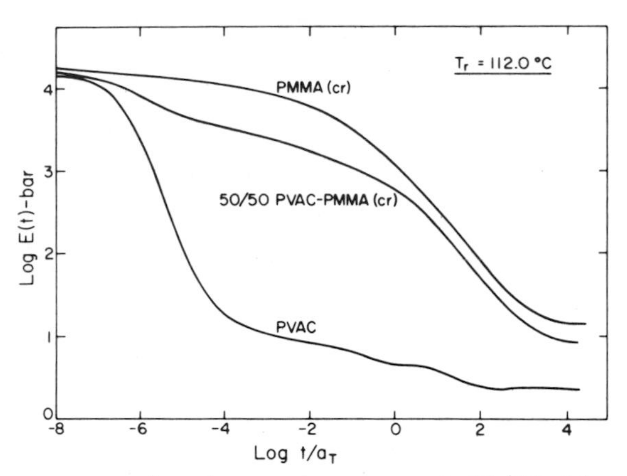

Fig. 5. Master curve for homopolymers and 50/50 blend at 112.0°C.

shifting, and the master curve calculated from the model for the 50/50 blend at the same temperature. For construction of the master curve, the same values of λ and φ were used in the appropriate time intervals as indicated above. A similarly constructed master curve, at 40.8°C, is shown in Fig. 6.

The empirically determined shift factors for the pure phases, at 112.0°C, are shown by the full circles in Fig. 7. For the PVAC homopolymer, shifting of the isothermal segments became difficult above 60°C and a WLF equation determined from the data between 40° and 60°C was used. This equation is

$$\log a_T = \frac{3.60\ (T - 112^\circ)}{112 + T - 112^\circ} \qquad T > 60^\circ \qquad (3)$$

For the PMMA(cr) homopolymer empirical shifting became difficult below 80°C. In this region we therefore resorted to an Arrhenius equation of the form

$$\log a_T = 19{,}000 \left(\frac{1}{T + 273^\circ} - \frac{1}{385}\right) \qquad T < 80^\circ \qquad (4)$$

This equation was determined from the data between 80° and 110°C.

Figure 7 shows the shift distances, $\log a_T(t)$, for the blend when t = 100 sec. These shift distances were found by translating the appropriate value of the blend modulus into coincidence on the master curve for the blend. The values of t = 100 and t = 3000 seconds are compared in Fig. 8 for the reference temperature of 112.0°C. These two values of time represent roughly the end points of the experimental window (cf. Fig. 1). Similar plots are shown

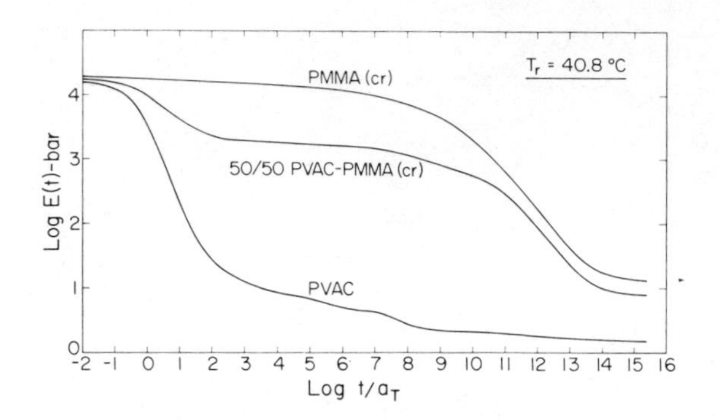

Fig. 6. Master curve for homopolymers and 50/50 blend at 40.8°C.

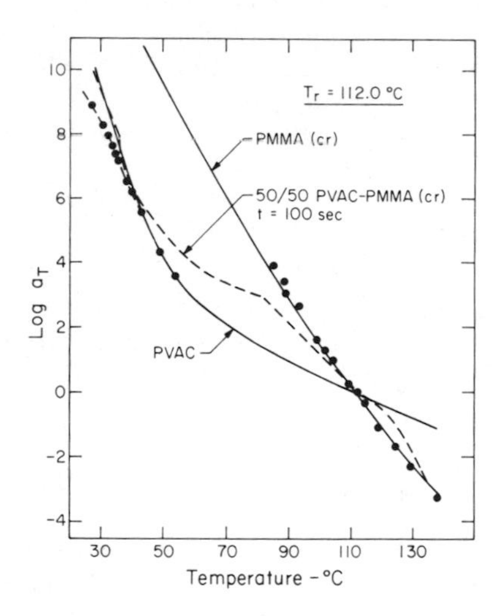

Fig. 7. Shift distances, log a_T, as function of temperature for homopolymers and 50/50 blend at t = 100 sec for reference temperature of 112.0°C.

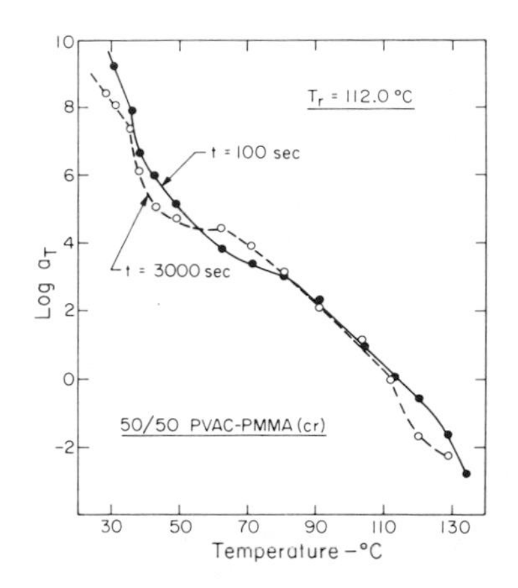

Fig. 8. Shift factors, log a_T, as function of temperature for blend at 100 and 3000 sec and reference temperature of 112.0°C.

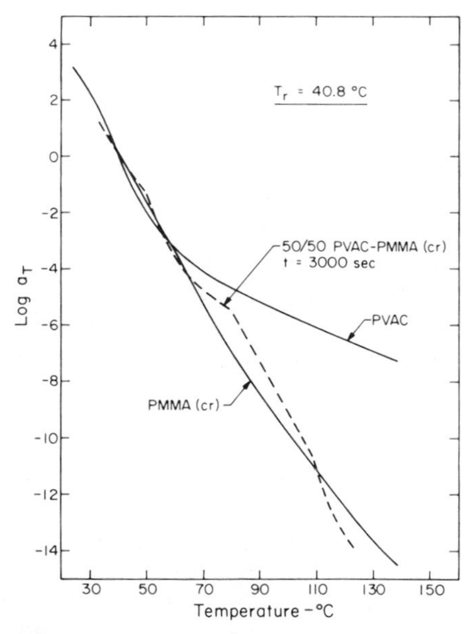

Fig. 9. Shift distances, log a_T, as function of temperature for homopolymers and 50/50 blend at t = 3000 sec for reference temperature of 40.8°C.

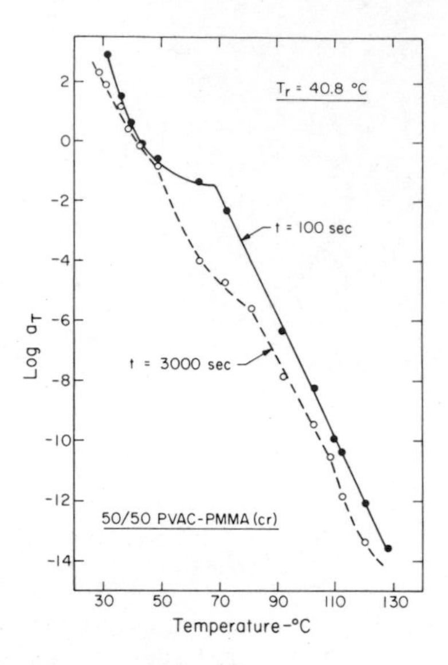

Fig. 10. Shift factors, log a_T, as function of temperature for blend at 100 and 3000 sec and reference temperature of 40.8°C.

in Figs. 9 and 10 for 40.8°C as the reference temperature. As anticipated from the earlier results [2] the shift distances are clearly seen to be functions of time. In addition, the slope of the shift factor curves as well as the slope of the master curves change with the reference temperature. With thermorheologically simple materials, a change in reference temperature merely results in a displacement of the WLF and master curves without a change in shape.

The plots shown in Figs. 8 and 10 share certain characteristics. The temperature dependence of the mechanical properties of the two-phase material changes from dominance by the softer phase at low temperatures to dominance by the harder phase at higher temperatures. At low temperatures one sees the typical Arrhenius dependence of the soft phase in the glassy state. This changes to a WLF dependence as the temperature increases. At a characteristic temperature (which depends both on time and the reference temperature), the WLF dependence of the soft phase changes into the Arrhenius dependence of the hard phase. It is interesting to note that the point-by-point shifting used here produced quite smooth plots of log $a_T(t)$ vs. T.

The change in shape of both the shift factor curves and the master curves with a change in reference temperature occurs most dramatically in the range between the two homopolymer transitions. This results because the slopes of the two homopolymer shift factors in the intertransition region are very different. Hence, a change in reference temperature results in a shift of the two curves by differing amounts. Below the T_g of the soft phase, however, both phases are glassy and the temperature dependencies can be approximated by Arrhenius equations. If the slopes are the same, a selection of a lower reference temperature would result in an equal shift of both homopolymer curves with no change in shape of the master curve. Similarly, above the T_g of the glassy phase, if the temperature dependence of both phases follows WLF behavior of equal slopes, an increase in reference temperature will only result in a shift of the master curve with no change in shape.

DISCUSSION

A general method has been developed to construct a master curve of any two-phase system, provided the properties of the components are known and the correct model can be found. This method must be used for all two-phase polymer systems regardless of whether empirical shifting appears to be possible or not. Empirical shifting can only yield one shift factor curve and one master curve for a particular reference temperature. A change in reference temperature would then merely result in a displacement of the master curve without a change in shape and therefore would not produce the true master curve representing the behavior at that temperature.

In our earlier work [2, 3] we generated, from an additive compliance model, short response curves for the times and temperatures of interest. Then, by comparing the time [2] at which the same compliance appears on the reference master curve, and taking the difference, we obtained a set of shift distances which was then applied to the data. This differs from our present approach in which the experimental data are directly shifted to the master curve. The earlier approach is valid when the behavior of the composite is determined predominantly by the shift factors of the homopolymers, i.e. when homopolymer shifts predominate in time intervals where the mechanical properties are changing rapidly, and thus the choice of model is not critical. This is essentially the case when the two phases of the composite exhibit glass transition temperatures very similar to those of the respective homopolymers. In the case of block copolymers the earlier approach may be preferable because it is not very sensitive to the precise nature of the mechanical response assumed for the two phases. This is difficult to deduce for a block copolymer while the information is readily available for a blend.

The study reported here shows that the temperature dependence of the mechanical properties of two-phase materials can, at the present state of our knowledge, be treated only on the basis of an appropriate model for combining the properties of the constituent phases. Once, however, a particular model has been assumed, the master curve at any temperature is automatically fixed. One can now obtain the dependence of the shift factors on temperature at a given isochronal time and for a given reference temperature as in Figs. 8 and 10, but one cannot determine shift factors to construct a master curve.

The shift factors determined here gave information on the dominance of each component of the blend as a function of time and temperature and thus provided insight into the behavior of two-phase materials. Dickie [8] has shown that the Takayanagi model is equivalent to Kerner's equation [8]. For spheres, the Halpin-Tsai equations [10] are, in turn, identical with the equation of Kerner. Hence, all three approaches are essentially identical. Our modification of making λ vary with temperature in the Takayanagi model is tantamount to letting Poisson's ratio become a function of temperature in the Kerner or Halpin-Tsai equations.

Models other than the Takayanagi approach can be used. The equations of Hashin and Shtrikman [9] give approximate upper and lower bounds for the moduli of composites when either the soft or hard phase is the continuous one. Nielsen [11] has modified the Halpin-Tsai equations to take into account the behavior of the composite during partial phase inversion when both phases may be continuous. The practical application of Nielsen's method restricted by the necessity of knowing the Poisson's ratios, maximum packing fractions ϕ_m, and the generalized Einstein coefficients [11] for the system under study. Furthermore, the theories are developed in a form which displays the concentration dependence of the moduli of the composite. The use of such equations to construct a master curve requires knowledge of the above mentioned variables as a function of time and temperature.

APPENDIX

The two variants of the Takayanagi model are shown as Model I and Model II in Fig. 11. Topologically, they correspond to a three-parameter Maxwell and Voigt model, respectively. Their equivalence can be demonstrated in essentially the same way in which one can show that the Maxwell and Voigt models are equivalent with an appropriate change of the parameters. We proceed as follows.

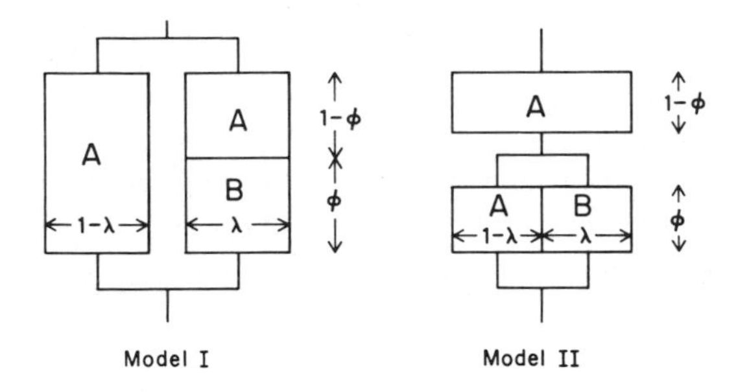

Fig. 11. Takayanagi Model I and II.

Model I gives the relaxation modulus, E(t, T), of the composite system as

$$E(t,\ T) = (1 - \lambda)\ E_A + \lambda \left(\frac{1 - \phi}{E_A} + \frac{\phi}{E_B} \right)^{-1} \tag{A-1}$$

where

$$\lambda\phi = V_B \tag{A-2}$$

and λ and ϕ are parameters relating to the degree of coupling of phase A and phase B, and V_B is the volume percent of component B. The relaxation modulus $E'(t,\ T)$ of the composite system is given by Model II as

$$E'(t,\ T) = \left[\frac{1 - \phi'}{E_A} + \frac{\phi'}{(1 - \lambda')\ E_A + \lambda E_B} \right]^{-1} \tag{A-3}$$

with

$$\lambda'\phi' = V_B \tag{A-4}$$

where λ' and ϕ' are analogous parameters and V_B is the same volume percent of B.

The following algebraic manipulations set the two equations, (A-1) and (A-2), equal to each other and finds the relationship between the mixing parameters which make the two models equivalent.

Simplifying Eq. (A-1) with the use of Eq. (A-2) and collecting similar terms gives

$$E(t, T) = \frac{[\lambda - V_B(1 - \lambda)]\ E_A\ E_B + (1 - \lambda)\ V_B\ E_A^2}{(\lambda - V_B)\ E_B + V_B\ E_A} \tag{A-5}$$

Similarly, Eq. (A-3) can be simplified by Eq. (A-4) to

$$E'(t, T) = \frac{(1 - \lambda')\ E_A^2 + \lambda'\ E_A\ E_B}{(1 - \lambda' + V_B)\ E_A + (\lambda' - V_B)\ E_B} \tag{A-6}$$

Setting Eq. (A-5) equal to Eq. (A-6) leads to

$$\begin{aligned}
&(1 - \lambda)(1 - \lambda' + V_B)\ V_B\ E_A^2 \\
&\quad + \left\{\left[(\lambda - V_B)(1 - \lambda') + \lambda' V_B\right] + 2V_B\left[\lambda(1 - \lambda' + V_B) - V_B\right]\right\} E_A E_B \\
&\quad + (\lambda - V_B + \lambda V_B)(\lambda' - V_B)\ E_B^2 \qquad \text{(A-7)} \\
&\quad = (1 - \lambda')\ V_B\ E_A^2 + \left[(\lambda - V_B)(1 - \lambda') + \lambda' V_B\right] E_A\ E_B \\
&\qquad\qquad + \lambda'(\lambda - V_B)\ E_B^2
\end{aligned}$$

But the coefficients of E_A^2, $E_A\ E_B$, and E_B^2 on each side must be equal. Hence we have

$$\lambda = \frac{V_B}{1 + V_B - \lambda'} \tag{A-8}$$

from which, with Eq. (A-2), Dickie's [12] equation

$$\lambda' = 1 + V_B - \phi \tag{A-9}$$

follows at once.

Equation (A-8) gives the relationship between the two independent and coupling parameters of Model I and Model II, i.e. between λ and λ'. The other parameters, ϕ and ϕ', follow from Eqs. (A-2) and (A-4).

Contentions that Model I or Model II fits a given set of data better [5, 7, 13] therefore result from the lack of recognition that the models are equivalent with the appropriate change in the model parameters.

ACKNOWLEDGMENT

The authors are indebted to Dr. D. G. Fesko for helpful suggestions. This work was supported by the National Science Foundation.

REFERENCES

1. J. D. Ferry, Viscoelastic Properties of Polymers, 2nd ed., Chapter 11; Wiley, 1970.

2. D. G. Fesko and N. W. Tschoegl, J. Polymer Sci., Part C, No. 35, pp. 51-69(1971).

3. D. G. Fesko and N. W. Tschoegl, submitted to Intern. J. Polymeric Mater.

4. C. K. Lim, R. E. Cohen, and N. W. Tschoegl, Adv. Chem. 99:397 (1971).

5. T. Horino, Y. Ogawa, T. Soen, and H. Kawai, J. Appl. Polym. Sci. 9:2261 (1965).

6. K. Fujino, Y. Ogawa, and H. Kawai, J. Appl. Polym. Sci. 8:2147 (1964).

7. M. Takayanagi, S. Uemura, and S. Minami, J. Polym. Sci. Part C, No. 5, pp. 113-122 (1964).

8. E. H. Kerner, Proc. Phys. Soc. (London) B 69:808 (1956).

9. Z. Hashin and S. Shtrikman, J. Mech. Phys. Solids 11:127(1963).

10. J. E. Ashton, J. H. Halpin, and P. H. Petit, Primer on Composite Materials: Analysis, Chapter 5, Technomics, Stamford, Conn. (1969).

11. L. E. Nielsen, Proc. 6th Int. Congress Rheol., 1972.

12. R. A. Dickie, J. App. Polym. Sci. 17:45 (1973).

13. M. Matsuo, T. K. Kwei, D. Klempner, and H. L. Frisch, Polymer Eng. and Sci. 10:327 (1972).

CONTRIBUTORS

H. E. Bair, Bell Telephone Laboratories, Incorporated, 600 Mountain Avenue, Murray Hill, New Jersey 07974

C. H. Bamford, Department of Inorganic, Physical and Industrial Chemistry, University of Liverpool, Liverpool L69 3BX, England

J. J. Charles, Institute of Polymer Science, The University of Akron, Akron, Ohio 44325

S. L. Cooper, Department of Chemical Engineering, University of Wisconsin, Madison, Wisconsin 53706

D. L. Davidson, Institute of Polymer Science, The University of Akron, Akron, Ohio 44325

A. A. Deanin, Plastics Department, Lowell Technological Institute, Lowell, Massachusetts 01854

R. D. Deanin, Plastics Department, Lowell Technological Institute, Lowell, Massachusetts 01854

A. A. Donatelli, Materials Research Center, Lehigh University, Bethlehem, Pennsylvania 18015

G. C. Eastmond, Department of Inorganic, Physical and Industrial Chemistry, University of Liverpool, Liverpool L69 3BX, England

G. M. Estes, Department of Chemical Engineering, University of Wisconsin, Madison, Wisconsin 53706; present address: Elastomer Chemicals Department, E. I. duPont de Nemours and Company, Wilmington, Delaware

L. J. Fetters, Institute of Polymer Science, The University of Akron, Akron, Ohio 44325

H. L. Frisch, State University of New York, Albany, New York 12207

K. C. Frisch, University of Detroit, Detroit, Michigan 48221

H. Ghiradella, State University of New York, Albany, New York 12207

E. Helfand, Bell Laboratories, Murray Hill, New Jersey 07974

G. Holden, Shell Development Company, Torrance, California 90509

D. Kaplan, Division of Chemistry and Chemical Engineering, California Institute of Technology, Pasadena, California 91109

J. P. Kennedy, Institute of Polymer Science, The University of Akron, Akron, Ohio 44325

J. F. Kenney, M & T Chemicals Incorporated, Rahway, New Jersey 07065

D. Klempner, University of Detroit, Detroit, Michigan 48221

G. Kraus, Research and Development Department, Phillips Petroleum Company, Bartlesville, Oklahoma 74004

S. Krause, Department of Chemistry, Rensselaer Polytechnic Institute, Troy, New York 12181

M. Matzner, Union Carbide Corporation, Bound Brook, New Jersey 08805

J. E. McGrath, Union Carbide Corporation, Bound Brook, New Jersey 08805

H. E. Railsback, Research and Development Department, Phillips Petroleum Company, Bartlesville, Oklahoma 74004

T. F. Reed, General Tire and Rubber Company, 2990 Gilchrist Road, Akron, Ohio 44329

L. M. Robeson, Union Carbide Corporation, Bound Brook, New Jersey 08805

R. W. Seymour, Department of Chemical Engineering, University of Wisconsin, Madison, Wisconsin 53706; present address: Tennessee Eastman Company, Kingsport, Tennessee

T. Sjoblom, Plastics Department, Lowell Technological Institute, Lowell, Massachusetts 01854

R. R. Smith, Institute of Polymer Science, The University of Akron, Akron, Ohio 44325

L. H. Sperling, Materials Research Center, Lehigh University, Bethlehem, Pennsylvania 18015

D. A. Thomas, Materials Research Center, Lehigh University, Bethlehem, Pennsylvania 18015

N. W. Tschoegl, Division of Chemistry and Chemical Engineering, California Institute of Technology, Pasadena, California 91109

R. G. Vadimsky, Bell Telephone Laboratories, Incorporated, 600 Mountain Avenue, Murray Hill, New Jersey 07974

J. L. Work, Research and Development Center, Armstrong Cork Company, Lancaster, Pennsylvania 17604

INDEX